Mining Innovation: Volume III

Mining Innovation: Volume III

Editor

Krzysztof Skrzypkowski

MDPI

Basel • Beijing • Wuhan • Barcelona • Belgrade • Novi Sad • Cluj • Manchester

Editor
Krzysztof Skrzypkowski
AGH University of Krakow
Krakow
Poland

Editorial Office
MDPI
St. Alban-Anlage 66
4052 Basel, Switzerland

This is a reprint of articles from the Special Issue published online in the open access journal *Energies* (ISSN 1996-1073) (available at: https://www.mdpi.com/journal/energies/special_issues/D86E1YG949).

For citation purposes, cite each article independently as indicated on the article page online and as indicated below:

Lastname, A.A.; Lastname, B.B. Article Title. *Journal Name* **Year**, *Volume Number*, Page Range.

ISBN 978-3-7258-0805-2 (Hbk)
ISBN 978-3-7258-0806-9 (PDF)
doi.org/10.3390/books978-3-7258-0806-9

Contents

About the Editor . . . . **vii**

Preface . . . . **ix**

Krzysztof Skrzypkowski, Krzysztof Zagórski, Anna Zagórska and Fhatuwani Sengani
Access to Deposits as a Stage of Mining Works
Reprinted from: *Energies* **2022**, *15*, 8740, doi:10.3390/en15228740 . . . . **1**

Krzysztof Skrzypkowski, René Gómez, Krzysztof Zagórski, Anna Zagórska and Roberto Gómez-Espina
Review of Underground Mining Methods in World-Class Base Metal Deposits: Experiences from Poland and Chile
Reprinted from: *Energies* **2023**, *16*, 148, doi:10.3390/en16010148 . . . . **17**

Gabriel Freire, Guillermo Ramirez, René Gómez, Krzysztof Skrzypkowski and Krzysztof Zagórski
Electro-Mechanical Modeling and Evaluation of Electric Load Haul Dump Based on Field Measurements
Reprinted from: *Energies* **2023**, *16*, 4399, doi:10.3390/en16114399 . . . . **41**

Zbigniew Burtan, Jerzy Cieślik, Dariusz Chlebowski, Paweł Piasecki and Krzysztof Gzik
Geomechanical and Technical Aspects of Torpedo Blasting under Seismic and Rockburst Hazard Conditions in Legnica–Glogow Copper District Mines
Reprinted from: *Energies* **2024**, *17*, 1174, doi:10.3390/en17051174 . . . . **58**

Yongkang Yang, Peipeng Gao, Chao Zhang and Chenlong Wang
Numerical Investigation of the Influence of Roof-Cutting Parameters on the Stability of Top Coal Gob-Side Entry Retaining by Roof Pre-Fracturing in Ultra-Thick Coal Seam
Reprinted from: *Energies* **2023**, *16*, 4788, doi:10.3390/en16124788 . . . . **75**

Jacek Feliks and Paweł Tomach
The Impact of Vibrating Screen Startup Time on Vibration Amplitude and Energy Consumption in Transient State
Reprinted from: *Energies* **2023**, *16*, 7129, doi:10.3390/en16207129 . . . . **95**

Krzysztof Krauze, Tomasz Wydro, Ryszard Klempka and Kamil Mucha
Application of an Analytical Model of a Belt Feeder for Assessing the Load and Stability of Its Structure
Reprinted from: *Energies* **2023**, *16*, 8111, doi:10.3390/en16248111 . . . . **116**

Krzysztof Kotwica, Grzegorz Stopka, Andrzej N. Wieczorek, Marek Kalita, Dominik Bałaga and Michał Siegmund
Development of Longwall Shearers' Haulage Systems as an Alternative to the Eicotrack System Used Nowadays
Reprinted from: *Energies* **2023**, *16*, 1402, doi:10.3390/en16031402 . . . . **133**

Paweł Baran, Stanisław Kozioł, Katarzyna Czerw, Adam Smoliński and Katarzyna Zarębska
Sorption–Dilatometric Properties of Coal from a High-Methane Mine in a CO_2 and CH_4 Atmosphere
Reprinted from: *Energies* **2023**, *16*, 1785, doi:10.3390/en16041785 . . . . **152**

Mariusz Kapusta and Krzysztof Skrzypkowski
Determination of the Salt-Dust Emission and the Efficiency of the Dedusting Installation in the Wieliczka Salt Mine
Reprinted from: *Energies* **2022**, *15*, 8122, doi:10.3390/en15218122 . . . . **166**

Marek Kęsek, Paweł Bogacz and Marcin Migza
Study on the Usefulness of Lean Management Tools and Techniques in Coal Mines in Poland
Reprinted from: *Energies* **2023**, *16*, 7240, doi:10.3390/en16217240 . . . . **180**

About the Editor

Krzysztof Skrzypkowski

Krzysztof Skrzypkowski is an associate professor of mining engineering at the AGH University of Krakow, Faculty of Civil Engineering and Resource Management, Department of Mining Engineering and Occupational Safety in Poland. Since 2021, he has been working as a University Professor. His scientific research concerns the stability of underground mine workings, particularly the selection of support for room and roadway excavations. In addition, his research interests focus on computer-aided design and determining the geotechnical parameters of backfill materials. As the head of the rock bolting laboratory, he performs model, numerical, and industrial tests on various cooperation mechanisms between the mining support and the rock mass. He is the author of over 160 scientific publications, several patents, and a utility model for monitoring and yielding rock bolt and steel arch support. As an academic lecturer, he works closely with scientific institutes, industries, and universities involved in exploiting raw materials from minerals.

Preface

Exploiting natural raw material deposits is closely related to new technologies that reduce operational losses while maintaining a high safety factor. The Special Issue of "Mining Innovation: Volume III" includes several articles concerning review, numerical, laboratory, and industrial research into underground mining. The issues discussed in the articles include: advantages and disadvantages of access to deposits as stage of mining works; characteristics of selected underground mining methods for world-class deposits of base metals; analysis of the operation of the electromechanical drive system through a dynamic model of the underground loader equipment using field data; original solutions for determining the areas of rock mass within a mining area where it is advisable to conduct torpedo blasting, and the experience of blasting in the Rudna copper mine in the LGOM in Poland; industrial research about the influence of roof-cutting borehole depth, borehole dip angle, mining height, and coal seam thickness on stability in an ultra-thick coal seam under 12 distinct mining conditions; laboratory research about trajectory and amplitude values during the startup of a vibrating screen, depending on the set startup time with reference into material enrichment processes is the screening process; analytical model of a typical belt feeder and its stability and forces in the supports in the mining industry; the results of field tests and their analysis and verification of operational correctness of newly developed haulage system of highly productive longwall systems; original laboratory tests with application of an Arduino microcontroller about sorption–dilatometric relationships of hard coal samples differing in vitrinite and inertinite content; innovative solution for monitoring salt-dust emissions in mining applications; the course and results of an original expert-mathematical study to assess the usefulness of specific lean management tools and techniques in hard coal mining. The solutions presented are innovative, developmental, and closely related to the geological and mining conditions. The obtained research results clearly confirm that the exploitation of mineral deposits is at a very high level of advancement; however, further research is still required using modern research laboratories and innovative verification techniques in industrial conditions.

Krzysztof Skrzypkowski
Editor

energies

Perspective

Access to Deposits as a Stage of Mining Works

Krzysztof Skrzypkowski [1,*], Krzysztof Zagórski [2], Anna Zagórska [3] and Fhatuwani Sengani [4]

1 Faculty of Civil Engineering and Resource Management, AGH University of Science and Technology, Mickiewicza 30 Av., 30-059 Kraków, Poland
2 Faculty of Mechanical Engineering and Robotics, AGH University of Science and Technology, Mickiewicza 30 Av., 30-059 Kraków, Poland
3 Research Centre in Kraków, Institute of Geological Sciences, Polish Academy of Science, Senacka 1, 31-002 Kraków, Poland
4 Department of Civil Engineering and Geomatics, Faculty of Engineering and the Built Environment, Durban University of Technology, P.O. Box 1334, Durban 4000, South Africa
* Correspondence: skrzypko@agh.edu.pl

Abstract: An extremely important role in the underground method of mining deposits is the type of access, which is the initial stage of the mining operations. The main feature of mineral deposits is their exhaustiveness and the inability to restore resources. This necessitates the rational management of deposit resources, especially the minimization of resource losses during exploitation. This article presents information on the mining area and methods of access the deposits. In particular, attention was paid to the advantages and disadvantages of access by means of an adit, decline, and a vertical and inclined shaft. Given the relationships among the various stages of mining works, it was found that the number of active levels depends on the volume of production and the adopted mining methods. In addition, attention was drawn to the fact that the access to deposits at increasing depth is related to the intensification of natural hazards that affect the access structure.

Keywords: mining area; access excavation; mining level; shaft; inclined shaft; decline; adit

Citation: Skrzypkowski, K.; Zagórski, K.; Zagórska, A.; Sengani, F. Access to Deposits as a Stage of Mining Works. *Energies* **2022**, *15*, 8740. https://doi.org/10.3390/en15228740

Academic Editor: Pavel A. Strizhak

Received: 21 October 2022
Accepted: 18 November 2022
Published: 21 November 2022

1. Introduction

Mining engineering is a field of science that includes work related to the exploration and recognition of a deposit and its access, preparation, exploitation and processing. More and more often, underground workings are used as hydrocarbon stores and for waste storage. The methods of obtaining raw materials can be divided into: underground; opencast; borehole; from the seabed [1]; from anthropogenic deposits [2]. It is worth mentioning that more and more raw materials (metals) are obtained from urban mining [3,4] and that a new field of science related to the exploitation of deposits in space is developing [5]. Among the methods mentioned, most of the workings are carried out in underground mining, and this trend will continue because shallow deposits, after reaching the maximum level of the floor, are transferred to underground mining [6,7]. An extremely important aspect when making a decision to continue exploitation is the model of the deposit [8] along with the estimation of the amount of resources that will be extracted to the ground surface by means of access workings. Hou et al. [9] stated that access layout optimalization can be treated as a network flow and should consider shortening the length of excavation and reducing the transportation distance. Musingwini [10] found that mine planning takes into account three phases: development layouts, production scheduling and equipment selection. Bołoz [11] noticed that improving the efficiency of the access workings execution is possible through the use of automation and robotization. Afum [12] determined that in order to make the deposit available, it is worth using computer-aided design, with the help of which it is possible to select the place of the deposit opening. Sirinanda et al. [13] pointed to the optimal access point, taking into account net present value.

Mining Area

The first step in setting up a mine in a place where there is a deposit that has been examined and considered suitable for favorable exploitation is to open access to the deposit, i.e., to make it available. Access to the deposit or part of it is understood to mean the performance of the main workings connecting the deposit with the ground surface. For each mine, a mining area is established, i.e., a space within which the entrepreneur is authorized to search for, identify and extract useful minerals covered by the concession. With small deposits, the mining area covers the entire deposit. The boundaries of the mining area are the lines on the surface and the vertical planes passing through them, reaching the depth of the deposit, taking into account its shape. The boundaries of mining areas are being adapted to the existing areas, taking into account adjustments according to natural conditions. In the case of deposits located over a large area, the mine field is sometimes defined by natural boundaries (for example: faults, outcrops, deposit deformations, protective pillars). The size of the area is determined so that the mine's useful life allows for the amortization of the costs incurred in its operation. Depending on the size of the deposit, one or more areas are allocated to individual mines for development. For horizontal deposits, the mining area has its width and length, while inclined deposits are characterized by width by extension and height: vertical and inclined. The size of the mining area is influenced by: the method of making the deposit available; number of beds and distance between them; deposit form, thickness, inclination, resources, tectonics, waterlogging, landform; the designed height of the level; projected extraction of minerals; service life; investment outlays and the period of mine construction. Increasing the mining area allows to reduce the depreciation per one ton of excavated material. Exceeding the optimal size of the mining area leads to an increase in transport and ventilation costs. When selecting the optimal size of the mining area, the sum of investment outlays and operating costs per one ton of extraction for which the value will be the lowest is taken into account.

The size of the mining area of a seam-type deposit can be determined by an analytical method [14]. According to the projected extraction and the method of access, with the size of the extraction level and the length of the mining area, the sum of capital expenditure and operating costs per one ton of extraction ($\sum R$) can be determined as a function of the number of levels or fields (n) and the length of the mining area (L) according to Equation (1):

$$\sum R = f(L, n) \tag{1}$$

where

L—length of the mining area;
n—number of levels.

In order to determine the extraction, the deposit utilization factor (η) according to Equation (2) and the dilution factor (ρ) according to Equation (3) should be taken into account:

$$\eta = (1 - 0.01a) \tag{2}$$

where

a—operational losses;

$$\rho = 0.001b \tag{3}$$

where

b—output dilution.

By inserting n = 1–3 and L = 100–1500 into Equations (4) and (5), respectively:

$$\frac{d\sum R}{dL} = 0 \tag{4}$$

$$\frac{d\sum R}{dn} = 0 \tag{5}$$

The number of mining levels corresponds to the current situation in underground hard coal and ore mining for Polish conditions. The length of the mining area was adopted due to the early stage of access works, where the deposit located very close to the shafts is cut.

For inclined and steep deposits, the length of the mining area is assumed with the number of mining levels corresponding to the depth of deposit exploitation (Figure 1).

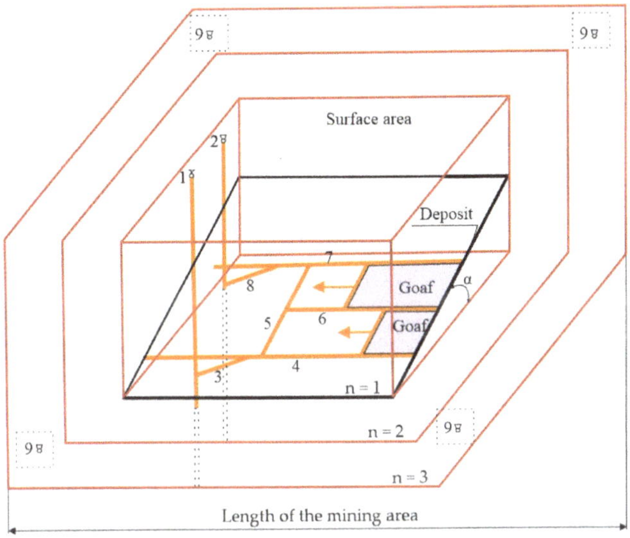

Figure 1. Dimensions of the mining area by the following method: n—number of operating mining levels; 1—output shaft; 2—ventilation shaft; 3—crosscut; 4—transportation gateroad; 5—inclined drift; 6—intermine gateroad; 7—ventilation gateroad; 8—ventilation crosscut; 9—ventilation shafts for expanding mining levels; α—inclination.

Individual mining areas are mined in parallel or in series, depending on: the degree of deposit recognition; the demand for a given raw material; the volume of extraction; the amount of capital expenditure and the period of mine construction. Deposits that are incompletely identified with very variable elements (veins) of deposition are made available in series. The areas of precious and rare metals are also exploited in series. With large-scale mining, mining areas can be operated in parallel. The areas of deposits that are insufficiently explored, with low resources and located at shallow depths, are exploited in series. The construction of the mine does not take long then, and the exploitation proceeds quickly. As the deposits are identified during exploitation in one area, the adjacent ones are opened. Inclined deposits running from the outcrop towards the fall are available in series. This direction enables gradual recognition of the deposit.

2. Ways of Access of Excavations

In underground mining, unit mines can be distinguished, serviced by two sets of shafts: mining-downhill and ventilation-material. A much more advanced organizational form is a combined mine consisting of several unit mines with independent movements but with a complex of mining shafts and a processing complex common to all mines. Combined mines are built when the mining areas are too large and hence the ventilation and transport roads are too long. Examples of combined mines are Mining Plants: Polkowice-Sieroszowice, Lubin and Rudna located in the Legnica-Głogów Copper District in Poland [15]. Each mine must have at least two connections to the ground surface [16]. One enters the mine with fresh air while the other one discharges exhausted air. The deposit can be accessed by: an adit; decline; vertical shaft; inclined shaft. The choice of the appropriate method of making the deposit available depends on: the topography; the shape of the bed; the thickness and angle of the bed; the size of the deposit and its depth; the nature of the rocks in which the deposit is located, especially the roof rocks (from the top to the surface); and the condition of the groundwater in the place where the deposit is to be made available. In underground mining, two basic access structures can be distinguished: deposit (Figure 1) and rock (Figure 2). In practical solutions, depending on the geological and mining situation, intermediate solutions are also used, with elements of both structures. In the case of low values of the strength parameters of the deposit and rocks directly surrounding the deposit or the occurrence of threats due to uplift of the floor, endogenous fires or rock bursts, it is advisable to limit the duration of the roadway workings in the deposit. In such cases, a rock structure is used, especially in hard rock mining conditions. The rock structure is characterized by a large number of excavations in the rock mass. However, this makes it possible to make the deposit available horizontally in many places and to separate independent exploitation areas (separate ventilation, crew movement, transport of excavated material, delivery of materials, supply of backfilling). In case of spontaneous combustion or roof falls, horizontal mining can be carried out with less disturbance than in the case of the deposit structure.

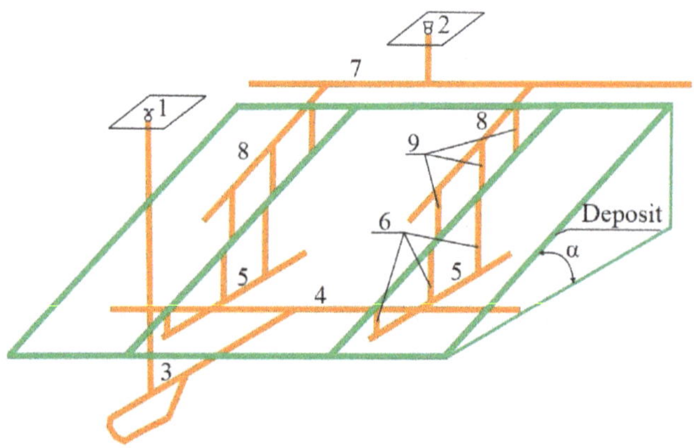

Figure 2. Rock access structure: 1—output shaft; 2—ventilation shaft; 3—transportation crosscut; 4—directional rock drift; 5—field rock drift; 6—output (mining) small shafts; 7—ventilation directional rock drift; 8—ventilation inclined drift; 9—ventilation small shafts.

2.1. Access by Means of Adit

The deposit is usually accessed by adits in mountainous or hilly areas. The adit usually opens access to the upper part of the deposit [17] (Figure 3a). The lower the adit on the slope of the hill is located, the greater part of the deposit is available. If the height of the hill is high, several adits can be made [18], (Figure 3b). The mouth of the adit should be above the highest possible water level in the valley during the period of spring thaws, heavy rain or prolonged rains [19]. Failure to do so may lead to the adit flooding. The adit site should be selected so that it is possible to build the necessary mining equipment on the surface, such as a processing plant, baths, warehouses and a convenient rail connection between the mine and the rail network. Depending on the arrangement of the layers and the deposit as well as the topography, the adits can be led perpendicularly to the direction of the deposit's extension or in the deposit along its length. Sometimes, the adits are carried out from the open pit (Figure 3c) to provide access to a part of the deposit lying under a thick overburden, the open pit mining of which is unprofitable, or to make separate deposits available.

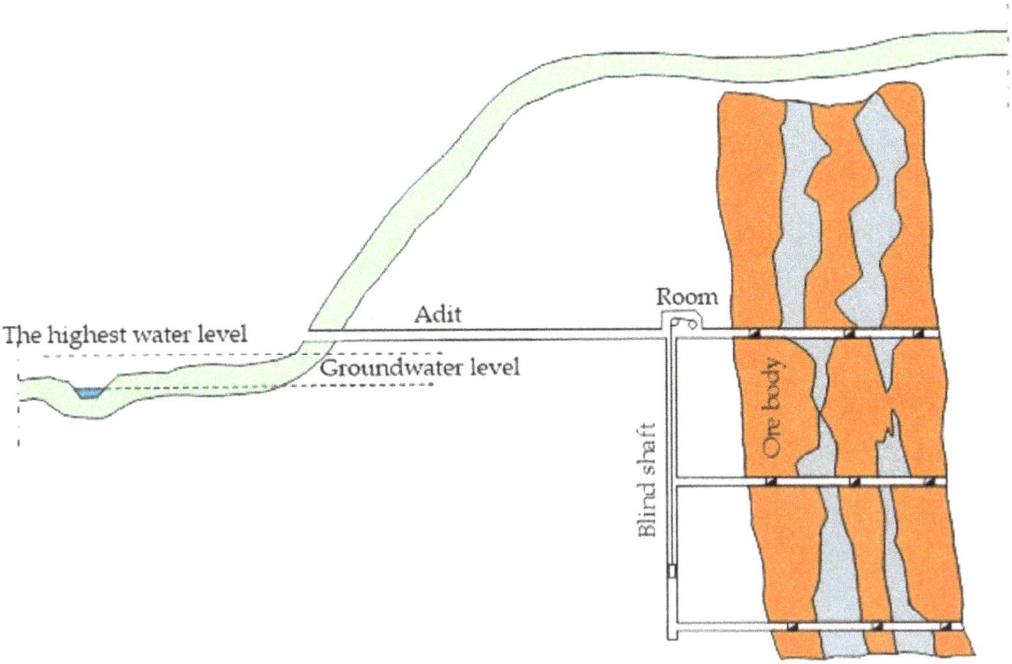

(a)

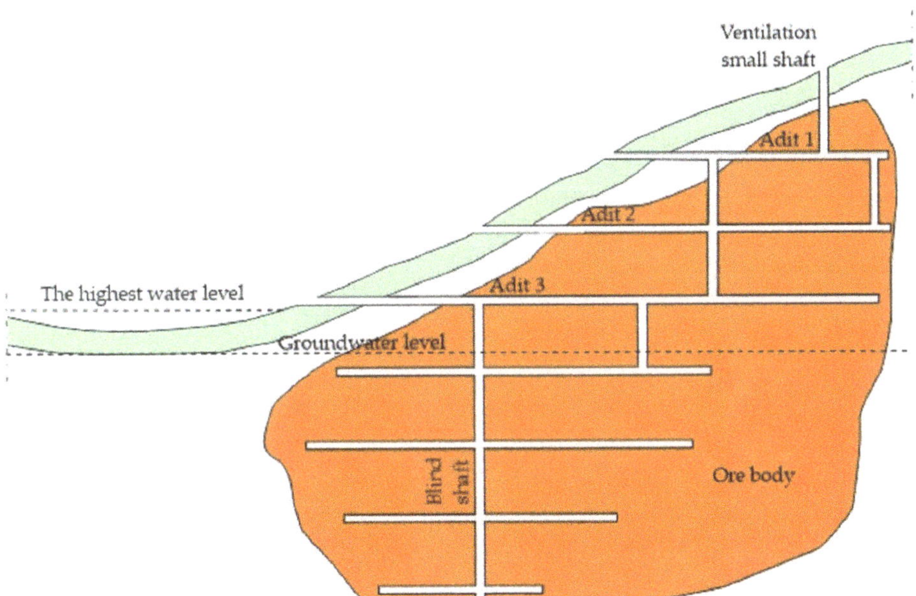

(b)

Figure 3. *Cont.*

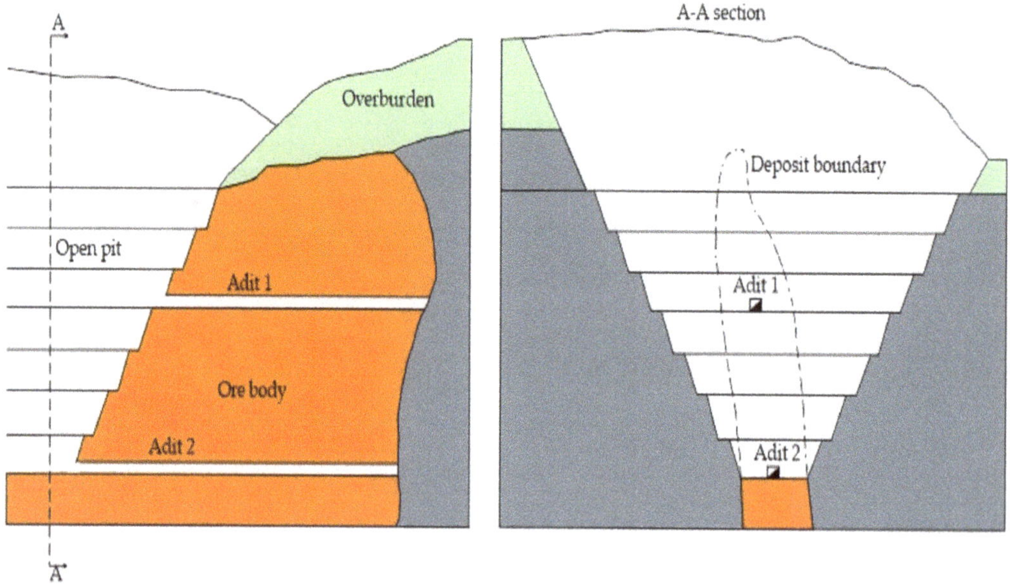

(c)

Figure 3. Access by means of adit: (**a**) Perpendicular to the strike; (**b**) Several adits; (**c**) From open pit mine.

The part of the deposit below the adit can be accessed with a blind shaft or a vertical shaft hollowed out from the ground surface. A blind shaft is used when the amount of the deposit below the adit is small. Then, the R room is constructed on the drift horizontal, in which the hoisting device is placed. The output is hoisted to the adit, and from there it is transported to the surface. When the resources of the deposit below the adit are significant, and therefore the daily output of the mine is sufficiently large, it is necessary to make larger rooms in order to place the hoisting device at the adit level. Then it may turn out to be more advantageous to hollow the shaft from the surface and install the hoist outside the mine. This solution also has the advantage that the machines on the surface do not wear as fast as underground, especially due to moisture. The output of the adit can be hoisted through the shaft or directly to the surface depending on the terrain conditions. If it is more convenient to locate the mine buildings at the mouth of the adit, the excavated material is hoisted to the adit, and from there it is transported to the surface. Then, usually above the adit, a container for the excavated material is made into which the material taken from the lower levels is poured. The adit becomes particularly important when the water flows into the mine, which saves the energy needed to transport it vertically to the surface. In addition, as a result of the construction of the adit, a part of the deposit lying above the adit is drained, which is a favorable phenomenon for mining.

2.2. *Access by Means of Vertical Shaft*

The most frequently used method of making deposits available is vertical shaft. The place where the shaft is established depends on the shape of the deposit, its size and angle of inclination as well as the topography [20]. It is not recommended to set up shafts in valleys near rivers and water reservoirs that would pose a risk of flooding the shaft. The shaft should be placed rather on a hill, from where it is easy to make rail and road connections with the main communication lines. Horizontal or low-slope on-board beds are accessible only through the vertical shaft, while incline beds require the excavation of stone drift from shaft to bed. If the inclined deck remains over a considerable space, a shaft is installed more or less in the center of the mine field (Figure 4a). Then the shaft passes through the roof rocks (part of the shaft marked with letters A–B), then through the deposit (part of the shaft marked with letters B–C) and then through the floor rocks (part of the shaft marked with letters C–D). The total length of all stone drifts in this case is the shortest. In order to protect the shaft and equipment on the surface, it is necessary to leave a part of the deposit around the shaft that cannot be exploited; this is the shaft's protective pillar. The thicker the bed and the greater the content of, for example, metals in the ore, the greater the losses due to leaving the protective pillar. Therefore, in order to reduce losses, shafts are sometimes located in places of irregularities in the deposit, such as of deformation, thinning or depletion. If the bed falls steeply, then it is more advantageous to locate the shaft outside the bed into roof or floor rocks (Figure 4b). With equally strong rocks in the floor and roof of the deposit, it is better to install a shaft in the floor rocks because it is easier to maintain. The pressure of the breaking rocks on the shaft lining due to the excavation of the deposit is lower in the floor than in the roof. In addition, the operation can be started earlier because the cross-sections of the upper levels are shorter. In order to avoid its destruction as a result of rock movement after selecting the deposit, if the shaft is located in the roof rocks, its distance from the deposit should be greater. The length of the stone drifts is then greater, so later it can start selecting the upper levels, and moreover, the costs of stone works increase significantly. Regardless of this, the movements of the rock mass can cause the stone drifts to collapse, or at least make them difficult to maintain. Therefore, the shaft is sinking out in the roof rocks only when it is impossible to locate it elsewhere, for example, if the floor of the deposit is covered with rocks that are not strong enough and water-bearing.

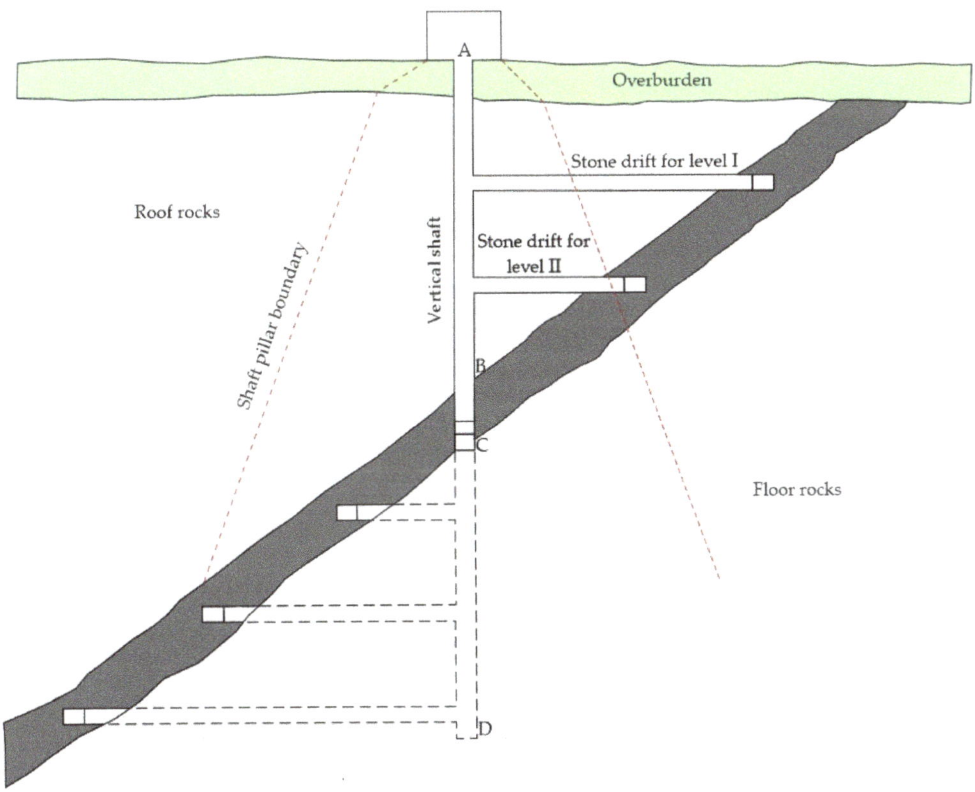

(a)

Figure 4. *Cont.*

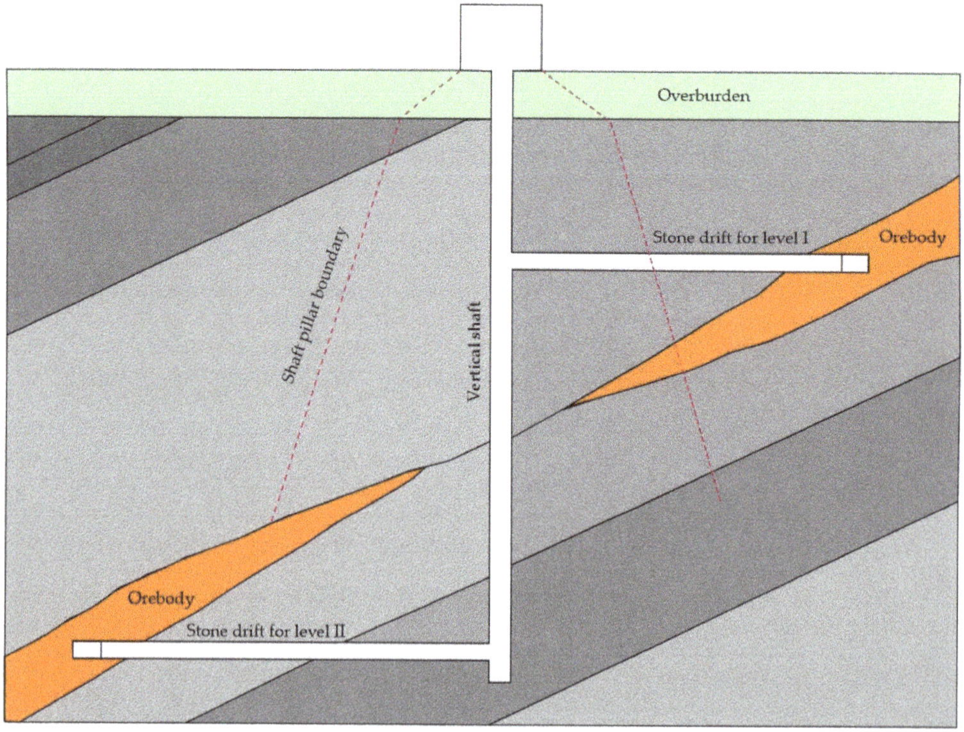

(b)

Figure 4. Access by means of vertical shaft: (**a**) With stone drifts; (**b**) Located in the place of thinning out; A,B,C,D—points in the rock mass from which the sections for the roof, deposit and floor are marked.

A vertical shaft with stone drifts is the most appropriate and most frequently used method of making available several decks or steeply lying veins, as well as nest beds consisting of one or more sockets [21]. If the nests are located at different depths, it is necessary to drive shafts from stone drifts. The small shafts should be driven from the lower level to the higher nest, so that the output can be transported. Transporting the output to a higher level is associated with higher costs and then the transport capacity of the small shaft is lower. Therefore, the output is only hoisted when the nest lies below the deepest mining level of the mine. To avoid or reduce losses in the protective pillar, the shafts should be located between the sockets or on the side of the deposit. The advantages of vertical shafts include:

- Keeping the shaft vertical is easier than inclined, as the rock mass pressure on the lining is less;
- The vertical shaft, which accesses the deposit at the same depth as the inclined shaft, has a shorter length;
- The permissible speed of hoisting of output in vertical shafts is higher than in inclined shafts, and therefore the transport capacity of the vertical shaft is higher;
- The rope wear in vertical shafts is smaller than in inclined shafts, where the rope wears very quickly due to friction between the rope and the pulleys;
- The cost of water drainage from the mine through a vertical shaft is lower (shorter pipeline length).

2.3. Access by Means of Inclined Shaft

Inclined shafts have an inclination of several to several dozen degrees [22–24]. They are sinking in a deposit (Figure 5a) or waste rocks, most often in floor rocks (Figure 4b). If the rock strength is high and the inclination is solid, then the shaft can be sunk out in the bed. With this method of driving a shaft, the operating costs are reduced, since the output is immediately obtained. Simultaneously with sinking the shaft, it is possible to start preparing the upper parts of the deposit for exploitation. On both sides of the shaft, there is a retaining pillar several dozen meters wide, depending on the strength of the rocks and the length of the shaft. It is associated with significant formation losses, especially in thick deposits. If the bed consists of several decks or veins, it is most advantageous to locate the shaft in the deepest and possibly thinnest deck or vein, since the losses in the protective pillars are then the lowest. With a variable inclination of the deposit, low rock strength or high thickness, it is more appropriate to locate the shaft in the floor rocks. Short stone drifts are driven from the shaft to the deposit (Figure 5b), usually alternately to the left and right of the shaft. This is to increase the vertical distance between the stone drifts. The length of the stone drifts on each level is therefore the same. The transport of the output in the shaft can be carried out in wagons or skips moving on the tracks. In order to prevent the output from falling out of the wagons, especially with a steep inclination of the shaft, cages with a horizontal platform can be used. The disadvantage of this solution is the requirement of a sufficiently large shaft height, which makes it much more advantageous to use a skip that has smaller dimensions for the same capacity.

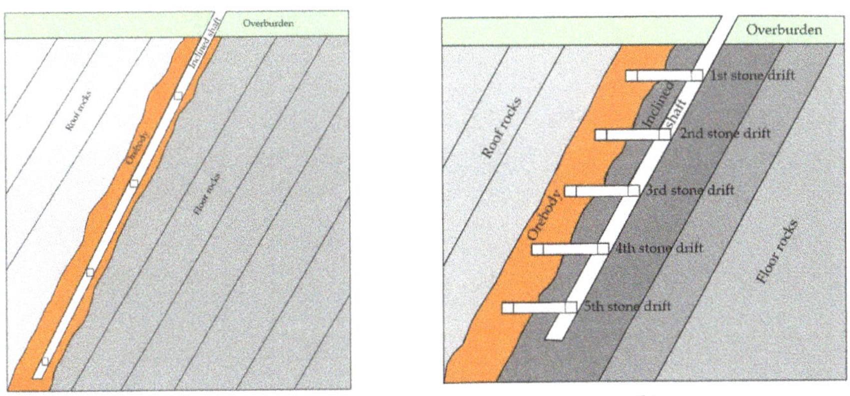

(a) (b)

Figure 5. Access by Means of Inclined Shaft: (**a**) In the deposit; (**b**) In the floor rocks.

The advantages of inclined shafts include:

- If the shaft is sinking in the deposit, the excavated material is obtained when the deposit is made available, and there is no need to drive stone drifts, which reduces the costs of stone works.
- In the case of driving a shaft in floor rocks, it avoids leaving a supporting pillar along the shaft, and the length of the stone drifts is shorter than in the case of a vertical shaft.
- Faster preparation of the deposit for exploitation.

The disadvantages of inclined shafts are in particular:

- The need to leave support pillars along the shaft, which are usually larger than the protective pillars at vertical shafts. If the deposit consists of several seams or veins, and the inclined shaft is sunk in one of the upper seams, it is necessary to leave a protective pillar in the lower seams (or veins), which significantly increases the loss of useful mineral.
- When sinking a shaft in a deposit with a variable inclination, either take back rock or change the direction of the shaft according to the inclination of the deposit. The

transport of the output in a shaft with a variable inclination encounters considerable difficulties and reduces its extraction capacity.

2.4. Access by Means of Decline

A characteristic feature of the decline is the direction of driving from top to bottom, while the transport of the output is carried out from the opposite direction (Figure 6a). Very often, ore mines use this type of access, because self-propelled machines can drive directly to the mine from the surface, eliminating the process associated with their time-consuming assembly and disassembly. In the deposits located at greater depths, spiral ramps are also driven (Figure 6b) [25]. The inclination of the decline when self-propelled machines are used is up to 8° [26]. Moreover, the use of self-propelled haulage trucks is adequate with the maximum daily production volume of the mine at the level of 3000 tons [27]. Belt conveyors are used for larger production capacities of the mining plant. It is also worth noting that a belt conveyor may be installed in the decline; then the maximum angle of inclination of the excavation may be in the range from 15° to 25° depending on the natural repose angle of the transported material [27]. This applies to a situation where the entire cross-section of the decline is intended only for transporting the excavated material by a belt conveyor. In the case of mixed transport, when mining machines on a tire chassis move next to the belt conveyor in the decline, then the angle of inclination is about 12°.

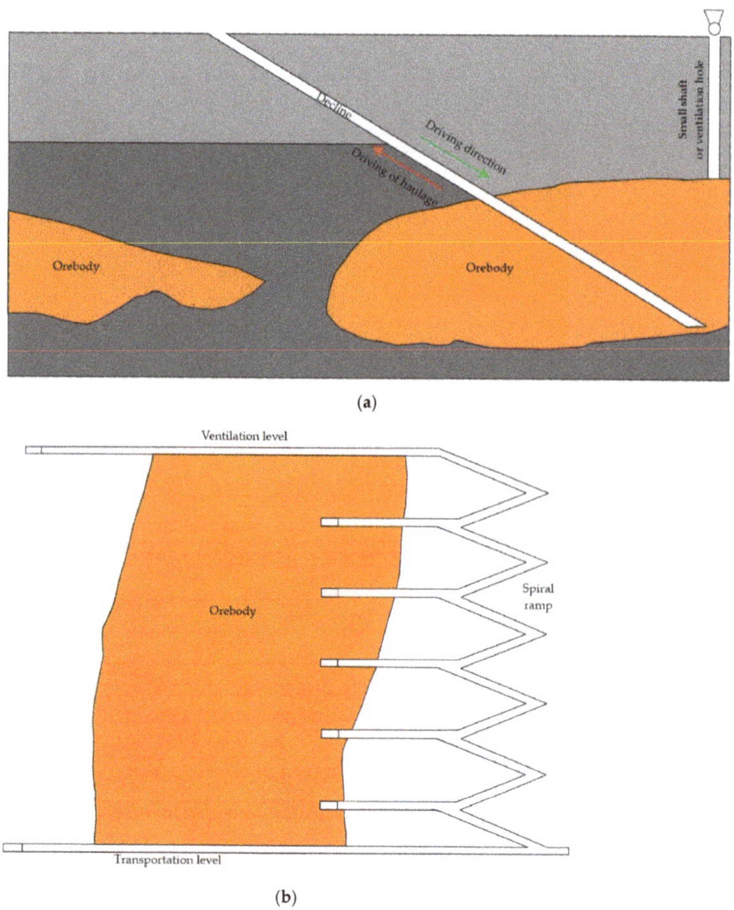

Figure 6. Access by means of decline: (**a**) With constant inclination; (**b**) Spiral ramp.

The advantages of inclined shafts include:

- The possibility of driving the excavation with the use of full-section machines (TBM) [28];
- Direct inspection of the excavation and equipment during the opening and operational works (for the vertical shaft, time should be allocated for its inspection);
- Easier movement of the mining crew between the levels (without the use of a hoisting machine);
- In the case of using a belt conveyor in a decline, high efficiency can be obtained (for a conveyor with a belt width of 1400 mm, up to 38.000 t/day can be obtained) [29].

The disadvantages of inclined shafts are in particular:

- The length of the decline providing access to a specific level and is several times greater than that of a vertical shaft;
- More resources are trapped in the decline protection pillar;
- In the case of transporting people and materials by suspended monorails installed in the decline, their efficiency is much lower compared with vertical transport in the shaft;
- Because the materials of the conveyor belt are flame retardant, it means that they can only be installed in the exhaust air stream. In addition, the greater length of the decline and, as a rule, a smaller cross-section in relation to the shafts, means several times greater ventilation resistance. Since the relative speed of the flowing air and the moving belt should be a maximum of 5 m/s, which allows to avoid excessive dust lifting, it means that the transporting decline cannot be used as the main ventilation excavation at the same time [30].

3. Relationship the Various Stages of Mining Works

The main stages of mining works: access, preparatory and exploitation of the deposit should be related to each other in time so that the previous stage overtakes the next one. Usually access overlaps with preparation in the same way that preparation overlaps with exploitation. The most appropriate ratio of the different stages is one where the upper mining level is exploited to the right extent, the lower mining level is carrying out preparatory work and the third mining level is made available. Often, depending on the geological and mining conditions, exploitation is carried out on several mining levels in order to obtain proper extraction. The advance notice for preparatory works should be determined taking into account: the recognition of the conditions and variable parameters of the deposit and the introduction of efficient mining methods. The advance in time of making the deposit available and preparing it for exploitation can be determined by the Equation (6):

$$t_o = \frac{t_e}{t_{a+d}} \tag{6}$$

where

t_o—time advance indicator;
t_e—horizontal exploitation time, (days);
t_{a+d}—time to access and prepare the level, (days).

Therefore, if the mining is carried out on the first level, the preparatory works on the second level, and the access works on the third level, then t_{a+d} is equal to the sum of the preparation time for the second level and the access of the third level. For the proper functioning of the mine, the value of the index $t_o > 1$, therefore $t_e > t_{a+d}$. The access of deposits and the division into levels is related to the performance of two characteristic workings. The lower boundary of the level is a drift made along the length of the deposit, also known as the main transporting drift, which connects to the mining shaft. The upper drift, called the ventilation drift, connects to the ventilation shaft. In fact, the level is a part of the resource bounded by two horizontal planes: The lower plane is the transport and drainage level, and the upper plane is the ventilation and material level. The basic drift of the transporting level is connected to the drift of the ventilation level by a transporting inclined gallery, which is used to transport the excavated material and, at the same time, to close the air circulation of the extraction and ventilation shaft. The boundaries of the level on the strike side are the boundaries of the deposit area. The number of active levels depends on the volume of production and the adopted mining methods. When dividing the deposit into horizontal lines, it is possible to use a depth chart of resources with the depth of the deposit on the vertical axis and the size of the resources on the horizontal axis. By projecting the amount of resources horizontally onto the vertical axis, the vertical height of the level is obtained. The height of the level is determined by the following factors: the degree of recognition of the deposit, with less-explored and complex deposits having a lower height; the physical and mechanical properties of rocks, specifically, in rocks of low strength the height is lower; the angle of inclination, that is, the smaller the angle of inclination, the smaller the height of the level; the amount of investment outlays per one ton of excavated material; and the period needed to access and prepare the deposit for exploitation. With the height of the level, the resources of the deposit increase, and the expenditure on the preparation of preparatory excavations per one ton of excavated material decreases. On the other hand, the costs of ventilation and support maintenance are rising. The cost of delivering materials for mining excavations is also increasing. The time of accessing and preparing longer workings is extended. Additionally, the costs of haulage and dewatering increase slightly with an increase in the height of the level. Lower heights are beneficial in thin seams and veins and in irregular and tectonically disturbed deposits. The height of H_l [31] according to the production capacity of the shaft and the intensity of exploitation works can be determined according to the Equation (7):

$$H_l = \frac{P \cdot \sin\alpha \cdot (1-\rho)}{n \cdot L \cdot m \cdot \gamma \cdot \eta},\ (\mathrm{m}) \tag{7}$$

where:

P—annual mining capacity of the shaft, (tons);
α—the angle of the deck, (°);
ρ—output dilution coefficient;
n—number of operating wings (parts);
L—annual progress of exploitation works along strike, (m);
m—deck thickness, (m);
γ—volume weight of the output, (t/m^3);
η—deposit utilization factor.

According to the minimum expenditure per one ton of extraction, the height of the H_l level can be determined analytically [31] according to the Equation (8):

$$H_l = \sqrt{\frac{(V \cdot K_0 + L_k \cdot K_k + L_e \cdot R_e) \cdot (1-\rho)}{0.5 \cdot (e + b + \Delta) \cdot S \cdot \eta \cdot \gamma}},\ (\mathrm{m}) \tag{8}$$

where:

V—volume of workings around shaft, (m^3);
K_o—cost of making 1 m^3 of a rock excavation;
L_k—length of the preparatory excavation of the level, (m);
K_k—unit cost of 1 m of the drift;
S—horizontal projection of the deposit, (m^2);
e—haulage cost of 1 ton of excavated material;
b—cost of dewatering per 1 ton of extraction;
Δ—increase in operating costs per 1 ton of excavated material (output) with an increase in the height of the level by 1 m.

Assuming that at the lower level, access and preparatory works are carried out in advance of the operational works at the higher level, then according to the time of making available and preparing the H_l level, Equation (9) can be used:

$$H_l > \frac{P \cdot w \cdot t \cdot (1-\rho)}{S \cdot \gamma \cdot \eta}, \text{ (m)} \tag{9}$$

where:
w—advance factor according to Equation (10):

$$w > \frac{t_o}{t}, \text{ (m)} \tag{10}$$

where:
t_o—time of exploitation;
t—time for access and preparatory the lower mining level.

The access of deposits with increasing depth is complex because with increasing depth, a number of natural factors that hinder underground work are more intense. The manifestations of the negative impact of depth on the conditions and results for access have different severity, and therefore the related problems occur in different frequencies. These problems, although some of them overlap very strongly, could be divided into four main groups, some of which are major in terms of solving them, and the other are secondary. The first group includes problems related to the determination of the size and nature of the stresses on the perimeter of the excavations and the pressure phenomena occurring during the excavation of the deposit in the protective pillars. Primary pressure in the rock mass increases proportionally to the depth [32], affecting the size and nature of stresses around mining excavations, as well as the structure of the mine and mining technology. This group includes: the selection of appropriately strength rocks in which it is most advantageous to drive excavations; the selection of the appropriate cross section and the method of supporting the workings as well as the selection of the appropriate structure of the mine and mining methods in terms of monitoring the symptoms of rock mass pressure and natural hazards. The second group of issues includes securing proper climatic working conditions. The primary temperature of rocks increases with depth [33]. The increase in the temperature of the mine air with increasing depth is caused not only by the more intense absorption of rock mass but also by air compression in the deep inlet shafts. Both these factors, along with the increased oxidation of, for example, coal dust suspended in the air, act to worsen the climatic conditions for mining crews. The importance of proper climatic working conditions in deep mines is especially important when the primary rock temperature is higher than the air temperature permitted by mining regulations. Profitability of mechanization of mining works requires replacement of compressed air favorable for physiological working conditions with electricity. The electrification of a mine increases the temperature of the mine air as a result of the change from electrical energy to thermal energy. The improvement of climatic conditions is realized by concentration of mining works and simultaneous application of intensive ventilation. In order to obtain intensive ventilation, ensuring proper climatic working conditions, it is necessary to: determine the correct location of the inlet and outlet shafts in relation to the exploitation fields; proper location of fresh air inflow and used air discharge roads in the mine as well as transport and ventilation roads in the exploitation field itself; determining the appropriate cross-sections and the type of excavation support; determination of air speed and cooling intensity as well as selection of appropriate parameters of the main fans. The third group includes problems related to the greater gas capacity of coal seams lying at great depths. As the depth increases, the gas content in the coal seams and in the surrounding rocks increases [34], which makes access and exploitation difficult and complicated. Excessive amount of methane emitted into mining excavations is a factor that inhibits the progress of works in both access, preparatory and operational excavations, and therefore has a decisive influence on the method and intensity of ventilation. The intensity of gas evolution can reach such an extent that it can manifest itself in the form of outbursts of gases and rocks [35,36]. The fourth group of problems is related to the negative impact of depth on the broadly understood transport of output, people, materials and water: dewatering. This group is relatively the easiest in terms of their technical solution. The advantages of modern vertical transport solutions to a large extent compensate for the negative effects of the increased depth and are relatively well understood from a technical and economic point of view.

4. Conclusions

The deposit structure, including access, preparatory and exploitation excavations is characterized by a small number of driven workings in the rocks, low costs and a short time to access individual parts of the deposit. The mining process (transport, ventilation), however, requires a long period of keeping the gateroads in the deposit. In the case of shallow deposits, the most popular method of access will still be declines. The use of haul trucks in the decline offers flexibility in terms of variable geometric parameters of excavations and complex mine spatial plans. However, in the case of the transport of the output with haulage trucks in the decline, there is a threshold above which the introduction of additional self-propelled machines will result in excessive traffic congestion and planning difficulties. A certain solution is the use of a belt conveyor in the decline, which is mainly characterized by small, occupied spaces as well as a small amount of service and the possibility of central control and automation. Of course, it should be remembered that often the output should be pre-crushed in order to obtain the appropriate granulation and to avoid spilling it out during transport. The choice of the place where a deposit is made available depends on the distribution of resources in the deposit. As the costs of transporting the excavated material depend on the length of the excavations, the transport roads should be shortened, especially for those parts of the deposit where the resources are the highest. Each underground mine must have at least two serviceable connections of the underground workings with the ground surface, one for fresh air supply (intake shaft) and one for exhaust air discharge (exhaust shaft). Accessibility should ensure access to the lowest part of the deposit intended for exploitation due to drainage and gravity transport. Due to the time of accessing and maintaining workings, effort should be made to shorten their length while taking into account the physical and mechanical parameters of the rocks in which their construction is planned. The selection of the access method and the justification of their parameters usually are based on comparison of different rational variants of access. Such a comparison is based on economic analyses (cost effectiveness analyses) which takes into account cash flows over that time—net present value (NPV). It's worth noting that the timing of cash flows has a significant importance for the present value of an investment and economical effectiveness of the method of access. As a result a choice of an variant of access cannot base only on operating costs per 1 ton.

The coexistence of natural hazards and the multitude of geological and mining factors contribute to the search for more effective methods of mining and haulage of the output. Taking into account the increasing temperature of the rock mass and the intensification of dynamic hazards, future mining works should focus on the full automation of driving and support for workings with a reduction in the number of miners in particularly hazardous zones. Future access works should be based on the more effective use of computer-aided design of mining excavations, both on the plane of the highest deposit utilization factor and taking into account the time of excavation work.

Author Contributions: Conceptualization, K.S., K.Z., A.Z. and F.S.; methodology, K.S., K.Z., A.Z. and F.S.; software, K.S., K.Z. and A.Z.; validation, K.S., K.Z. and A.Z.; formal analysis, K.S., K.Z., A.Z. and F.S.; investigation, K.S., K.Z., A.Z. and F.S.; resources, K.S., K.Z., A.Z. and F.S.; data curation, K.S., K.Z., A.Z. and F.S.; writing—original draft preparation, K.S., K.Z. and A.Z.; writing—review and editing, K.S., K.Z., A.Z. and F.S.; visualization, K.S., K.Z., A.Z. and F.S.; supervision, K.S., K.Z., A.Z. and F.S.; project administration, K.S., K.Z. and A.Z.; funding acquisition, K.S., K.Z. and A.Z. All authors have read and agreed to the published version of the manuscript.

Funding: This research received no external funding.

Data Availability Statement: The data presented in this study are new and have not been previously published.

Conflicts of Interest: The authors declare no conflict of interest.

References

1. Leng, D.; Shao, S.; Xie, Y.; Wang, H.; Liu, G. A brief review of recent progress on deep sea mining vehicle. *Ocean Eng.* **2021**, *228*, 108565. [CrossRef]
2. Winterstetter, A.; Heuss-Assbichler, S.; Stegemann, J.; Kral, U.; Wäger, P.; Osmani, M.; Rechberger, H. The role of anthropogenic resource classification in supporting the transition to a circular economy. *J. Clean. Prod.* **2021**, *297*, 126753. [CrossRef]
3. Pietrzyk, S.; Tora, B. Processing of Non-Ferrous Metals Secondary Raw Materials in Poland–Trends, Opportunities and Threats. *J. Pol. Miner. Eng. Soc.* **2017**, *18*, 81–92. [CrossRef]
4. Tejaswini, M.S.S.R.; Pathak, P.; Gupta, D.K. Sustainable approach for valorization of solid wastes as a secondary resource through urban mining. *J. Environ. Manag.* **2022**, *319*, 115727. [CrossRef] [PubMed]
5. Xu, F.; Su, J. Towards a legal regime of benefits sharing for space mining: With some experience from the Area. *Resour. Policy* **2022**, *76*, 102627. [CrossRef]
6. Flores, G.; Catalan, A. A transition from a large open pit into a novel "macroblock variant" block caving geometry at Chuquicamata mine, Codelco Chile. *J. Rock Mech. Geotech. Eng.* **2019**, *11*, 549–561. [CrossRef]
7. Ross, I.T.; Stewart, C.A. Issues with transitioning from open pits to underground caving mines. In *MassMin 2020: Proceedings of the Eighth International Conference and Exhibition on Mass Mining, Virtual, 4–8 October 2020*; Castro, R., Báez, F., Suzuki, K., Eds.; University of Chile: Santiago, Chile, 2020; pp. 221–238. [CrossRef]
8. Skrzypkowski, K. Determination of the Backfilling Time for the Zinc and Lead Ore Deposits with Application of the BackfillCAD Model. *Energies* **2021**, *14*, 3186. [CrossRef]
9. Hou, J.; Li, G.; Hu, N. Optimization of underground mine access layout and production scheduling. *Miner. Resour. Manag.* **2020**, *26*, 87–108.
10. Musingwini, C. Optimization in underground mineplanning-developments and opportunities. *J. S. Afr. Inst. Min. Metall.* **2016**, *116*, 809–820. [CrossRef]
11. Bołoz, Ł.; Biały, W. Automation and Robotization of Underground Mining in Poland. *Appl. Sci.* **2020**, *10*, 7221. [CrossRef]
12. Afum, B.O.; Ben-Awuah, E. A Review of Models and Algorithms for Surface-Underground Mining Options and Transitions Optimization: Some Lessons Learnt and the Way Forward. *Mining* **2021**, *1*, 112–134. [CrossRef]
13. Sirinanda, K.; Brazil, M.; Grossman, P.; Rubinstein, H.; Thomas, D. Optimally locating a junction point for an underground mine to maximise the net present value. *ANZIAM J.* **2014**, *55*, 315–328. [CrossRef]
14. Takuski, S. *The Technique of Underground Mining of ore Deposits*; University script no. 769; AGH University of Science and Technology Publishing House: Kraków, Poland, 1980; p. 92. (In Polish)
15. Skrzypkowski, K.; Korzeniowski, W.; Zagórski, K.; Zagórska, A. Adjustment of the Yielding System of Mechanical Rock Bolts for Room and Pillar Mining Method in Stratified Rock Mass. *Energies* **2020**, *13*, 2082. [CrossRef]
16. Nie, X.; Wei, X.; Li, X.; Lu, C. Heat Treatment and Ventilation Optimization in a Deep Mine. *Adv. Civ. Eng.* **2018**, *2018*, 1529490. [CrossRef]
17. Yardimci, A.; Karpuz, C. Shortest path optimization of haul road design in underground mines using an evolutionary algorithm. *Appl. Soft Comput.* **2019**, *83*, 105668. [CrossRef]
18. Zeqiri, K.; Ibishi, G.; Shabani, M.; Kortnik, K.; Bilir, M.E.; Geniş, M.; Yavuz, M.; Hetemi, M.; Bacak, G. Preliminary support design for underground mine adit, Artana mine, Kosovo. *Min. Sci.* **2021**, *28*, 141–159. [CrossRef]
19. Galdón, J.M.; Rey, J.; Martínez, J.; Hidalgo, M.C. Application of geophysical prospecting techniques to evaluate geological-mining heritage: The Sinapismo mine (La Carolina, Southern Spain). *Eng. Geol.* **2017**, *218*, 152–161. [CrossRef]
20. Bakhtavar, E.; Yousefi, S.; Jafarpour, A. Evaluation of shaft locations in underground mines: Fuzzy multi-objective optimization by ratio analysis with fuzzy cognitive map weights. *J. S. Afr. Inst. Min. Metall.* **2019**, *119*, 855–864. [CrossRef]
21. Costa, L.; Silva, J.; Lima, H. Analysis of options of production and access ways in underground mines. *Min. REM Int. Eng. J.* **2018**, *70*, 237–242. [CrossRef]
22. Wu, F.; Qin, Y.; Xu, H.; Zhang, F.; Chu, X. Numerical Simulation of Deformation and Failure Mechanism of Main Inclined Shaft in Yuxi Coal Mine, China. *Appl. Sci.* **2022**, *12*, 5531. [CrossRef]
23. Yu, W.; Wang, W.; Wu, G.; Yu, X.; Peng, W. Three Zones and Support Technique for Large Section Incline Shaft Crossing Goaf. *Geotech. Geol. Eng.* **2017**, *35*, 1921–1931. [CrossRef]
24. Krishna, P. Analysis on the Dynamics of Burst Debris Flood at the Inclined Pressure-Shaft of Svandalsflona Hydropower Project, Norway. *Rock Mech. Rock Eng.* **2014**, *47*, 923–932.
25. Haviland, D.; Marshall, J. Fundamental behaviours of production traffic in underground mine haulage ramps. *Int. J. Min. Sci. Technol.* **2015**, *25*, 7–14. [CrossRef]
26. Rupprecht, S.M. Mine development–access to deposit. *J. S. Afr. Inst. Min. Metall.* **2012**, *2012*, 101–120.

27. Wilson, R.B.; Willis, R.P.H.; Du Plessis, A.G. Considerations in the choice of primary access and transportation options in platinum mines. In *International Platinum Conference: Platinum Adding Value*; The South African Institute of Mining and Metallurgy: Johannesburg, South Africa, 2004; pp. 269–274.
28. Zheng, Y.L.; Zhang, Q.B.; Zhao, J. Challenges and opportunities of using tunnel boring machines in mining. *Tunn. Undergr. Space Technol.* **2016**, *57*, 287–299. [CrossRef]
29. Czaja, P.; Kamiński, P. The possibility of access to copper ore deposits on Fore-Sudetic Monocline with a inclined shaft. *Cuprum* **2015**, *3*, 19–35.
30. Jakubowski, M. A comparison of technical-economical aspects of primary access to the orebody via the vertical shaft and the decline. *Cuprum* **2014**, *4*, 73–92.
31. Takuski, S.; Maciejasz, Z.; Gajoch, K.; Piechota, S.; Wziętek, B. *Mining*; University script no. 723; AGH University of Science and Technology Publishing House: Kraków, Poland, 1980; p. 64. (In Polish)
32. Wagner, H. Deep Mining: A Rock Engineering Challenge. *Rock Mech. Rock Eng.* **2019**, *52*, 1417–1446. [CrossRef]
33. Zhu, S.; Wu, S.; Cheng, J.; Li, S.; Li, M. An Underground Air-Route Temperature Prediction Model for Ultra-Deep Coal Mines. *Minerals* **2015**, *5*, 527–545. [CrossRef]
34. Dreger, M.; Kędzior, S. Methane emissions against the background of natural and mining conditions in the Budryk and Pniówek mines in the Upper Silesian Coal Basin (Poland). *Environ. Earth Sci.* **2021**, *80*, 746. [CrossRef]
35. Wierzbicki, M.; Młynarczuk, M. Structural aspects of gas and dolomite outburst in Rudna copper mine, Poland. *Int. J. Rock Mech. Min. Sci.* **2013**, *57*, 113–118. [CrossRef]
36. Wierzbicki, M.; Skoczylas, N. The outburst risk as a function of the methane capacity and firmness of a coal seam. *Arch. Min. Sci.* **2014**, *59*, 1023–1031. [CrossRef]

energies

MDPI

Review

Review of Underground Mining Methods in World-Class Base Metal Deposits: Experiences from Poland and Chile

Krzysztof Skrzypkowski [1,*], René Gómez [2,3], Krzysztof Zagórski [4], Anna Zagórska [5] and Roberto Gómez-Espina [2]

1 Faculty of Civil Engineering and Resource Management, AGH University of Science and Technology, Mickiewicza 30 Av., 30-059 Kraków, Poland
2 Faculty of Engineering, Universidad de Concepción, Concepción 4030000, Chile
3 Advanced Mining Technology Center, Santiago 4070371, Chile
4 Faculty of Mechanical Engineering and Robotics, AGH University of Science and Technology, Mickiewicza 30 Av., 30-059 Kraków, Poland
5 Research Centre in Kraków, Institute of Geological Sciences, Polish Academy of Science, Senacka 1, 31-002 Kraków, Poland
* Correspondence: skrzypko@agh.edu.pl

Abstract: There are several massive deposits around the world with different geological characteristics. Thus, different mining methods and strategies are applied based on the particularity of each method and mine experience. Particularly, in this work, we review and summarize the underground exploitation of some world-class base metal deposits based on Poland and Chilean experiences. Here, the main geological and mining parameters of Poland and Chilean mines applied in massive deposits are reported and analyzed. In Poland, mainly room and pillar methods (and variants) have been applied in massive deposits. Here, back-filling is required to maintain the mine's stability due to the large deposit size and open areas. In Chile, the block caving method is commonly used in massive underground deposits where less development is required. Here, the cave is naturally filled with broken material and a large subsidence zone is generated. In this review, it has been observed that different underground methods and strategies can be effectively used in massive deposits. Some parameters that influenced the method selection are mainly related to rock mechanics, ore recovery and dilution, subsidence zone, extraction rate, and mining experience. Here, key mining variables and parameters such as productivity, support, and equipment, as well as various issues related to the world-class deposit are studied. Additionally, a comparison between both experiences is presented, highlighting the main geological and mining parameters. This study can be used as a reference to evaluate the different option of underground mining methods to be applied in future massive mine projects with similar geological characteristics.

Keywords: block caving; massive deposit; mine design; room and pillar; underground mining; world-class deposit

Citation: Skrzypkowski, K.; Gómez, R.; Zagórski, K.; Zagórska, A.; Gómez-Espina, R. Review of Underground Mining Methods in World-Class Base Metal Deposits: Experiences from Poland and Chile. *Energies* **2023**, *16*, 148. https://doi.org/10.3390/en16010148

Academic Editor: Sergey Zhironkin

Received: 14 November 2022
Revised: 9 December 2022
Accepted: 19 December 2022
Published: 23 December 2022

1. Introduction

The technical and economic progress in the exploitation of ore deposits consists of the introduction of modern technologies, mechanization, and automation of processes while meeting the requirements of occupational safety and protection of the deposit and the human environment. In the mining process, in terms of deposit protection, underground mining methods are of particular importance, for which the increase in extraction, labor efficiency, and lower extraction costs should be accompanied by a reduction in operating losses. Mass mining is frequently applied worldwide, related commonly to low-grade deposits, large depths, water tables, and high capital costs. In particular, in underground mining, these deposits are related to several challenges that have to do with the mining method, ventilation, and drainage requirements, mine support, and mine planning. Mass mining is defined as that with production greater than 10 kt/day or 3 Mt/year [1]. Mass-mining methods are commonly applied to low-grade, high-tonnage deposits, so the size and shape of the deposit are relevant characteristics in determining the suitability of a mass-mining deposit [2].

The characteristics of the rock mass are of vital importance to determine the most suitable mining method based on the stress deformations that occur according to the chosen method [3]. In addition, it is necessary to select the extractive method considering the operational costs [4]. To face these challenges, a methodology is proposed based on an algorithm to choose the optimal method for the development of deposits, considering the geological, technical, and operational factors of the deposit [5].

Ladinig et al. [2] mainly identify two typologies of deposits favorable for these exploitation techniques: narrow tabular deposits of large area extent, as might be the case with some coal deposits and thick tabular deposits and massive deposits, characterized by a large extension in all three spatial directions. In general, this last category adjusts quite well to the so-called world-class deposits. Although the typologies and formation processes of the deposit that this term encompasses are varied, these are characterized by being large-tonnage and volumetrically extensive deposits. These characteristics are more remarkable in the deposits of base metals compared to those of precious metals, presenting, in addition to higher tonnages, higher grades.

Singer [6] defined world-class mineral deposits as the upper 10% of all deposits in terms of the content of a given metal. The term world-class deposit is an informal term applied to ore deposits with an exceptionally large tonnage of economically recoverable ore [7]. This author points out that world-class deposits can be divided into giant and supergiant, with an ore metal content in a deposit/metal greater than 1×10^{11} t and 10^{12} t of average crust material, respectively. For example, in the case of gold, the resource in a giant deposit is equal to or greater than 100 t Au, and supergiants are deposits that contain equal to or greater than 1200 t [8]. For copper, supergiant deposits are defined as those with more than 24 Mt Cu, and giants as having more than 2 Mt Cu; 2400 t of silver, 1.7 Mt of zinc, or 1 Mt of lead are required to be considered a world-class deposit [6]. Although this definition and the limits it uses are relatively old, taking into account the advancement of exploration techniques and the continuous discovery of new large, mineralized ore bodies, these ranges are still used today to categorize deposits as world-class, for example in the works of [9–12], among others. Among the world-class deposits of base metals, there are different typologies according to their genesis, ores, lithologies, etc. Porphyry copper deposits stand out for their tonnage, these being the world's principal source of Cu and Mo [13]. Another important type of deposit that occasionally gives rise to world-class mineralization is a sediment-hosted stratiform copper deposit. There is a low number of known stratiform copper deposits, but with very attractive Cu and Ag grades and tonnages [14]. Other types of deposits such as the manto type, Sedimentary Exhalative deposits (SEDEX), Volcanogenic Massive Sulphide (VMS), iron oxide copper gold (IOCG), and Mississippi Valley Type (MVT) are also main sources of base metals that occasionally generate massive ore classifiable as world-class [15].

Different criteria can be used to define a world-class deposit. However, in this work, we focus on the underground mine activities applied in large deposits using Poland and Chilean large deposits as the example. A review of current underground mines within massive deposits is a useful tool to identify the main methodologies used nowadays that have shown good results and to identify the lessons learned in these environments. Therefore, the purpose of this study is to analyze different underground mines in Poland and Chile focusing on key parameters such as productivity, support, and equipment, as well as main issues.

2. Chilean and Polish World-Class Base Metal Deposits

In Chile, the most abundant deposits and the largest tonnages correspond to porphyry copper deposits. Porphyry deposits occur in arc-related settings of various ages throughout the world. However, giant systems are restricted to only a few mineral provinces and periods [16]. In accordance with reference [11] among the 31 largest porphyry copper deposits discovered to date, 14 fall under the supergiant class (Chuquicamata, El Teniente, Los Bronces, Escondida, Los Pelambres, Collahuasi, Pebble Copper, Safford, Morenci/Metcalf, Continental/Butte, Almalyk, Grasberg, Oyu Tolgoi, and Bingham) of which 6 are in Chile. The porphyry deposits are characterized by low-grade copper, gold, and/or molybdenum mineralization developed within and around a porphyritic intrusive complex, where vein stockworks and hydrothermal breccias are common [16].

The iron oxide copper gold deposits (IOCG) in north-central Chile form part of the Andean IOA-IOCG belt, which extends from immediately north of Santiago to north of Antofagasta. These IOCG deposits are commonly spatially associated with or hosted in faults that form part of the Atacama fault system [17]. Among all the deposits associated with this belt (El Espino, Montecristo, Mantoverde, etc.), the Candelaria–Punta del Cobre district is the most important due to its size. It is within the range of world-class deposits. The majority of the IOCG mineralization in the district is hosted in the upper part of the lower andesite unit and the overlying volcanic-sedimentary and dacite units, all within the Punta del Cobre Formation [17]. Mineralization is hosted in fault zones, breccias, and specific lithologies.

Another type of common deposit in Chile that can reach large tonnages, although they do not reach the magnitudes of porphyries, are the stratabound deposits, known as manto type. These deposits are the third source of Chilean copper production after the porphyry copper deposits and the IOCG deposits [18]. Similar deposits in North America are named "volcanic red-bed" [19]. Within this manto-type located in Chile are the deposits of Papomomo [20], Mantos Blancos, Cerro Negro, and El Soldado, among others. They are mainly Cu–Ag mineralized bodies that are hosted in Upper Jurassic to Lower Cretaceous volcano sedimentary sequences along the Chilean coast from center to north [21]. They originate in back-arc extensional basins associated with calc–alkaline volcanic belts in the continental crust on an active convergent margin. The manto-type deposits in Chile present very similar characteristics, with comparable geometries, mineralization controls, and mineral paragenesis. They are considered epigenetic, hydrothermal, or metamorphic fluid origins are suggested, and albite is the most common alteration. The main controls are lithologic (permeable strata), structural (extensional faults), or intrusion-related [22]. Ore appears in andesitic lavas and occasionally in volcanoclastic and sedimentary rocks, where it is commonly disseminated and associated with zones with organic matter. Pyrobitumen has been described in stratabound deposits in Chile, such as El Soldado [23] and Papomomo [22]. These manto-type Cu deposits tend to display relatively high grades with variable quantities of Ag and Au as a by-product [24].

Poland is a country with abundant mineral resources ranging from coal and copper to other raw materials, such as rock salt or zinc, and lead ores. Regarding copper, the most important source of this metal produced in the world in addition to porphyry copper deposits involves sediment-hosted stratiform copper deposits [25,26]. These copper deposits are an important, economically attractive, world-class mineral deposit type, traditionally represented by supergiants such as the Kupferschiefer of north-central Europe and the Copperbelt of Central Africa [14]. In the New Copper district, located on the Polish side of Kupferschiefer, resources were identified in 2018 totaling 32.62 Mt Cu and 97,938 kt Ag [27]. The Lubin, Polkowice-Sieroszowice and Rudna mines are in this district. The principal host rock in the Kupferschiefer deposit is shale, with 2 to 10% Cu content, and a thickness of up to 20 m [28]. Structurally it is located on a monocline dipping gently to the north-east (less than 12°).

The morphology and distribution of mineralization in these deposits are markedly different from that of the porphyries. Brown [14] affirms that sediment-hosted stratiform copper deposits tend to be high-tonnage deposits because of their wide lateral extents along preferred stratigraphic units, and their copper grades frequently surpass those of porphyry copper ores. In addition, they may contain significant amounts of other highly desirable metals such as silver and cobalt.

The Upper Silesian ore district in south-central Poland is an important producer of zinc, lead, and silver [28]. Five clusters of orebodies have been discovered: the eastern Olkusz area (Pomorzany and Olkusz), the southern Chrzanów area (Trzebionka mine), the western Bytom area, and two northern clusters in the Zawiercie and Klucze area [29]. These deposits have been classified as Mississippi Valley Type [22,29,30]. The ore forms replacements, cavity fillings, linings, veins, and mineralized breccias [31]. The lower and upper units of the sequence enclosing the ores are marly or argillaceous sediments. The main ore minerals include sphalerite, galena, pyrite, and marcasite, accompanied by the gangue minerals dolomite, calcite, barite, chalcedony, and quartz. According to Sass-Gustkiewicz and Kwiecinska [29], the host rock of the sulfide ores is a coarse-crystalline, ore-bearing dolomite that tends to occur in the form of extensive, roughly tabular bodies in Triassic sequences or bodies of various shapes in Devonian carbonates. The orebodies in Triassic rocks occupy various positions within the ore-bearing dolomite and are tabular, lenticular, or nestlike, except for a single, large, chimney-like body hosted in Roethian carbonates. The Devonian orebodies occur as vertical, sub-vertical, and/or horizontal forms. The vertical bodies contain abundant mineralized breccias, and higher in the sequence, grade into horizontally disposed ores in both the Paleozoic and Triassic rocks [32]. The total extractable metal content of the district is estimated at 30 Mt [28].

3. Chilean Study Cases

In this section, three Chilean underground mines that are currently in different stages are reviewed. These mines are exploited using caving methods (block and inclined). The caving methods are commonly applied in massive underground deposits, highlighting the use of block and sub-level caving. Block caving is an underground method commonly applied in Chile [33], used in important copper deposits such as the El Teniente mine, Salvador mine, Andina mine, and currently in transition from open pit mining, the Chuquicamata underground mine is also included [34]. The Chuquicamata underground mine transition to block cave mining is a consequence of the deepening of reserves [35], which also could occur in other Chilean massive deposits (actually open pit mines), such as in the Escondida mine, Collahuasi mine, Los Bronces mine, and Radomiro Tomic mine. Additionally, the Papomono mine is a new underground mine project that will be exploited by the Inclined Caving method [36].

Block caving is an underground method applied in massive ore deposits. This method relies on gravity and induced mining stresses to cave the orebody which is known as caved rock [37,38]. The caved or fragmented rocks are extracted using gravity by drawing them from drawpoints located at the production level. The current challenges of this method, as well as other underground methods, are related to deeper deposits, hard rock, high-stress fields, higher production requirements, and uncertainties in ground conditions [39–41].

3.1. Copper Exploitation in the Coquimbo Region

The Papomono mine is in the Coquimbo Region (north-central, Chile). The Deposit consists mainly of a sequence of andesite interbedded with pyroclastic rocks of the Quebrada Marquesa Formation, N 40° W strike, dipping 15° to the SW. The deposit is 450 m long and about 80 m wide.

Papomono mine reserves were calculated to be 2.68 Mt at 1.22% of recoverable Cu [36]. In Papomono, the favorable lithological horizons that served as fluid conduits were strongly affected by a pervasive to sparse albitic alteration [22]. The lithologies have a thickness that ranges from 2 to 50 m, up to 500 m strike extension and length along dip reaching 600 m. In addition to the most common stratiform geometry, these deposits can present lenses, breccia pipes, veins, or irregular shapes, sometimes all together in the same deposit [22].

3.1.1. Underground Mining Method

The underground mine has two main access points, one located at 1429 m.a.s.l. and another located at 1315 m.a.s.l. The selected mining method is Inclined Caving, which is currently under construction. The deposit inclination and poor rock strength are some of the main reasons for the selected underground method. The production level is divided into four sub-level spaced 11 m in height and 13 m in width, with extraction drifts of 4 × 4 m. Here, a total of 97 drawpoints were defined. A total of 5173 m of horizontal development and 182 m of vertical excavations (including ventilation and ore pass requirements) are required. Figure 1 shows an example of an inclined (front) caving design.

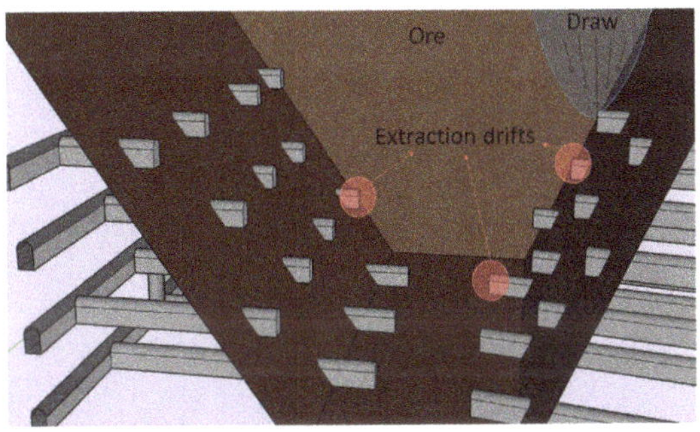

Figure 1. Example of inclined caving design, isometric view.

In this mine, there is no undercut level. Then, drawbells are drilled and blasted to start the caving. A small hydraulic radius of 9 m is estimated to start the caving process. This parameter was estimated using an MRMR of 20. In this method, the draw strategy is key to ensuring interaction between drawpoints, increase ore recovery, and delay dilution. Here, the drawpoint spacing is a critical parameter to ensure drawpoint interaction (flow zones) and production pillar stability.

3.1.2. Mining Support

The support requirements used in this project included the following: helical bolts (2.6 m length; 22 mm diameter), self-drilling bolts, electro-welded mesh (C-196) and brained steel mesh, internal anchoring with grouting, resin and expansion pin, shotcrete with and without fiber (150 mm), and reticulated frames. The support design varies depending on the rock mass quality of each sector. Figure 2 shows an example of the support design applied in a drift.

3.1.3. Equipment

In the current construction phase, there has been different mining equipment such as one dumper, two arm jumbos, one arm jumbo, one Simba, four mixer trucks, two LHD of 6 yd^3, two robotshots (equipment used for shotcrete projection), and one backhoe with a hammer for scaling. The extraction equipment requirement is currently under study. These probably will include one or two LHD due to the low production rate and the trucking system to move the ore out of the mine.

3.2. Copper and Molybdenum Exploitation in the Antofagasta Region

The Chuquicamata deposit is in the foothills of the Atacama Desert, west of the modern volcanic arc of the Andes Mountains. The deposit is of the porphyry copper type and is related to intrusive magmatism of Eocene—Oligocene age (41–31 Ma), a porphyry copper belt that extends for about 1400 km, from 18° S to 31° S [42,43]. The orebody geometry is 3000 m north–south, 350 to 800 m east–west and the drilling campaign suggested the existence of more than 900 m of mineable ore from the final pit bottom [44]. An important geological feature is the west fault observed in Figure 3. This fault separates the waste on the western side from the ore body.

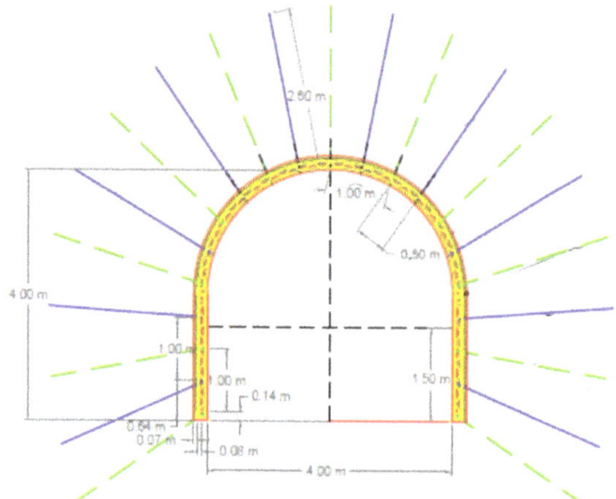

Figure 2. Example of drift support used in Papomono mine, including Helical bolts (blue), Shotcrete (red), and frames (yellow).

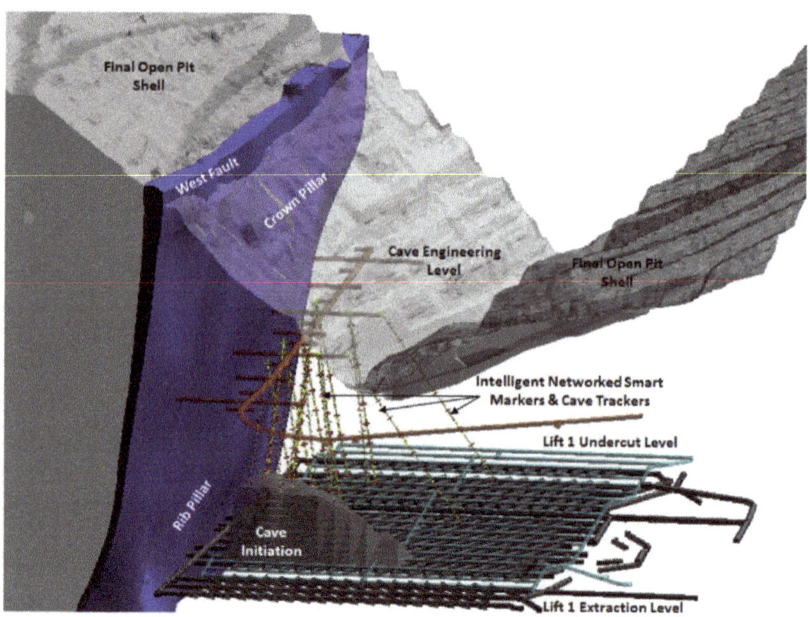

Figure 3. Chuquicamata transition project and west fault [44], isometric view.

3.2.1. Underground Mining Method

This project is a transition from an open pit project that was transitioned to an underground mine due to the open pit depth (1.1 km). The underground method selection was block caving with macroblocks applying the El Teniente extraction level layout [44,45]. The ore column heights are 216 m for level 1, 432 m for level 2, and 216 m for level 3 [44]. The ore extraction already started at the first macroblocks (in the center of level 1) while level 1 continues to be under development, at the time of this writing.

In this project, preconditioning through hydraulic fractures and confined blasting is applied in the rock mass over the production level to improve the caving process. The holes were drilled from some drawbells, the undercut level, and a hydro-fracturing level. The undercutting level is west–east oriented, with galleries of 4.2 m × 4.4 m. This size is defined by the driller equipment and ventilation ducts.

In Figure 4, the production level development is shown where three macroblocks are completed (N2N3, N1S1, and S2S3). Different extraction layout spacing is used based on the macroblocks (32 × 16 m and 32 × 20 m [46]). For example, two macroblocks (N2/N3) include the following: 300 confinement walls, 278 drawpoints, 1278 Norwegian frames, 15 IE frames, 9689 m of horizontal development at the production level, and 4694 m of development at the undercut level. The production drifts are 5.0 × 4.5 m, by applying the El Teniente layout.

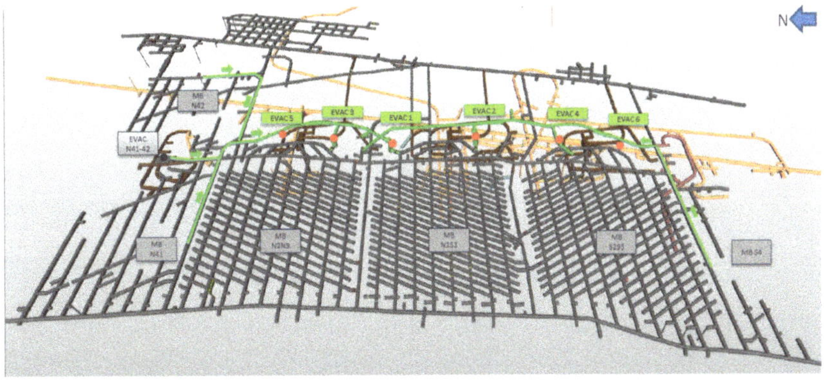

Figure 4. Macroblocks development in level 1, Chuquicamata underground mine [47].

3.2.2. Mining Support

The supporting element and structures used in this project are helical bolts, cable bolts, shotcrete, steel frames (H type), and confinement walls. Confinement walls are used to reinforce pillars, which consist of a group of concrete blocks reinforced with steel bars with the purpose of maintaining the pillar shape and assisting with the resistance of load above the production level [46]. Figure 5 shows examples of support construction at the production level (level 1).

Figure 5. Production and extraction drift support constructions [47].

3.2.3. Equipment

Construction equipment used includes Jumbo and Simba drillers, backhoes, excavators, blind hole, raise borer, loaders, scoops (LHD), dumpers, mixer trucks, robotshot, trucks of explosives, and forklifts. In production, LHD of 7 yd^3 that dump in a primary crusher connected to an ore pass are currently being used.

3.3. Copper and Molybdenum Exploitation in the O'Higgins Region

The El Teniente mine is one of the largest underground copper mines in the world with more than 3000 km of drifts. It is in the Andes range (Farellones formation) in the central zone of Chile, about 70 km from Santiago, Chile. El Teniente is a porphyry Cu–Mo deposit, anomalous in size, with over 94 million tonnes (Mt) of contained fine copper, where the principal host rocks are andesites [48]. The primary copper ore is very competent and massive, with almost no open discontinuities [49]. The mine is subdivided into smaller production units that are located around a sterile pipe. The deposit is formed by three main lithologies called El Teniente Mafic Complex (CMET), Felsic Complex (SC), and Brecha Braden Complex (Pipe). Figure 6 shows the sterile pipe (green) around which different exploitation mines are located.

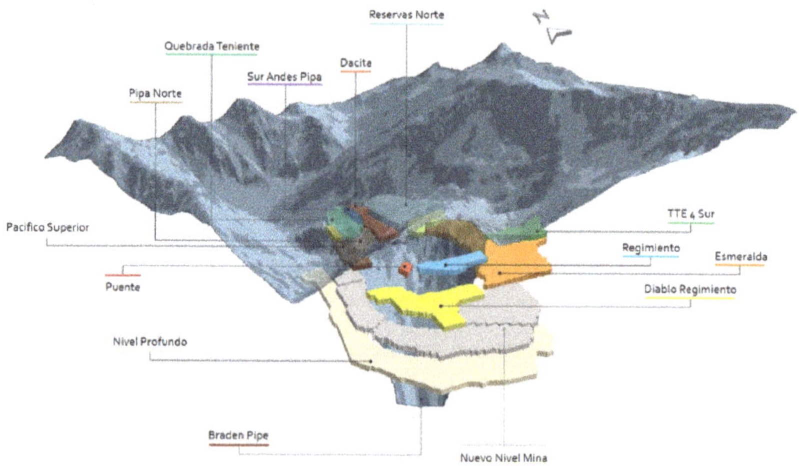

Figure 6. Location of different exploitation levels around Pipa Braden [50].

3.3.1. Underground Mining Method

The underground mines or sectors are exploited using the panel caving method with conventional undercutting and advanced undercutting using crinkle cut, among other techniques, to initiate the caving. The ore is collected at drawpoints by LHDs, which dump the ore in ore passes to an intermediate transport level operated by trucks system. El Teniente also has an open pit operation called Rajo Sur, located between 2730 and 3240 m.a.s.l. Nowadays, the new mine level (Figure 6 bottom) has been constructed and divided into different mines or sectors. In the El Teniente mine, there are different extraction layouts geometries, and dimensions. However, the El Teniente layout type is currently used in all sectors.

Figure 7 shows an example of different levels required to apply a panel caving method in a large deposit. The first level is used to blast the deposit base and start the caving. At the production level, the ore is drawn using LHD that dumps the material to ore passes connected with the haulage level. From the undercutting and production levels, preconditioning in the rock mass is commonly applied. At the haulage level, the ore is commonly transported using underground mining trucks. The ventilation level consists of a layout of galleries connected by chimneys to the other levels.

3.3.2. Mining Support

During the construction of the (originally called) new mine level in El Teniente, several rock bursts occurred due to the competent rock and high-stress field. Thus, a precondition was applied during drifts development using hydraulic fracturing and distress blasting to decrease stress on the working front. In addition, a strong fortification was considered, highlighting a minor use of shotcrete (due to the fall of concrete blocks) and an intensive used of bolts, cable bolts, double galvanized mesh, shotcrete with fiber, and confining walls in production level pillars. Additionally, the use of mechanized and tele-commanded equipment increased during development. These measures increased the safety during development but also the cycle time.

Figure 8 shows an example of the support used during the new mine level construction. The bolt layout is shown in blue, while the cable bolt layout is shown in pink. Bolts and cable bolt adhered to rock mass with concrete and additives. Two shotcrete layers are applied by dividing by two meshes. The bolts are pressed between both shotcrete layers.

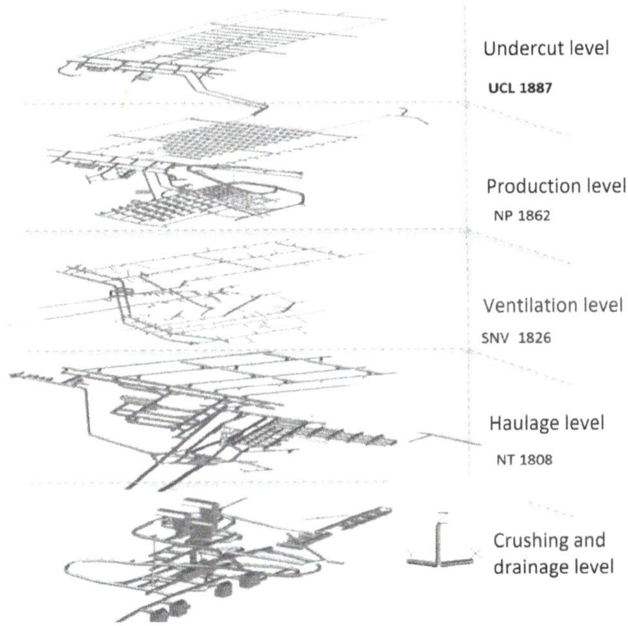

Figure 7. Isometric view of the initial new level mine project, El Teniente [51].

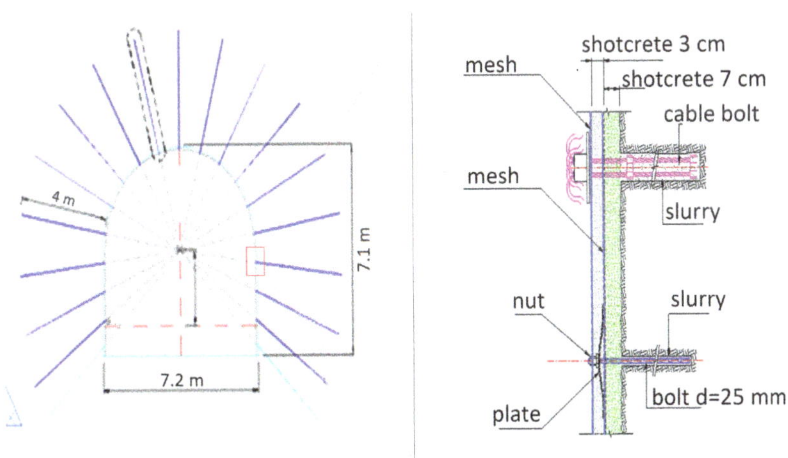

Figure 8. Example of support used in high seismic risk [52].

3.3.3. Equipment

Diverse mining equipment is used in this mine during construction and ore extraction. At the production level, LHD of 7 yd^3 (~9.7 tonnes) CAT R1600H are commonly used, with a productivity of around of 3000 tpd. In addition, other LHD models are used such as the Sandvik lh410 and lh517 models, and the Caterpillar R1600H model. The haulage level (Teniente 7) is an intermediate haulage level where underground trucks are used. These trucks have commonly 40 t or 80 t of capacity (Volvo and Supra 012H model, respectively). Then, the ore is transported by ore pass to the railway located in El Teniente 8 level. In the new levels, conveyor belt will also be used in the new haulage level.

4. Poland Study Cases

Deposits with a thickness of more than 7 m are considered thick, this applies to deposits located in Poland. Currently, the exploitation of thick copper ore deposits is carried out only in the Legnica–Głogów Copper district at Rudna mine owned by KGHM Polska Miedź S.A. A characteristic feature of the deposit is its horizontal deposition with a small angle of inclination, about 8°. Exploitation is carried out at a depth of more than 1200 m in the sediment-hosted stratiform copper deposits, Kupferschiefer-type.

The article also presents the methods of exploitation of thick zinc and lead ore deposits, which were used in the Trzebionka mine (Chrzanów region) until 2009 and in the Olkusz mine (Olkusz region) until 2020. The deposits in the form of nests and lenses were deposited at shallow depths of 80 to 200 m. Both mines are already closed due to resource depletion. Nevertheless, the presented mining methods indicate their evolution, especially in terms of the future exploitation of zinc and lead ore deposits in the Zawiercie region, which were identified by drilling holes from the ground surface and for which it is possible to protect buildings with the use of hydraulic backfilling.

Due to the fact that copper ore deposits occur at a depth of less than 1000 m, they are accessible only by means of vertical shafts. On the other hand, deposits of zinc and lead ores lying at a small depth (about 100 m) are accessible by means of a decline and vertical shafts. Regardless of the access method, one excavation is used to supply fresh air and the other to discharge used air. The output from the exploitation field was transported to the surface by means of a vertical shaft. An exemplary transport route is shown in Figure 9.

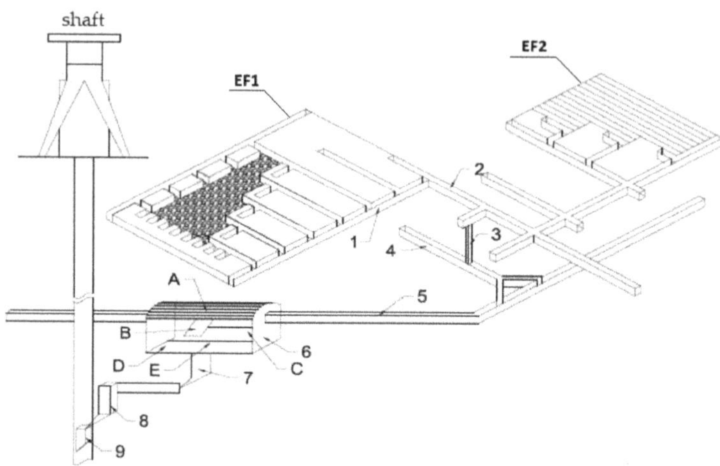

Figure 9. The way of transporting the excavated material from the exploitation field to the surface: EF1—exploitation field no 1; EF2—exploitation field no 2; 1—haulage room; 2—transportation drift; 3—discharge (chute) small shaft; 4—field rock drift; 5—main rock drift; 6—dump's room; 7—storage reservoir; 8—measure reservoir; 9—skip hoist; A—passable dump; B—inclined plane; C—steel membered conveyor; D—jaw crusher; E—belt conveyors.

4.1. Zinc and Lead Exploitation in the Chrzanów Region

In Poland, in the Chrzanów region, until 2009, various variants of room and pillar mining methods were used to select thick deposits, with the mining of the deposit with rooms to the full thickness and the liquidation of the room with the use of hydraulic backfilling. The contoured area with exploration workings was divided into two mining fields (Figure 10).

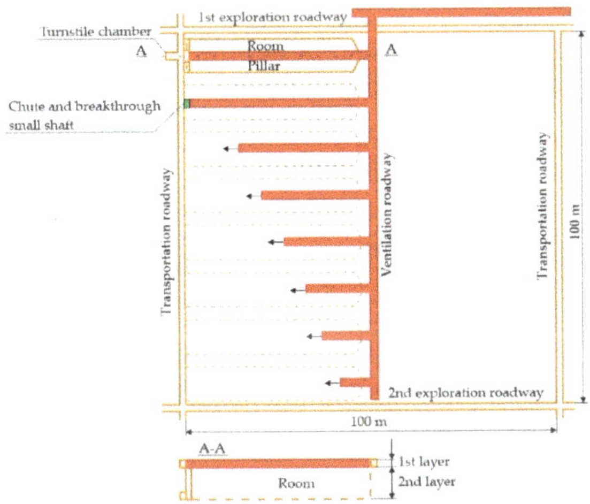

Figure 10. Scheme of preparation of a thick orebody for exploitation.

In the mining field, rooms with a width of 10 m to 16 m and a length of up to 70 m were designed, as well as inter-room pillars with a width of 3 m to 4 m. In the deposit up to 10 m thick, the deposit with the contour of the designed room was selected by dividing it into two layers. First, the 3 m thick roofing layer was exploited, and then the bottom layer was as well. The preparation of the rooms for exploitation consisted of the construction of a transportation roadway on the bottom of the deposit, located under the exploration roadway, over-room roadways located in the axes of the designed rooms under the roof of the deposit, and chute and breakthrough small shafts connecting the transporting roadway with the over-room roadway (Figure 11).

In the first stage of exploitation, the over-room roadway was widened to the width of the designed room. The deposit in the first layer was made with short blast holes. The output was placed in the chute and breakthrough small shaft. The roof of the workings was secured along the entire length with the rock bolt support with expansion heads or with an application of resin cartridges. The length of the bolt support was at least 1.6 m. The distance between the bolts did not exceed 1.2 m. The mining of the second layer began with widening the small shaft chute to the width of the future room. The widening was performed with short blast holes, leaving a 3 m wide retaining barrier pillar from the side of the transporting roadway. Then the bottom layer was selected with the use of long blast holes drilled from the top layer. During the exploitation of the bottom layer, 4 m wide and 4 m deep cut-rooms were made in the sidewall of the room from the side of the orebody. The distance between the cut-rooms was 8 m. The cut-rooms were crossed by the designed inter-room pillar, reducing operational losses. Finally, there were unselected rectangular pillars with dimensions of 4 × 8 m between the rooms.

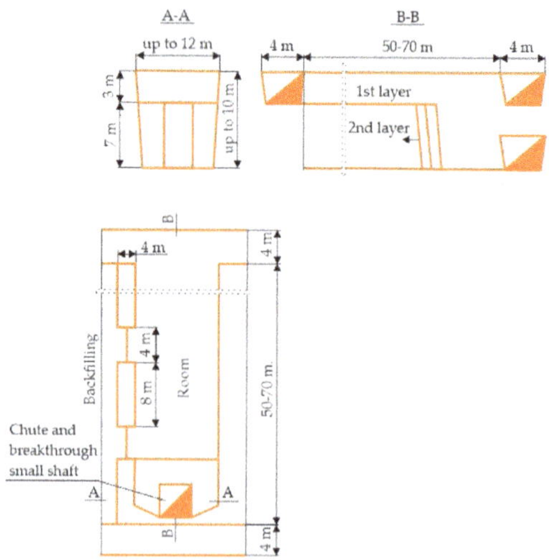

Figure 11. Room and pillar mining method in an orebody with a thickness of up to 10 m.

Exploitation works were carried out simultaneously in several rooms, usually in three rooms: the first room—excavating the roof layer; the second room—selecting the bottom layer and cut-rooms in the pillar; the third room—building backfilling wooden dams in the cut-rooms in the pillar. The rooms were shaped into a trapezoid as a result of the deviation of the sidewalls towards the orebody at an angle of about 5°. The deposit was mined with the use of explosives such as dynamite, ammonite, and saletrol. The haulage of the excavated material was carried out with the use of loading and hauling trucks. Wooden dams and backfilling cloth were used to enclose the room. The room was supplemented in two stages. In the first stage, the room was extended to half its height. Complementing the filling to the roof took place while filling the next room. In the ore deposits up to 15 m thick, layers up to 6 m thick were separated in the room, which were mined from the room to the bottom of the room (in the case of stable rooms) or in the direction from the floor to the roof (in the case of less stable roofs). In the case of mining layers in the top-down direction, the filling of the rooms took place after selecting the orebody in the room [53]. In the case of the reverse order of selecting the layers, after mining the layer, backfilling was performed, and the next higher layer was selected along the backfilling. The scheme of mining a deposit in a 15 m high room using three layers in the top-down direction is shown in Figure 12.

4.2. Zinc and Lead Exploitation in the Olkusz Region

Room and pillar mining methods with the bottom discharge of the output was used in the Olkusz region in the 1970s to extract a thick deposit of zinc and lead ores with a thickness of 9 m to 15 m [54]. The exploitation was conducted in two stages. In the first stage, rooms with a width of 10 m to 12 m were selected in the mining field and continuous pillars 7 m wide were left between the rooms (Figure 13). In the second stage, the inter-room pillars in the mining field were mined. It was assumed that after the simultaneous mining of the pillars in the selection field, the excavated material would be ejected to the rooms and the roof rocks would fall. The material thrown into the rooms was to be recovered through discharge openings made under the rooms. The elementary field was contoured with exploration roadways in a 100 × 100 m grid, which was divided into two fields, and the rooms were perpendicular to the longer edge of the field. Preparatory works in the field were performed on two or three levels: under the orebody, under the roof of the orebody and above the orebody. Above the roof of the orebody (about 5 m), over-pillar roadways were made along the axis of the designed pillars. They were intended for drilling blast holes in the pillars and mining the pillars between the rooms. This concept was quickly abandoned and the pillar roadways were replaced with pillar roadways made at the same level as the haulage roadways. Under the roof of the orebody, over-room roadways were driven, located in the axis of the designed rooms or the corner of the room. These roadways were connected by a system of preparatory workings with a ventilation level.

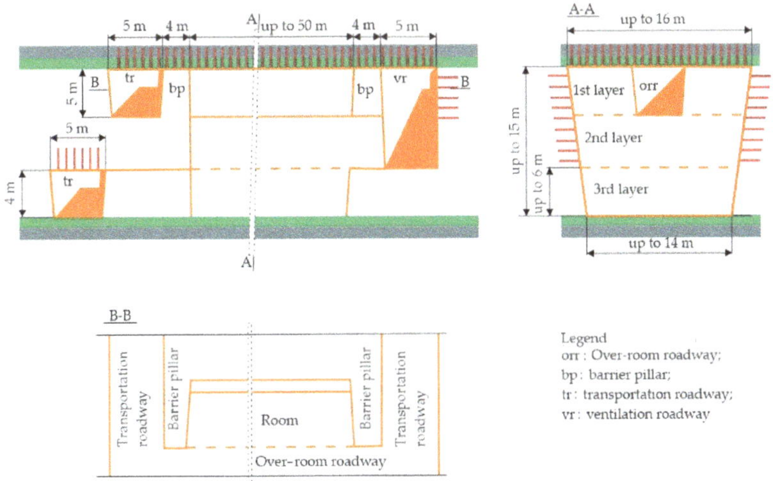

Figure 12. Room and pillar mining method in an orebody with a thickness of more than 10 m.

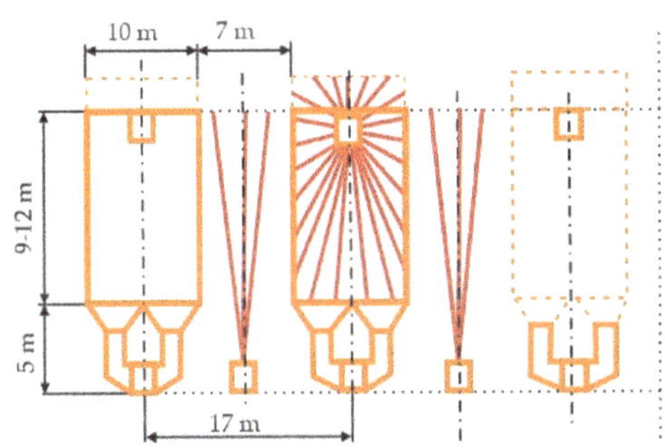

Figure 13. Room and pillar mining method with bottom discharge. Blasting holes in red.

They were used to drill long cutting holes and discharge the used air. About 5.5 m below the bottom of the orebody, there was a transporting level. At this level, in the axes of the future rooms, haulage roadways were driven, and in the axes of the pillars, roadways for excavating the pillars were also driven. The haulage roadways were driven at intervals of about 17 m. From the haulage roadways, the excavations were driven and used to discharge the output. These were funnels made at intervals of 5.5 m, arranged in a checkerboard pattern. The diameter of the funnels in the upper part was 5 m and they undercut the bottom of the room on about 70% of the surface. A breakthrough small shaft was constructed at the beginning of each room. It was an extension of the first funnel and was connected with the over-room roadway using a short cut-off roadway, made perpendicularly along the axis of the room.

In the haulage roadways, scraper loaders were installed in order to place the output pouring out of the openings on the transporting roadway. The mining in the room was started until the breakthrough was completed, i.e., the small shaft was extended to the full width of the room. The breakthrough was made with long blast holes and it was a compensation space in the case of the material blasted from the first advance (the distance between the fans of the holes, was 2.5 m). The diameter of the blast hole in the fan was 70 mm, and the length of the holes was from 6 m to 15 m. The total length of the holes in the fan was from 96 m to 115 m.

For the blasting, saletrol was used and injected into the holes using pneumatic injectors. The maximum mass of the explosive in the fan was 350 kg. One or two fans were blasted simultaneously. The output rolled down to the discharged chutes, which undercut the bottom of the room. The first stage was related to the mining of the orebody in the rooms. After mining the orebody in the selection field, the mining of the inter-room pillars was started, which was the second stage. For this purpose, one row of long blast holes was drilled from the over-pillar roadways and all the pillars were mined at the same time. After blasting, only part of the excavated material was thrown toward the rooms. Most of the output remained between the rooms and it could not be recovered through the discharge openings made for the rooms.

In addition, the planned caving of the roof rocks did not take place. Maintaining such a large underground void was dangerous. The aforementioned technological defects, compared to the planned design assumptions, showed that the operating losses in this method were very high—up to 40% and due to the second stage of operation, the system was uneconomical. In the following fields, various attempts were made to improve the technology in the second stage of operation. In some fields, all the pillars were abandoned and only every second pillar in the field was mined. The attempts at improvement did not bring the expected economic effects, which prompted the discontinuation of this exploitation mining method.

Later, successful attempts were made with the help of a sublevel caving method. This method was used in the Olkusz region in Poland to extract zinc and lead ore deposits, the roof of which was disturbed by the earlier exploitation of the higher layers, and to extract the remains of the ore [55]. The sublevel caving method could also be used in a deposit that had not been affected by previous mining operations when the roof rocks showed a high tendency to collapse. The method could not be used in the case of watered layers in the overburden or the upper layers. In the case of a thick orebody, 7 m thick layers were separated in horizontal planes, and 8 m wide blocks in vertical planes (Figure 14).

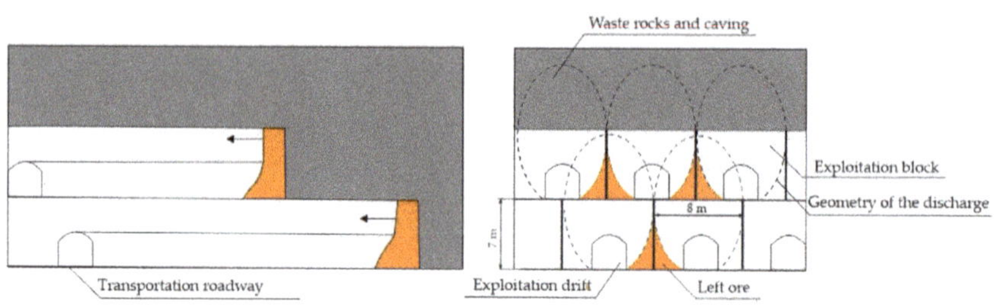

Figure 14. Sublevel caving used in the Olkusz–Pomorzany mine in Poland.

The layers were mined with blocks/stopes exploited from top to bottom. In the axis of each block, mining drifts with dimensions of 3 × 3 m were driven on the floor of the layer, and they were protected using rock bolt support. In the lower layers, the blocks were moved by half their length between the axes of the drifts, which ensured the recovery of the ore left in the leftovers between the mining drifts in the upper layer. The length of the exploitation blocks did not exceed 100 m.

The blast holes were drilled from the exploitation drift in a fan arrangement. The diameter of the holes ranged from 50 mm to 70 mm, and the distance between the fans ranged from 1 m to 1.8 m. About 12 blast holes were drilled in one fan. The ore was blasted with saletrol, which was loaded into the holes using a pneumatic injector. The blasted material rolled down to the roadways and was loaded and set aside with self-propelled machines—loaders. The main advantages of the method were the ability to safely select the deposit under the roof of rocks damaged by previous mining; the possibility of recovering the ore left in the collapsed space of the upper layers; the ability to safely extract the ore left in the inter-room pillars. The disadvantages of the method were the dilution of the ore with rocks permeating from the caving space; enormous devastation of the surface, as the exploitation was carried out at shallow depths, about 60 m below the ground surface. The above-mentioned disadvantages resulted in abandoning the exploitation of zinc and lead ore deposits using the sublevel caving method. From 1990 to 2020, thick deposits were mined from the bottom up, with horizontal layers 5–6 m high. In the first layer, a room and pillar method was always used, while the subsequent layers were exploited with a strip method with the liquidation of post-mining space with the use of hydraulic backfilling [56,57].

4.3. *Copper Exploitation in the Legnica–Głogów Copper District*

Diversified strength parameters of rocks in the mining gate led to the development of a two-layer method of extracting thick copper ore deposits in the Legnica–Głogów Copper district in Poland, with the use of hydraulic backfilling. In the thick orebody, two layers are separated—the roofing layer with a thickness of 4 m to 5 m and the bottom layer with the remaining thickness of the deposit. The preparatory workings for both layers are made in the roofing layer, in which the roof of the workings is made of carbonate layers with high strength parameters. This enables the stability of these excavations to be maintained despite the large depth of exploitation and the occurrence of induced seismicity. Attempts to make and maintain the preparatory workings in the bottom layer, which are usually sandstones with low strength parameters, failed. Both layers are mined at a certain time in advance of each other. One preparatory excavation made in the bottom layer is the inclined drift (ramp). This inclined drift connects the preparatory workings in the roofing layer with the bottom of the rooms in the lower layer. It is used to haul the output from the rooms of the lower layer [58].

The room and pillar method with a two-layer deposit extraction (proper name Rudna 5) was used in an orebody with a thickness of up to 15 m and an inclination angle of up to 8°. The preparatory works consisted of the construction of double roadways and inclined drifts in the roof layer separating the exploitation fields with the following dimensions: field width from 150 m to 200 m; the length of the field was about 800 m, depending on the geological and mining conditions. At the same time, mining was carried out in at least two fields of exploitation (Figure 15).

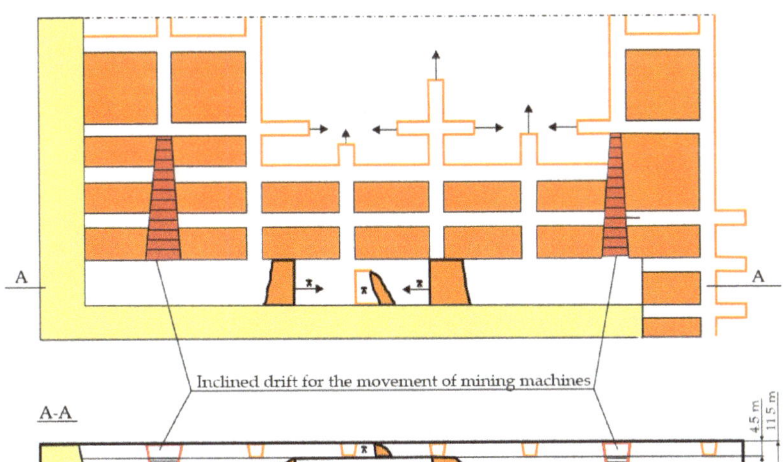

Figure 15. Room and pillar mining method using two-layer orebody exploitation.

First, the roof layer with a thickness of about 4.5 m was selected [59]. It included dolomites, shales, and sandstones. In this layer, 28 × 10.5 m pillars were separated by room workings. The longer side of the pillar was parallel to the goal line. The rooms were about 7 m wide at the roof, and their sidewalls were inclined towards the whole body at an angle of about 10°. There were three rows of pillars ahead of time. Preparation for the exploitation of the bottom layer consisted of making inclined drifts from the top layer, giving them a slope that enabled the movement of mobile machines (from 5° to 8°). Flooring down the inclined drifts was performed continuously, as the work front progresses. The selection of the orebody in both layers took place simultaneously with the maintenance of appropriate advance. In the top layer, the last row of pillars was selected and the maximum width of the exposed roof was 17.5 m.

The selection of pillars proceeded on both sides from the inclined drifts toward the center of the field. The mining of the bottom layer followed the selected pillars. A distance of 12 m to 36 m was maintained between the face in both layers due to the blasting works and the movement of machines. The lower layer was cut with long blast holes drilled from the upper layer. After mining both layers, a room about 15 m high, 17.5 m wide at the roof, and 150 m to 200 m long was created. The drilling of blast holes and loading and hauling of the excavated material were performed with typical self-propelled machines. Full column resin bolt support (2.6 m) was used, installed in a 1 × 1 m net bolting. In the upper layer, the workings had a trapezoidal shape, while the inclined drifts and room workings oval shape. The rooms in the upper layer were ventilated until they were connected using duct ventilation. In the perpendicular rooms of the upper layer, backfilling wooden dams were built at the edge of the room. In the inclined drifts for the movement of the mining machines, wooden dams were built on the sand, after the room was backfilled up to the floor of the upper layer (Figure 16). When designing the excavation at a distance of 15 m from the stream of air flow generated by the main fan, they are ventilated by diffusion. Excavations above 15 m are ventilated with separate fans. The amount of fresh air introduced must ensure the appropriate composition, such as O_2—min. 19%; CO_2—max. 1.0% (10,000 ppm); CO—max. 0.0026% (26 ppm); H_2S—max. 0.0007% (7 ppm); SO_2—max. 0.000075% (0.75 ppm); NO_2—max. 0.00026% (2.6 ppm).

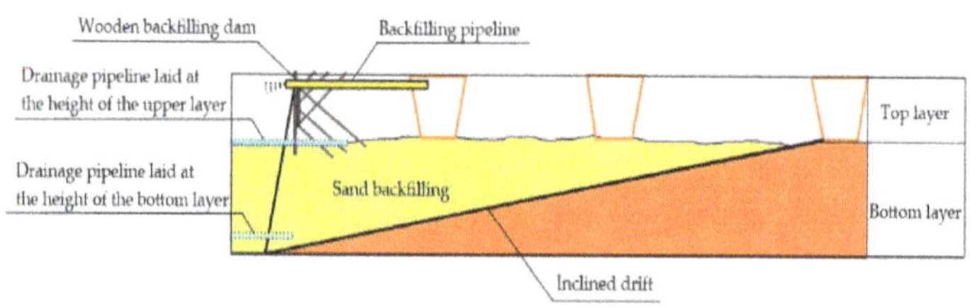

Figure 16. Backfilling in the inclined drift.

Mining of copper ore deposits in the area of LGOM is carried out in conditions of high seismic activity of the rock mass. Therefore, in the blasting metric, apart from breakage, mining, and contouring holes, destress relief holes are often made, the length of which are usually twice as long as the mining holes (Figure 17). Moreover, the diameter of the destress blasting holes is almost twice as large. For example, for 30 holes, the explosive load is 84 kg, and with destress holes with a diameter of 0.089 m, the maximum explosive load is 116 kg.

The backfilling water was discharged to the field sedimentation reservoirs through pipe channels (worn pipes with a diameter of 500 mm) laid in the backfilling at the level of the bottom of the upper layer and through perforated tubing and wrapped with a backfilling cloth laid on the bottom of the lower layer. Pipe channels were arranged in each upper layer room and the end layer rooms of the bottom layer. The backfilling pipeline was installed in the rooms of the upper layer. After backfilling the post-exploitation area, the sand was removed from the inclined drifts (picked up with loading trucks) and the cycle was repeated in the field. Exploitation was performed simultaneously in at least two fields of exploitation. The times of selection, liquidation, and preparatory works in both fields were selected in such a way as to ensure the continuity of the mining process.

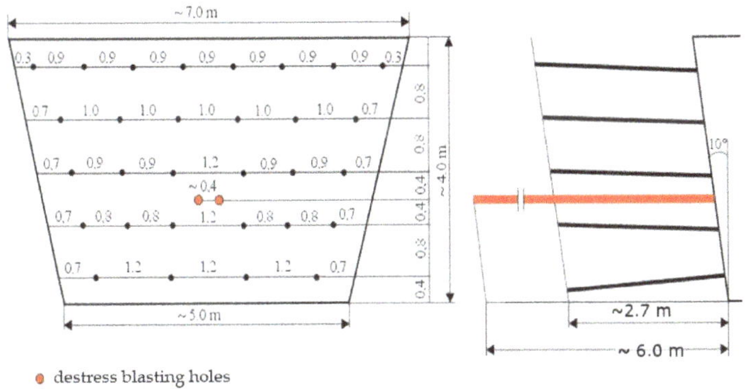

Figure 17. Arrangement of blasting holes in the ore forehead.

The backfilling that is used in mining excavations is either dry or hydraulic. In the case of dry backfilling, the rock material comes from preparatory and mining excavations and is delivered to the space selected by mine loaders. The hydraulic backfilling, on the other hand, is transported using pipelines with a diameter of 0.150 m and 0.185 m. Backfilling material may be non-combustible and non-toxic solids. Filling material can also be used: sand; gravel; waste: slags; gang; industrial waste. It is worth noting that the maximum dimension of grains should not be greater than 0.06 m; the content of particles of size less than 0.1 mm are at most 20%; and the compressibility at pressure 15 MPa is at most 15%.

Water flowing into mine workings is discharged to the surface through drainage devices and special workings included in the mine drainage. In the vast majority of cases, water runoff occurs automatically, by gravity, through transport headings in the direction from the mining fields to the main mine drainage system. Due to the drainage of water, it is advisable to drive a drift and roadways with an inclination of up to 3‰ towards the shafts, and in the case of more polluted water up to 5‰. The whole of the main drainage workings (Figure 18) consists of water reservoirs at the pump chamber; drifts and inclined drifts supplying water to reservoirs; pump chambers; suction wells and water channels—connecting water reservoirs with water wells; an access roadway connecting the pump chamber with the bottom shaft; a pipe channel connecting the pump chamber with the shaft and used to accommodate water pipelines and to ventilate the chamber.

Water reservoirs are mining excavations in the vicinity of the pump chamber of the main drainage, designed to collect underground water from the natural inflow and from the hydraulic backfilling. In Poland, the capacity of active water reservoirs should be sufficient for at least a 12-h supply of water from natural inflow and backfilling to the excavations, where main drainage systems are capable of discharging the highest daily water inflow in less than 20 h. The pump chamber is a mining excavation with built-in pump units for pumping water directly to the surface or to a higher level of the mine. The chamber is located parallel to water reservoirs and most often between water reservoirs. The size of the pump chamber depends primarily on the number of installed pump units. In Poland, pump chambers of the main drainage at levels with water inflow over 1 m^3/min are equipped with at least three pumps. If the pump chamber is equipped with pump units, the number of pumps in these units shall be at least: five pumps with two pumps working in a group; seven pumps with three pumps working in a group. The following numbers of pumps should be in full operational readiness at all times in the pump chamber of the main drainage: two pumps, in chambers equipped with three pumps; two sets of pumps, in chambers equipped with sets of pumps. The devices, together with the main drainage systems, should enable the discharge of the highest daily water supply in no more than 20 h.

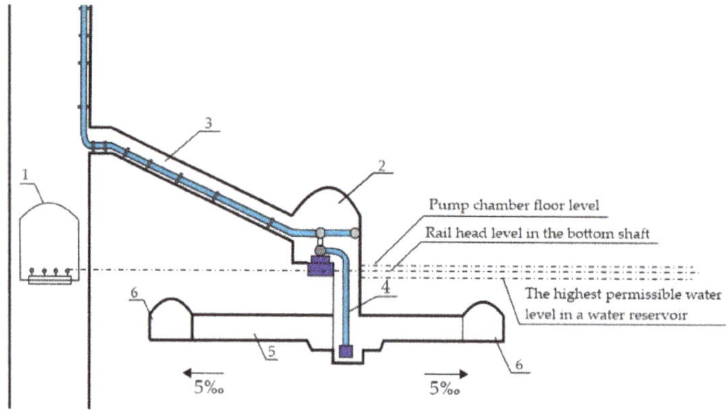

Figure 18. Main drainage workings: 1—bottom shaft; 2—pump chamber; 3—pipe channel; 4—pump suction well; 5—water channels; 6—water reservoirs.

During the 1990s, due to the increasing seismic risk and rock bursts, attempts were made to select a thick deposit with the use of pillars operating in the post-destruction phase, located perpendicular to the operational front line. The positive results of industrial research confirmed the legitimacy of the introduced geometric changes and made it possible to develop a one-stage method with the liquidation of the space selected with a hydraulic filling (the proper name of the mining system "Rudna 7") [60]. With this method, ore deposits with a thickness of up to 18 m and an inclination angle of up to 8° can be mined. The roof conditions should allow the use of the rock bolt support. It is recommended that this method be used when rocks with increased strength parameters and high rock burst tendency are present in the exploitation gate. In the first phase of excavation works in the mining field, a 6 m to 7 m wide room is cut into the upper layer, up to 5 m high, separating technological pillars with dimensions of 7–9 × 16–38 m (Figure 19).

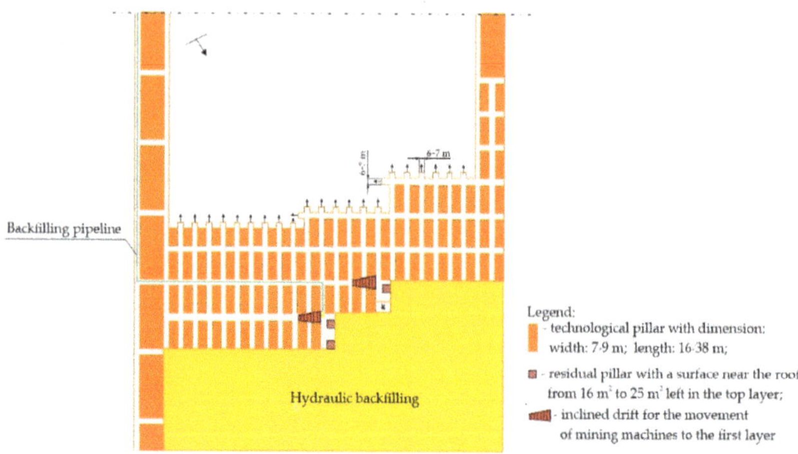

Figure 19. A two-layer room and pillar mining method with a hydraulic backfill in the case of the exploitation gate conditions made of rocks strongly prone to rock bursts.

There are no less than two rows of technological pillars in the front. The width of the rooms in the upper layer may be increased to 10 m as a result of induced seismicity and falling rocks from the sidewalls. Before mining the bottom layer, a pillar adjacent to the backfill is partially mined, leaving residual pillars with a roof area ranging from 16 m^2 to 25 m^2. Under favorable roof conditions, it is permissible to select the pillar completely, without leaving any residual pillars. To the bottom of the layer in an elementary field with one room and one technological pillar, there are inclined drifts with an inclination from 5° to 8° that allow the movement of mobile machines. The inclined drifts are used in successive elementary fields, but this requires constant flooring as the front of works progresses. Then the bottom layer in the field is mined. The orebody is processed with explosives. Blasting holes in the lower layer are drilled from the floor of the lower layer (horizontal holes) or the floor of the upper layer (vertical holes).

The roofs of the excavations and the sidewalls are secured using rock bolt support. After the bottom layer is mined in the elementary field, the filling dams are built in the room leading to the field and the inclined drift (after partial backfilling), and the post-exploitation space is filled with hydraulic backfill. The backfill water is discharged through pipe collectors or workings fenced off in the filling to field settlers. The mining of the deposit in the next field may take place after the entire scope of liquidation works in the previous field has been completed. All these works are performed with the use of self-propelled machines: blast holes are drilled with self-propelled drill trucks; loading explosives such as saletrol or emulsion is carried out with blasting vehicles, while the loading of the cartridge materials is carried out manually; the output is loaded and set aside from the fronts with the use of loading and hauling trucks or cooperating with box haulage vehicles; the filling material brought in during the selection works is loaded and put away into the post-exploitation voids using loading and hauling trucks; mechanical and resin rock bolts with a length of 1.6 m to 2.6 m are made with the use of self-propelled machines; ripping off the roof and sidewalls is performed mechanically with the use of trucks equipped with cutters.

5. Discussion

Mining in massive deposits has various challenges, mainly related to great depth, hard rock, in-rush related to water, and mine stability. Additionally, the mine development in these environments requires the planning and optimization of several activities to improve the construction times, mining times, and support requirements. This work has summarized different underground mining methods applied to massive deposits located in Chile and Poland.

Underground mining of ore deposits in Chile and Poland is diversified due to the geological conditions and the exploitation methods used. For the ore deposits located in the Legnica–Głogów Copper district in Poland, copper and silver are mined together, as well as zinc and lead (Chrzanow and Olkuski regions). A comparison of the basic geological and mining parameters is presented in Table 1. Currently, the only exploitation of ore deposits in Poland is carried out in the Legnica–Głogów Copper district. Mining of Cu–Ag ores in this region in 2021 amounted to 30 million tonnes of ore with a content of 1.48% Cu and 50.73 g/t Ag, containing 0.443 million tonnes of metallic copper and 1.522 million tonnes of silver. Exploitation of zinc and lead ore deposits in the Chrzanów region (Trzebionka mine) and in the Olkusz region (Olkusz–Pomorzany mine) was completed in 2009 and 2020, respectively. The annual output of zinc and lead ores in 2009 amounted to 2.349 million tonnes, including 0.089 million tonnes of zinc and 0.034 million tonnes of lead. In 2020, it was 1.435 million tonnes, including 0.043 million tonnes of zinc and 0.018 million tonnes of lead. The content of zinc and lead in the ore was 2.6% and 1.3%, respectively.

On the other hand, the deposit in Chine are in different stages. The deposit located in the Coquimbo region, is under development, and is being designed to mine 2000 tpd of copper ore with a content of 1.22% Cu containing 2680 million tonnes of cooper. The deposit located in the Antofagasta region is an open pit to underground transition project (Table 1). Mining of Cu–Mo ores in this region has 1700 million tonnes of ore with a content of 0.7% Cu and 450 ppm Mo. The deposit located in the O'Higgins region is a large block caving project (Table 1) with more than 4500 km of drifts mined from 1905 to date. Mining of Cu–Mo ores in this region has 2020 million tonnes of ore with a content of 0.86% Cu, containing several underground exploitation areas including an open pit.

According to reference [2], mass mining exploitation methods must consider the mechanical behavior of the rocks and the depth of the deposits. Caving is more difficult to initiate in competent rock masses and cave progression is typically slower. Furthermore, the rock mechanics aspects in deep cave mines are more relevant due to the increasing primary stress magnitudes. There is no common characteristic regarding the quality of the rock in world-class deposits since their typology and genesis are varied and, therefore, the host rocks and their characteristics are also varied. Particularly, the porphyry deposits present their ore generally associated with granodiorites and andesites, on which different types of alteration due to hydrothermal fluids are superimposed. This alteration implies the presence of abundant clays, which decrease the resistance of the rock. The depth at which the ore appears depends largely on the degree of erosion suffered by the rock mass, although the location of the magmas that give rise to the hydrothermal fluids is around 2–3 km and even more superficial. In any case, given the large dimensions of the mineralized bodies, it is common for at least part of them to be found at a considerable depth for their exploitation.

The mine development in these environments required the planning and optimization of several activities to improve the construction times and support requirements. This work summarized some of the activities for different mine deposits and underground mine methods and designs. In particular, the caving method required minor development compared with the deposit volume to be mined. However, the caving process implies high induced stress that must be included in stability. Additionally, the crater of subsidence generated in the surface due to caving involves an extensive zone. In Chile, a few studies have evaluated the impact of refilling this zone [61], but to date they have not been applied. Some of the current tendencies in the Block Caving method are focused on decreasing the development required and proposed improvement in mine design [62–64]. For example, the last caving projects used wide space extraction level layouts as in the Chuquicamata underground mine [44] and the Diamante sector of the El Teniente mine [65]. Additionally, in Cadia East in Australia, the undercut level has been omitted, so the caving was started directly from the extraction level [62]. On the other hand, the Polish mines described in massive deposits here use more selective underground methods. In these methods, the development and mining activities are commonly carried out at the same time (as well as caving propagation and undercutting). The mine planning then involved several challenges. Additionally, a back-filling process must be used to control the mine stability in thick deposits with these selective methods due to the large open areas.

For the underground copper ore mines in Poland in the area of LGOM in the coming years, three fundamental issues will arise with which the mining staff will have to look for and improve solutions. The first refers to the climatic hazard associated with the primary temperature of the rock mass. The primary temperature of the rock mass at the level of the mining excavations (Zechstein bottom level) ranges from 34 °C to 47 °C, with an average of 39 °C. On the other hand, the geothermal gradient ranges from 2.5 °C to 3 °C for every 100 m. Considering the greater depth of exploitation, the primary temperature of the rocks will increase, which will make it necessary to ensure the appropriate temperature for the mining crew by cooling the air temperature. The second issue is closely related to the decreasing thickness of the selected deposit. For thin deposits below 2 m, mines currently use split mining methods, trying to minimize dilution and operational losses. Further reduction of the exploitation gate will be related to the introduction of low-profile machines while maintaining the current level of extraction. The third issue relates to the risk of rock bursts. The occurring tremors have a value of 10^9 J. Therefore, mines are constantly improving the methods of rock bolt support, capable of absorbing as much dynamic load energy as possible. In addition, in order to minimize the risk of the crew staying in hazardous zones, methods for the remote control of mining operations are being developed, including blasting and bolt hole drilling, as well as loading and haulage of ore output. Further research works will be closely related to innovative mining technologies for the increasing level of exploitation.

Table 1. Comparison of mining and geological parameters for underground mining of selected ore deposits (copper and silver, zinc and lead) in Chile and Poland.

Parameter	Unit	Chile		Poland	
		Copper + Molybdenum (Chuquicamata)	Copper + Molybdenum (El Teniente)	Copper + Silver	Zinc + Lead
Depth of exploitation	(m)	100–650 (current level 1)	~1150	800–1300	80–120
Annual production	(mln tonnes)	50.4 (in regimen)	48.6	30	1.435
Metal content	(%); (g/Mg for silver)	0.77 (copper); 450 (ppm Mo)	0.86 (copper)	1.28–2.3 (copper); 54.5–62.5 (silver)	2.6 (zinc); 1.3 (lead)
Deposit thickness	(m)	250	~1200 (porphyry diameter in surface)	Thin up to 2 m; medium form 2 to 7 m; Thick more than 7 m	up to 30 m
The form of the deposit		Massive	Massive with mineralized veins.	The deposit belongs to the type of stratoid deposits (pseudo-seam)	Lenses and nests
Deposit inclination	(°)	-	-	about 8	about 5
Access		Tunnels	Tunnels	Vertical shafts	Decline and vertical shafts
The shape of preparatory and mining excavations		Horseshoe	Horseshoe	Trapezoidal	Rectangular with a flat or oval roof
Natural hazards		Rock burst; collapses; spalling	Rck burst; collapses; mudrushes	Rock burst; water; rock and gas outburst; roof falls; climatic (primary temperature of rock mass)	Water
Mining methods		Block caving (Macroblocks)	Block/panel caving	Room and Pillar	Room and Pillar
Liquidation of post-mining space		Subsidence	Subsidence	Bending of roor layers with further caving; hydraulic and dry backfilling	Hydraulic backfilling
Type of explosives		ANFO and Emulsion	ANFO and emulsion	Cartridge (manually loaded); emulsion materials (mechanically loaded)	
Support type		Shotcrete, cement grout rock bolt support, mesh	Shotcrete, cement grout rock bolt support (helical and split set), mesh	Mechanical and resin (full column) rock bolt support.	Resin (full column) rock bolt support.
Additional support		Shotcrete with syntetic fiber, concrete walls, cable bolts, steel frames	Shotcrete over mesh, shotcrete with fibet, concrete walls, cable bolts, steel frames	Cable up to 8 m; wooden cribs; arch yieldig support; steel and wooden props; geosynthetic and steel grids	Wooden cribs; arch yieldig support; wooden props; steel grids

Author Contributions: Conceptualization, K.S. and R.G.; methodology, R.G., K.Z. and A.Z.; formal analysis, K.S., R.G. and K.Z.; investigation, K.Z., A.Z. and R.G.-E.; resources, K.S.; writing—original draft preparation, R.G., K.Z. and R.G.-E.; writing—review and editing, K.S. and A.Z.; visualization, R.G., K.Z. and A.Z.; supervision, K.S.; project administration, K.S. and R.G.; funding acquisition, K.S. All authors have read and agreed to the published version of the manuscript.

Funding: This research was partially funded by CONICYT/PIA Project AFB220002 and AGH University of Science and Technology in Poland, scientific subsidy under number: 16.16.100.215.

Data Availability Statement: Not applicable.

Conflicts of Interest: The authors declare no conflict of interest.

References

1. Nordlund, E. Preface of Mass Min 2008 Proceedings. In Proceedings of the 5th Conference and Exhibition on Mass Mining, Luleå, Sweden, 9–11 June 2008.
2. Ladinig, T.; Wimmer, M.; Wagner, H. Raise Caving: A Novel Mining Method for (Deep) Mass Mining. In Proceedings of the Fifth International Conference on Block and Sublevel Caving, Adelaide, SA, Australia, 30 August–1 September 2022; pp. 651–666.
3. Rybak, J.M.; Kongar-Syuryun, C.; Tyulyaeva, Y.; Khayrutdinov, A.M.; Akinshin, I. Geomechanical substantiation of parameters of technology for mining salt deposits with a backfill. *Min. Sci.* **2021**, *28*, 19–32. [CrossRef]
4. Khayrutdinov, A.M.; Kongar-Syuryun, C.B.; Kowalik, T.; Tyulyaeva, Y.S. Stress-strain behavior control in rock mass using different-strength backfill. *Min. Inf. Anal. Bull.* **2020**, *10*, 42–55. [CrossRef]
5. Kongar-Syuryun, C.; Ubysz, A.; Faradzhov, V. Models and algorithms of choice of development technology of deposits when selecting the composition of the backfilling mixture. *IOP Conf. Ser. Earth Environ. Sci.* **2021**, *684*, 012008. [CrossRef]
6. Singer, D.A. World class base and precious metal deposits; a quantitative analysis. *Econ. Geol.* **1995**, *90*, 88–104. [CrossRef]
7. Laznicka, P. Quantitative relationships among giant deposits of metals. *Econ. Geol.* **1999**, *94*, 455–473. [CrossRef]
8. Robert, F.; Poulsen, K.H. World-class Archean hydrothermal gold deposits: A Canadian perspective. *Geol. Soc. Aust. Abstr.* **1996**, *41*, 368.
9. Yang, Z.M.; Cooke, D.R. Porphyry copper deposits in China. In *Mineral Deposits of China, Society of Economic Geologists*; Zhaoshan, C., Richards, G., Eds.; Society of Economic Geologists: Littleton, CO, USA, 2019; pp. 133–187. Available online: https://eprints.utas.edu.au/32587/ (accessed on 18 December 2022).
10. Redwood, S.D. The mineral deposits of Panama: Arc metallogenesis on the trailing edge of the Caribbean large igneous province. *Boletin de la Soc. Geol. Mex.* **2020**, *72*, A130220. [CrossRef]
11. Tabelin, C.B.; Park, I.; Phengsaart, T.; Jeon, S.; Villacorte-Tabelin, M.; Alonzo, D.; Yoo, K.; Ito, M.; Hiroyoshi, N. Copper and critical metals production from porphyry ores and E-wastes: A review of resource availability, processing/recycling challenges, socio-environmental aspects, and sustainability issues. *Resour. Conserv. Recycl.* **2021**, *170*, 105610. [CrossRef]
12. Hao, H.; Park, J.-W.; Campbell, I.H. Role of magma differentiation depth in controlling the Au grade of giant porphyry deposits. *Earth Planet. Sci. Lett.* **2022**, *593*, 117640. [CrossRef]
13. Chiaradia, M.; Caricchi, L. Supergiant porphyry copper deposits are failed large eruptions. *Commun. Earth Environ.* **2022**, *3*, 117864. [CrossRef]
14. Brown, A.C. World-class sediment-hosted stratiform copper deposits: Characteristics, genetic concepts and metallotects. *Aust. J. Earth Sci.* **1997**, *44*, 317–328. [CrossRef]
15. Sillitoe, R.H. Copper provinces. *Soc. Econ. Geol.* **2012**, *16*, 1–18.
16. Cooke, D.R.; Hollings, P.; Walshe, J.L. Giant porphyry deposits: Characteristics, distribution, and tectonic controls. *Econ. Geol.* **2005**, *100*, 801–818. [CrossRef]
17. del Real, I.; Thompson, J.; Simon, A.C.; Reich, M. Geochemical and Isotopic Signature of Pyrite as a Proxy for Fluid Source and Evolution in the Candelaria-Punta del Cobre Iron Oxide Copper-Gold District, Chile. *Econ. Geol.* **2020**, *115*, 1493–1518. [CrossRef]
18. Zamora, A. *Geología Proyecto Las Luces*; Internal report; Compañía Minera Las Cenizas: Cabildo, Chile, 2011.
19. Cabral, A.R.; Beaudoin, G. Volcanic red-bed copper mineralisation related to submarine basalt alteration, Mont Alexandre, Quebec Appalachians, Canada. *Miner. Depos.* **2007**, *42*, 901–912. [CrossRef]
20. López, J. M Geología y Microtermometría del Yacimiento Estratoligado Don Gabriel. Comuna de Salamanca, Región de Coquimbo, Chile. Master's Thesis, Universidad de Concepción, Concepción, Chile, 2010.
21. Maksaev, V. Chilean Strata-Bound Cu-(Ag) Deposits: An Overview. In *Hydrothermal Iron Oxide Copper-Gold and Related Deposits: A Global Perspective*; Porter, T.M., Ed.; Publishers Group Canada: Vancouver, BC, Canada, 2002; pp. 185–205.
22. MTV. *Minera Tres Valles Copper Project, Salamanca, Coquimbo, Chile*; NI43–101 Technical Report; Minera Tres Valles Sprott Resource Holdings Inc.: Coquimbo, Chile, 2018.
23. Boric, R.; Holmgren, C.; Wilson, N.S.F. The Geology of the El Soldado Manto type Cu (Ag) Deposit, Central Chile. In *Hydrothermal Iron Oxide Copper-Gold and Related Deposits: A global Perspective*; Porter, T.M., Ed.; Publishers Group Canada: Vancouver, BC, Canada, 2002; pp. 163–184.
24. Movahednia, M.; Maghfouri, S.; Fazli, N.; Rastad, E.; Ghaderi, M.; González, F.J. Metallogeny of Manto-type Cu-(Ag) mineralization in Iran: Relationship with Neo-Tethyan evolution and implications for future exploration. *Ore Geol. Rev.* **2022**, *149*, 105064. [CrossRef]
25. Cox, D.P.; Lindsey, D.A.; Singer, D.A.; Moring, B.C.; Diggles, M.F. Sediment-Hosted Copper Deposits of the World: Deposit Models and Database. *Sci. Chang. World* **2003**. Available online: http://pubs.usgs.gov/of/2003/of03-107/ (accessed on 18 December 2022).
26. Hitzman, M.; Kirkham, R.; Broughton, D.; Thorson, J.; Selley, D. The Sediment-Hosted Stratiform Copper Ore System. *Econ. Geol.* **2005**, 609–642. [CrossRef]
27. Oszczepalski, S.; Speczik, S.; Zieliński, K.; Chmielewski, A. The Kupferschiefer Deposits and Prospects in SW Poland: Past, Present and Future. *Minerals* **2019**, *9*, 592. [CrossRef]
28. De Vos, W.; Batista, M.J.; Demetriades, A.; Duris, M.; Lexa, J.; Lis, J.; Marsina, K.; O'Connor, P.J. Metallogenic mineral provinces and world class ore deposits in Europe. *Geochem. Atlas Eur.* **2005**, *1*, 43–49.

29. Sass-Gustkiewicz, M.; Kwiecinska, B. Organic matter in the Upper Silesian (mississippi valley-type) deposits, Poland. *Econ. Geol.* **1999**, *94*, 981–992. [CrossRef]
30. Paradis, S.; Hannigan, P.; Dewing, K. Mississippi Valley-Type Lead-Zinc Deposits. In *Mineral Deposits of Canada: A Synthesis of Major Deposit-Types, District Metallogeny, the Evolution of Geological Provinces, and Exploration Methods*; Goodfellow, W.D., Ed.; Geological Association of Canada, Mineral Deposit Division: St. John's, NL, USA, 2007; pp. 185–203.
31. Szuwarzyński, M. Ore bodies in the Silesian-Cracow Zn-Pb ore district, Poland. *Pol. Geol. Inst. Spec. Pap.* **1996**, *154*, 9–24.
32. Kurek, S. Problems of modelling of Zn-Pb ores of the Mississippi Valley-type in the sediments of the Younger Paleozoic. *Geol. Q.* **1993**, *37*, 147–156.
33. Gómez, R.; Labbé, E. Una opción para el paso a la minería subterránea masiva: La actual metodología aplicada en Chile del "block caving". *Bol. Inst. Geol.* **2019**, *130*, 181–198. [CrossRef]
34. Araneda, O. Codelco: Present, Future and Excellence in Projects. In Proceedings of the Eighth International Conference & Exhibition on Mass Mining 2020, Virtual, 9–11 December 2020; pp. 1–9.
35. Arancibia, E.; Flores, G. Design for Underground Mining at Chuquicamata Orebody, Scoping Engineering Stage. In Proceedings of the International Conference & Exhibition on Mass Mining 2004, Santiago, Chile, 22–25 August 2004; pp. 603–608.
36. Villa, D.; Baraqui, J.; Maass, S. Papomono Mine Modelled as an Inclined Cave. In Proceedings of the Eighth International Conference & Exhibition on Mass Mining 2020, Virtual, 9–11 December 2020; pp. 182–194.
37. Laubscher, D.H. Cave mining-the state of the art. *J. South. Afr. Inst. Min. Metall.* **1994**, *94*, 279–293.
38. Edwin, E.T.; Brown, T. *Block Caving Geomechanics*, 2nd ed.; Julius Kruttschnitt Mineral Research Centre: Indooroopilly, QL, Australia, 2007.
39. Flores, G. Future Challenges and Why Cave Mining Must Change. In Proceedings of the 3rd International Symposium on Block and Sublevel Caving, Santiago, Chile, 5–6 June 2014; pp. 23–52.
40. Fairhurst, C. Some challenges of deep mining. *Engineering* **2017**, 3, 527–537. [CrossRef]
41. Tshisens, J.N.; Moss, A.; Sullivan, M.; Yuniar, A.; Casten, T.; Zimmer, C. Organising for the Successful Management of Complex Underground Caving Mines. In Proceedings of the Fifth International Conference on Block and Sublevel Caving, Adelaide, SA, Australia, 30 August–1 September 2022; pp. 81–94.
42. Sillitoe, R. Epochs of intrusion-related copper mineralization in the Andes. *J. South Am. Earth Sci.* **1988**, *1*, 89–108. [CrossRef]
43. Barra, F.; Alcota, H.; Rivera, S.; Valencia, V.; Munizaga, F.; Maksaev, V. Timing and formation of porphyry Cu–Mo mineralization in the Chuquicamata district, northern Chile: New constraints from the Toki cluster. *Miner. Deposita* **2013**, *48*, 629–651. [CrossRef]
44. Flores, G.; Catalan, A. A transition from a large open pit into a novel "macroblock variant" block caving geometry at Chuquicamata mine, Codelco Chile. *J. Rock Mech. Geotech. Eng.* **2018**, *11*, 549–561. [CrossRef]
45. Arancibia, E.; Carrasco, F.; Fuentes, G.; Guarda, J. Design of Extraction Layout for the Chuquicamata Underground Mine Project. In Proceedings of the International Conference & Exhibition on Mass Mining 2008, Luleå Sweden, 9–11 June 2008; pp. 3–14.
46. Mühlenbrock, K.; Silva, D.; Navarrete, R.; Mardonez, L. Construction of the First Macro Block at the Chuquicamata Underground Project. In Proceedings of the Eighth International Conference & Exhibition on Mass Mining, the Eighth International Conference & Exhibition on Mass Mining 2020, Virtual, 9–11 December 2020; pp. 127–139.
47. Mülhembrock, K. Preparación Minera GRMD. *Concepción* **2022**, *47*.
48. Cannell, J.; Cooke, D.R.; Walshe, J.L.; Stein, H. Geology, Mineralization, Alteration, and Structural Evolution of the El Teniente Porphyry Cu-Mo Deposit. *Econ. Geol.* **2005**, *100*, 979–1003. [CrossRef]
49. Brzovic, A.; Villaescusa, E. Rock mass characterization and assessment of block-forming geological discontinuities during caving of primary copper ore at the El Teniente mine, Chile. *Int. J. Rock Mech. Min. Sci.* **2007**, *44*, 565–583. [CrossRef]
50. Castro, R.; Garces, D.; Brzovic, A.; Armijo, F. Quantifying Wet Muck Entry Risk for Long-term Planning in Block Caving. *Rock Mech. Rock Eng.* **2018**, *51*, 2965–2978. [CrossRef]
51. Pozo, R.; Díaz, G.; Medina, O. Centro Integrado de Construcción en Obras de Minería Subterránea, Proyecto Nuevo Nivel Mina El Teniente. In Proceedings of the UMining 2016, 1er Congreso Internacional en Minería Subterránea, Santiago, Chile, 19–21 October 2016; pp. 299–312.
52. Ritter, C. Effect of Constructability in Underground Mining under High Stress Conditions. Ph.D. Thesis, Universidad de Concepción, Concepción, Chile, 2018. (In Spanish).
53. Kurek, T. *Perspectives of Development of Zn-Pb ore Exploitation in the Safety Pillars of 'Trzebionka' Mine'*; AGH: Kraków, Poland, 1987. (In Polish)
54. Piechota, S. Basic principles of the exploitation of thick ore deposits. *Ores Non Ferr. Met.* **1973**, *18*, 172–177. (In Polish)
55. Roszkowski, J.; Piechota, S. About some problems of the exploitation of zinc-lead ore deposits. *Ores Non Ferr. Met.* **1979**, *24*, 472–478.
56. Skrzypkowski, K. Decreasing Mining Losses for the Room and Pillar Method by Replacing the Inter-Room Pillars by the Construction of Wooden Cribs Filled with Waste Rocks. *Energies* **2020**, *13*, 3564. [CrossRef]
57. Skrzypkowski, K. Comparative Analysis of the Mining Cribs Models Filled with Gangue. *Energies* **2020**, *13*, 5290. [CrossRef]
58. Piechota, S. Development of copper ore mining methods in LGOM region. *Bull. Polish Geol. Inst.* **2007**, *423*, 23–43. (In Polish)
59. Piechota, S.; Stopyra, M.; Poborska-Młynarska, K. *Systemy Podziemnej Eksploatacji Złóż Węgla Kamiennego, Rud i Soli*; AGH: Kraków, Poland, 2009. (In Polish)
60. Burta, J. *Eksploatacja Złoża Rud Miedzi w Warunkach Zagrożenia Tąpaniami i Zawałami*; KGHM Cuprum sp. z o.o. Centrum Badawczo—Rozwojowe: Wrocław, Poland, 2010.

61. Armijo, F.; Irribarra, S.; Castro, R. Experimental Study of Fines Migration for Caving Mines. In Proceedings of the 3rd International Symposium on Block and Sublevel Caving, Santiago, Chile, 5–6 June 2014; pp. 356–362.
62. Paredes, P.; Popa, I.; Kline, P.; Sykes, M. Undercutless caving at Newcrest: Towards the Next Generation of Cave Mining. In Proceedings of the Eighth International Conference & Exhibition on Mass Mining 2020, Virtual, 9–11 December 2020; pp. 166–181.
63. Le-Feaux, R.; Castro, R.; Cortez, D.; Gómez, R.; Silva, D. A hybrid extraction level layout design for block caving. *Min. Technol.* **2021**, *131*, 51–65. [CrossRef]
64. Paredes, P. Evaluation of the Effect of Wider-Spaced Layouts in Recovery for High Column Block Caves. In Proceedings of the Caving 2022: Fifth International Conference on Block and Sublevel Caving, Adelaide, SA, Australia, 30 August–1 September 2022; pp. 393–416.
65. Cornejo, J.; Meléndez, M.; Lasagna, G.; Bustamante, D.; Castro, R.; Arancibia, L.; Gúzman, D. Design at Drawpoint Spacing at Diamante Project—El Teniente Mine. In Proceedings of the Eighth International Conference & Exhibition on Mass Mining 2020, Virtual, 9–11 December 2020; pp. 761–772.

energies MDPI

Article

Electro-Mechanical Modeling and Evaluation of Electric Load Haul Dump Based on Field Measurements

Gabriel Freire [1], Guillermo Ramirez [1], René Gómez [2,3,*], Krzysztof Skrzypkowski [4,*] and Krzysztof Zagórski [5]

1 Centro de Energía, Faculty of Engineering, Universidad Católica de la Santísima Concepción, Concepción 4090541, Chile; guillermoramirez@ucsc.cl (G.R.)
2 Faculty of Engineering, Universidad de Concepción, Concepción 4030000, Chile
3 Advanced Mining Technology Center, University of Chile, Santiago 8370451, Chile
4 Faculty of Civil Engineering and Resource Management, AGH University of Science and Technology, Mickiewicza 30 Av., 30-059 Kraków, Poland
5 Faculty of Mechanical Engineering and Robotics, AGH University of Science and Technology, Mickiewicza 30 Av., 30-059 Kraków, Poland; zagkrzys@agh.edu.pl
* Correspondence: regomez@udec.cl (R.G.); skrzypko@agh.edu.pl (K.S.)

Abstract: In underground mining, conventional loader equipment uses diesel as a power source, implying different drawbacks, such as combustion gases, low visibility, worker's health problems, and high ventilation requirements. Thus, hybrid and electric loaders are being developed by the main industry suppliers who prefer clean technology. In this study, we analyzed the performance of an electro-mechanical powertrain through a dynamic model of underground-loader equipment using field data. This electric LHD model was compared to a diesel loader under the same operational conditions. For the case study, the results showed that the proposed electro-mechanical model, considering 14 tons of capacity, consumed 86.8 kWh, representing 60.5% less energy than the diesel loader with similar speed and torque characteristics. Thus, the proposed methodology is a valuable tool for operators, process engineers, and decision-makers, allowing an energy-efficiency evaluation for electric LHD adoption, based on the current operational data available for conventional equipment.

Keywords: electro-mobility; electromechanical; diesel; LHD; underground mining

Citation: Freire, G.; Ramirez, G.; Gómez, R.; Skrzypkowski, K.; Zagórski, K. Electro-Mechanical Modeling and Evaluation of Electric Load Haul Dump Based on Field Measurements. *Energies* **2023**, *16*, 4399. https://doi.org/10.3390/en16114399

Academic Editor: Sergey Zhironkin

Received: 24 April 2023
Revised: 26 May 2023
Accepted: 27 May 2023
Published: 30 May 2023

1. Introduction

In underground mining, the Load-Haul-Dump (LHD) or scoops, generate a significant impact in mining extraction by promoting the mechanization of operations with a versatile and highly productive machine [1]. Consequently, the manufacturers have been improving characteristics, such as higher powers and capacities, in this equipment. Thus, the main providers of LHDs, such as CAT, Atlas Copco, Sandvick, and Komatsu, have begun to offer hybrid and/or electric alternatives in addition to conventional diesel models. The main advantage of hybrid and full-electric LHDs is the ability to reduce and/or eliminate polluting gas emissions [2,3]. Additionally, these loaders generate benefits in terms of production efficiency, versatility, and maintenance, by reducing maintenance time and costs [4].

LHDs can be classified into three types based on their power source: Diesel, Hybrid, and Electric. The diesel or conventional LHDs have a diesel engine that provides mechanical energy to the entire powertrain. The main disadvantage of using these conventional LHDs in underground mining is that they require a robust ventilation system for the extraction of gases from diesel combustion that affect visibility and also turn out to be toxic to people, which generates eventual occupational-health problems [5–7]. Hybrid LHDs combine a diesel engine coupled to an electric generator that produces energy to feed the reluctance motors attached to each wheel, reducing the consumption of diesel by up to 20% compared to conventional LHDs [8]. Electric LHDs commonly use an induction motor for traction and in some cases include an additional induction motor for the hydraulic bucket drive.

Currently, conventional LHDs represent around 74% of the models available, with a load capacity that can vary between 2 and 21 tons. Hybrid LHDs account for approximately 5% of the market, ranging in load capacity between 18 and 22 tons. Electric LHDs cover around 21% of the market, with load capacities between 3.5 and 25 tons. For power supply, electric LHDs are fed by a trailing cable or battery pack. In the first case, the feeding cable is extended and contracted according to displacement requirements by a reel at the back. On the other hand, battery LHDs are more versatile but have limited autonomy, which is determined by the battery pack size. Some examples of these types of LHDs are the Scooptram ST7 Battery of Atlas Copco with a load capacity of up to 10 tons [9], the R1700 XE LHD from Caterpillar [10], and model LH518B from Sandvik [11]. Additionally, the current battery-operated LHDs do not generate major differences in equipment size.

Electric LHDs have the potential to be powered by clean energy sources, enabling the reduction of the carbon footprint, as well as lower operational costs [4,12–14]. On the other hand, one of the main disadvantages of an electric LHD powered by trailing cables is the potential for constant deterioration, requiring frequent maintenance, in addition to the limitations on displacements presented by the cable length [15]. In the case of a battery-powered LHD, productivity is affected by regular battery replacement due to the limited battery autonomy, in addition to the operational requirements of a specially conditioned space for battery storage and charge.

There are different motor types used by electric LHDs. The electric LHD models LH514E and LH625IE, developed by Sandvik use three-phase squirrel-cage induction motors for traction, hydraulic steering and bucket pumps, and ventilation purposes. Other providers, such as Komatsu, use switched reluctance motors considering their small moment of inertia and high dynamic response even at low speed [16]. Considering the great torque capacity at low speeds, permanent magnet motors have been incorporated into electric LHDs during the last years as in the case of Sandvik LH625IE.

Schultz and Huard [17] compared five types of motors (induction, salient pole permanent magnets, non-salient pole permanent magnets, switched reluctance, and synchronous reluctance) and highlight that the induction motor will not disappear due to its low cost, reliability, and robustness. Moreover, alternatives such as permanent magnet motors (salient and non-salient pole) and reluctance motors have complex controls and must deal with torque ripple and efficiency optimization at different speeds [18].

Considering its proven robustness and extended documentation, the vector-control method is one of the most used for both induction motors and synchronous motors [19,20]. Theoretically, the vector control of an induction motor seeks to model the asynchronous machine as a simple and versatile DC machine, with independent control loops for flux, speed, and current in *qd* coordinates, facilitating the use of linear controllers and dynamic performance analysis.

Based on a literature review, the reluctance motor is more efficient than an induction machine; however, there is a lack of evidence regarding the use of a reluctance motor for full-electric LHDs except for the case of hybrid options (Komatsu). Thus, in the present work, an electro-mechanical model for a full-electric LHD was developed by integrating the dynamic model of the powertrain driven by induction motors, a vector- control scheme for speed control, and a dynamic load model as a function of the mechanical inertias, forces, and friction coefficients during operation. Then, the proposed model was analyzed considering the load torque and speed profiles obtained from the field measurements of a diesel LHD to finally compare their results in terms of energy consumption during the operational cycle under evaluation. The obtained results allowed for the proposed electric LHD model to be validated by visualizing a consistent behavior in terms of developed speed and power when compared with the diesel LHD field curves. In this way, the proposed methodology represents a valuable tool for operators and process engineers to project the energy requirements required for the adoption of battery or trailing cable electric LHDs, based on current and historic operational data of diesel LHDs.

2. Methods

The methodology used in this study follows similar steps described in [21], adapted to model a full-electric LHD, starting with the definition of the dynamic load model. This load model consists mainly of obtaining the total equivalent LHD inertia considering all the inertial components present in the powertrain. These inertial components depend on geometries and masses of tires, rims, motor, and gearboxes added to the inertial components associated with the translational motion, which depends on the fixed mass of the LHD structure, and the variable mass of the payload carried in the LHD bucket.

Once the inertial components have been defined, the next step is to calculate the dynamic load torque reflected in the electric motor side, which depends on a variety of mechanical forces, including grade force, rolling-resistance force, aerodynamic-drag force, and rotating-friction torque. Each of these expressions has specific parameters that, once integrated into a single equation, allow for the complete dynamic model of the LHD load torque to be calculated.

The third stage consists of the definition of equivalent motor drives present in the LHD powertrain, considering the induction motor with vector-control dynamic models based on the nameplate power data of the operational LHD diesel engine. Then, the integration of all the components previously described allows for the construction of the electro-mechanical LHD model, which is finally evaluated using field records of typical work cycles determined by the study case.

Study Case

For this study, the field data were collected from a diesel LHD working in an underground copper mine located in the central zone of Chile. The data analyzed considered the operation of two diesel LHDs models with load capacities of 5.3 and 9.9 m^3 with a limited maximum velocity of 20 km/h due to the mine's internal safety standards. All the loaders were equipped with a data acquisition system to monitor and record the operating variables in a centralized monitoring software systems during various periods. This data included fuel consumption curves, traveled distance, LHD velocity, engine speed, torque, power, acceleration, and accelerator-pedal position. This study focused on the 5.3 m^3 (7 yd^3) model, and Figure 1 shows an example of the variables measured during 15 min of operation. This figure includes four time segments labeled with literals R, S, T, and U. It should be noted that all the variables are depicted in per unit (p.u.) considering the base values shown in Table 1.

Table 1. Base values are used for per-unit variables representation.

LHD Parameters	Value	Unit
Engine power	275	kW
Engine torque	1905	Nm
Engine speed	2100	rpm
LHD velocity	15	Km/h

The four time segments marked in Figure 1 show characteristic behaviors related to specific operational conditions. The time segment (R) shows the case where the operator of the diesel LHD fully presses the throttle, increasing the engine speed, torque, and power while the LHD velocity is decreased. This behavior is characteristic of the bucket operation for load scooping in the extraction zone where one-step scooping demands high torque and power [22]. The time segment (S) is representative of the operational stops during which the operators must wait to access the dumping zone (ore pass) or must wait until the return route is available for safe transit. During this period, the engine speed stays in idle mode (around 750 rpm), as seen in Figure 1b. Segment (T) represents similar behavior to that described for (R) but with more intense acceleration movements needed to shake the bucket during dumping or to accommodate material for later loading into the bucket. The last time segment (U) is nearly equal to segment (R), with high engine acceleration in a short time resulting in a low LHD velocity. Figure 1d shows that the distance traveled by the LHD during the 15-min window is 1.52 km. It should be noted that the distance traveled between the end of (R) and the beginning of (S) is approximately 190 m, being very similar to the distance between the end of (U) and the next stop stage, demonstrating that the case studied has a cyclical behavior commonly found in LHD mining routes.

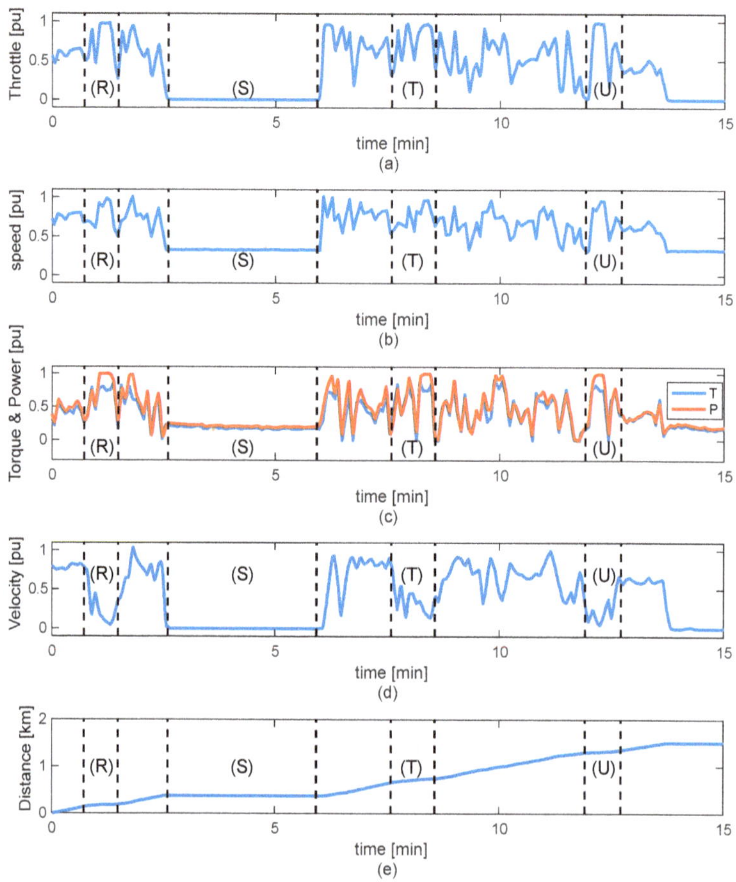

Figure 1. Operational curves of a diesel LHD: (**a**) throttle position; (**b**) engine speed; (**c**) engine torque and power; (**d**) LHD velocity; and (**e**) traveled distance.

Then, the extended dataset of variables shown in Figure 1, was used to evaluate the proposed full-electric LHD model based on field measurements, considering a maximum capacity of 14 tons.

3. Full-Electric LHD Model

Figure 2 shows a general schematic of the proposed full-electric LHD model. The model is composed of an electric drive that includes an induction motor controlled by an indirect-vector-control topology [23]. The electric drive generates the power required to feed the powertrain composed of the mechanical pieces of the LHD that deliver power to the drive wheels. The powertrain characterization enables the implementation of the dynamic load model to provide both the total inertia of the electric LHD and load torque.

Total inertia is required for tuning of the control law of the electric drive while load torque acts as a disturbance input to the electric drive, defining the acceleration characteristics of the LHD.

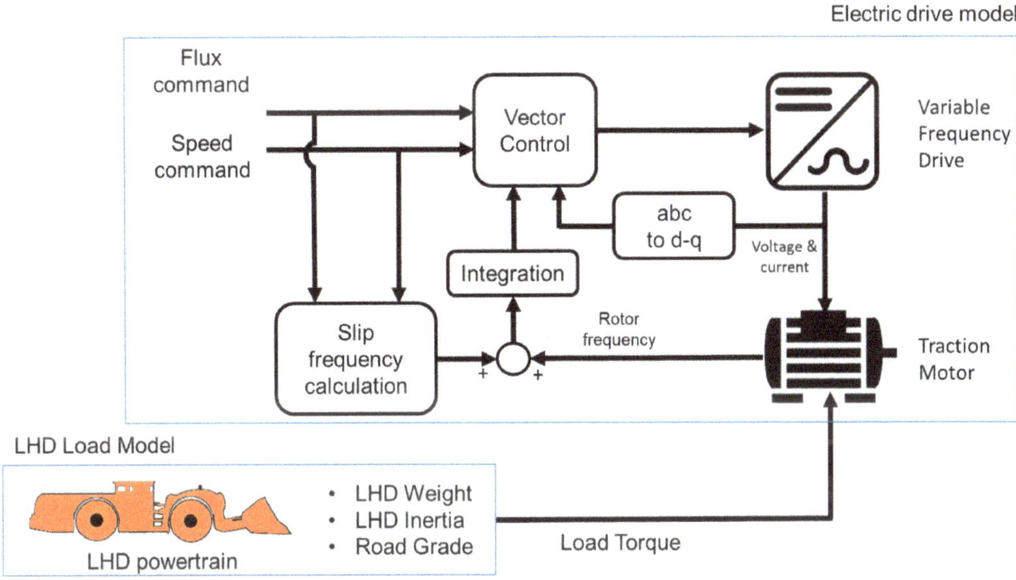

Figure 2. Full-electric LHD general schematic.

The first stage of the full-electric LHD modeling starts with defining and analyzing the load model, considering the mechanical parameters, forces, and drag coefficients during operation.

3.1. LHD Load Model

This study was based on a commercial full-electric LHD (Sandvik, LH514-E), whose main technical specifications are presented in Table 2. Figure 3 shows the free-body diagram of all the forces acting over the LHD and the main wheel dimensions used to calculate the wheel and rim inertia.

Table 2. Technical specification of LHD model LH514-E.

LHD Parameters	Value	Unit
Load Capacity	14	t
Load LHD weight	52.5	t
Power of traction motor	132	kW
Power of hydraulic pump	45	kW
Power of ventilation system	3	kW
Total power	180	kW
Tire	25.5 × 25 L5S 36-ply	
Gears	4	

3.1.1. Inertial Components

In this section all the inertial components of the powertrain are calculated, considering the moments of inertia of the tires, rims, LHD structure, and LHD load [21]. The LHD model uses 26.5 × 25 L5S 36-ply tires, as indicated in Table 2. In this nomenclature, the first number refers to tire width, and "L5S" refers to the fact that it has an extra-deep tread which is ideal for working in very aggressive rocky areas. The last term (36 ply) refers to the thickness of the rim in inches (in). Figure 3C shows the approximate size of the LHD tire, whose mass is around 450 kg, while Figure 3D shows the approximate size of the tire rim in cm.

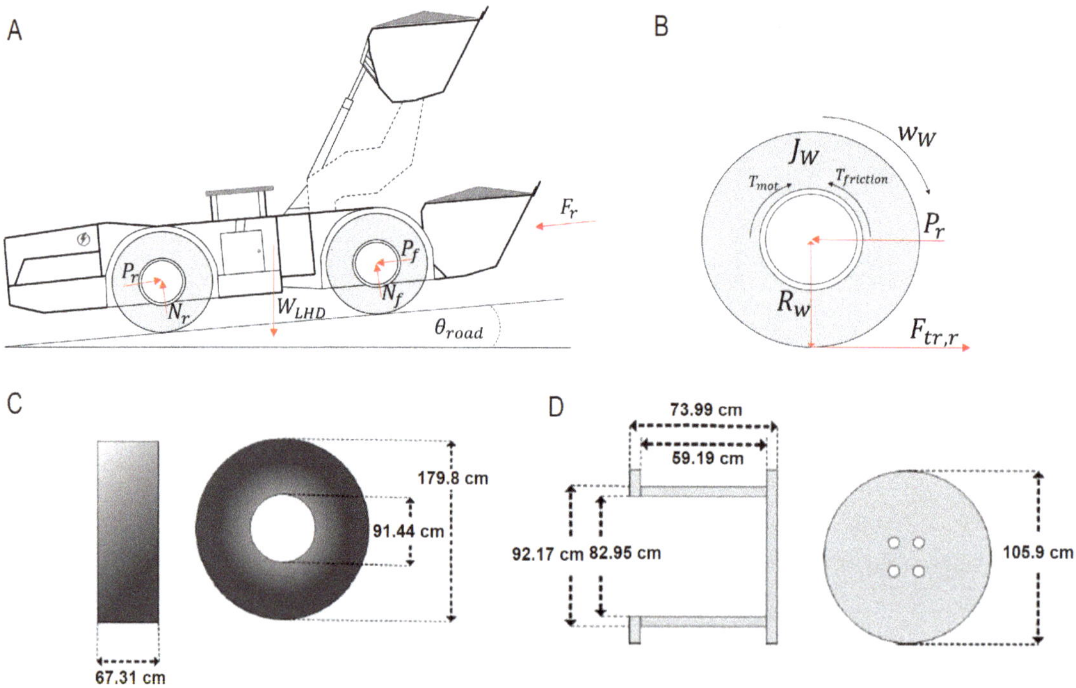

Figure 3. (**A**) LHD free-diagram body; (**B**) traction wheels; (**C**) tire dimensions; (**D**) rim dimensions.

The inertia of the tire can be calculated based on the tire dimensions and then dividing the tire into three bodies: two hollow discs and one cylindrical shell (*cs* subscript).

Thus, the tire inertia is calculated by the sum of three components as follows:

$$A_{disc1} = \pi\left(\left(\frac{De}{2}\right)^2 - \left(\frac{Di}{2}\right)^2\right) = A_{disc2} \tag{1}$$

$$A_{cs} = \pi DeW \tag{2}$$

$$A_{tire} = A_{disc1} + A_{disc2} + A_{cs} \tag{3}$$

where A_{disc1} and A_{disc2} are the areas of tire discs (m^2), De is the external diameter of the tire (m), Di is the internal diameter of the tire (m), A_{sc} is the area of the cylindrical shell (m^2), W is the tire width (m), and A_{tire} is the total area of the tire (m^2). Then, with the mass of the tire already known, the tire inertia is determined as follows.

$$m_{disc1} = \frac{A_{disc1} m_{tire}}{A_{tire}} = m_{disc2} \tag{4}$$

$$m_{cs} = \frac{A_{cs} m_{tire}}{A_{tire}} \tag{5}$$

$$J_{tire} = 2\left(\frac{1}{2}m_{disc1}\left(\left(\frac{De}{2}\right)^2 + \left(\frac{Di}{2}\right)^2\right)\right) + m_{cs}\left(\frac{De}{2}\right)^2 = 296.6165 \text{ kgm}^2 \quad (6)$$

where m_{disc1} and m_{disc2} are the tire disc masses (kg), m_{tire} is the tire mass (450 kg), and J_{tire} is the total inertia of the tire (kgm^2). Then, the approximate inertia of the rim is estimated, considering an iron density of 7874 kg/m^3 which results in a rim mass of 1302.6 kg.

With the mass of the rim and its dimensions, the inertia of the rim is determined considering three bodies: a hollow disk, a solid disk, and a hollow cylinder. The volume of the bodies is calculated as follows,

$$V_{hd} = \pi\left(\frac{de}{2}\right)^2\left(\frac{l_e - l_i}{2}\right) - \pi\left(\frac{di}{2}\right)^2\left(\frac{l_e - l_i}{2}\right) \quad (7)$$

$$V_{sd} = \pi\left(\frac{de}{2}\right)^2\left(\frac{l_e - l_i}{2}\right) \quad (8)$$

$$V_{hc} = \pi\left(\frac{de}{2}\right)^2 l_e - \pi\left(\frac{di}{2}\right)^2 l_i \quad (9)$$

where V_{hd}, V_{sd}, and V_{hc} are the volumes of the hollow disc, solid disc, and hollow cylinder, respectively (m^3), de and di are the external and internal diameters of the rim (m), and l_e and l_i are the external and the internal lengths of the rim (m). Then, the mass of the tire with the mass of the rim is determined considering that:

$$m_{hd} = V_{hd}\rho_{iron} \quad (10)$$

$$m_{sd} = V_{sd}\rho_{iron} \quad (11)$$

$$m_{hc} = V_{hc}\rho_{iron} \quad (12)$$

where m_{hd}, m_{sd}, and m_{hc} are the masses of the hollow disc, solid disc, and hollow cylinder, respectively (m), and ρ_{iron} is the iron density (kg/m^3), then,

$$J_{hd} = \frac{1}{2}m_{hd}\left(\left(\frac{de}{2}\right)^2 + \left(\frac{di}{2}\right)^2\right) \quad (13)$$

$$J_{sd} = \frac{1}{2}m_{sd}\left(\frac{de}{2}\right)^2 \quad (14)$$

$$J_{hc} = \frac{1}{2}m_{hc}\left(\left(\frac{de}{2}\right)^2 + \left(\frac{di}{2}\right)^2\right) \quad (15)$$

$$J_{rim} = J_{hd} + J_{sd} + J_{hc} = 230.4043 \text{ kgm}^2 \quad (16)$$

where J_{hd}, J_{sd}, and J_{hc} are the inertia of the hollow disc, solid disc, and hollow cylinder, respectively (kgm^2), and J_{rim} is the total inertia of the rim. Subsequently, the inertia related to the translational movement of the LHD is estimated as:

$$J_{tran} = \left(\frac{R_w}{n_{gear}}\right)^2 m_{LHD} + \left(\frac{R_w}{n_{gear}}\right)^2 m_{load} \quad (17)$$

where R_w is the tire radius (m), n_{gear} is the gear ratio (dimensionless), m_{LHD} and m_{load} are the mass of the LHD and the material loaded, respectively (kg). The first term of the equation represents the inertia of the LHD without a load. Finally, the total inertia of the LHD is the sum of the translational inertia (J_{tran}), the motor inertia (J_{mot}), and the inertia of each tire and rim (J_{tire}, J_{rim}).

$$J_{LHD} = J_{mot} + n_{gear}^{-2}(4J_{tire} + 4J_{rim}) + J_{tran} \tag{18}$$

where J_{LHD} is the total inertia of the LHD (kgm^2). Then, the inertia of the LHD under no load conditions is 62.33 kgm^2, while the inertia of the LHD considering a full load is 83.55 kgm^2. In both cases, a planetary gearbox with a gear ratio n_{gear} = 23.08 is used, considering that the motor rotates at 1396 rpm when the LHD velocity reaches 20.5 km/h.

3.1.2. Load Torque

According to [21] the equation that models the load torque by all the forces that act on a motorized vehicle with wheels depends on the air force, drag coefficients, the mass of the LHD, rolling coefficients, frontal area of the LHD, gravity, and road inclination. The following expression represents the load torque of the LHD:

$$T_{LHD} = \frac{1}{n_{gear}} R_w \left(\frac{1}{2}\rho_a (v_a + v_{tire})^2 C_{drag} A_{LHD} + (m_{LHD} + m_{load}) g sin\left(\frac{\theta_{road}}{100}\right) + u_{rr}(m_{LHD} + m_{load}) g cos\left(\frac{\theta_{road}}{100}\right)\right) \tag{19}$$

Here, T_{LHD} is the load torque reflected in the LHD motor (Nm) where ρ_a is the air density, v_a is the air velocity (m/s), v_{tire} is the tire velocity (m/s), C_{drag} is the drag coefficient (dimensionless), A_{LHD} is the frontal area of the LHD (m^2), m_{LHD} is the LHD mass (kg), m_{load} is the mass of the material carried (kg), g is the gravity force (m/s^2), u_{rr} is the rolling coefficient (dimensionless), and θ_{road} is the road inclination (%).

In the load torque equation, the air coefficient can be considered constant at an atmospheric pressure of 101 kPa and a temperature of 20 °C. The aerodynamic drag coefficient is considered to be 1.95, according to [24]. The frontal area of the LHD can be obtained from its dimensions as described in [25], considering a height of 2.545 m and a width of 2.897 m. The rolling coefficient used is the average value of those exposed in [26], which depends on the tire pressure and the LHD load. The inclination of the road will be equal to zero considering a horizontal route. Table 3 shows the values used for the load torque of the LHD.

Table 3. Summary of parameters used to calculate the load torque.

Parameter	Value	Unit
ρ_a	1.2038	kg/m^3
v_a	1.3667	m/s
C_{drag}	1.95	-
u_{rr}	0.03	-
A_{LHD}	7.3729	m^2
g	9.81	m/s^2
θ_{road}	0	%

To observe the influence of the mass of the material carried in the bucket of the LHD Figure 4 shows the load torque obtained considering different load masses as a function of the traveled distance, considering a flat route and a rolling coefficient of 0.03. Initially, the bucket is empty (0 t before loading), then after 5 km, the load is incremented to 7 t (loaded) until the 10 km where the bucket is unloaded remaining with 0.14 t. Then, after 15 km, the bucket is fully loaded with 14 t. Figure 4 shows that the load torque is slightly higher than 500 Nm when no load is carried in the bucket, while for full load conditions, the load torque reaches around 685 Nm.

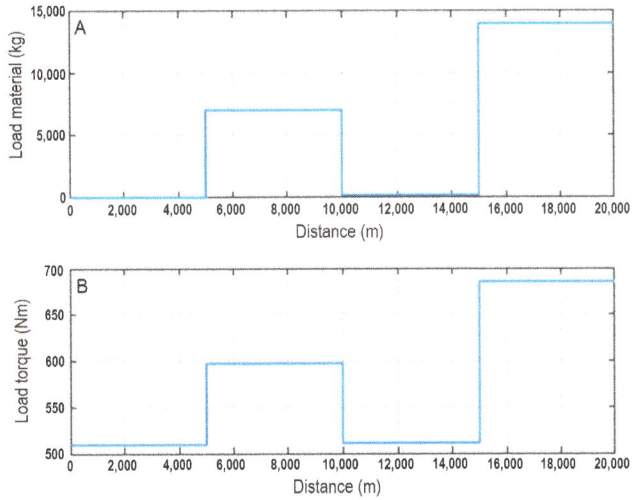

Figure 4. (**A**) Material loaded in the bucket; and (**B**) load torque response.

According to equation 19, the LHD speed is another variable that influences the load torque. Figure 5 shows the load torque calculated when considering different motor speeds, which results in different LHD velocities. In the figure, the load torque is 664 Nm when the LHD operates at zero speed considering the effect of a constant wind speed of 2.78 m/s added to the torque component associated with the LHD mass with no load in the bucket. Then, the motor speed is incremented in successive ramps followed by steady speeds that, for instance, reach 272.5 rpm at 1000 m and 1396 rpm at 10,000 m. The load torque developed at those points is 666.9 Nm and 686 Nm, respectively, showing a lower variation when compared with the effect of mass increments (Figure 4) over the load torque.

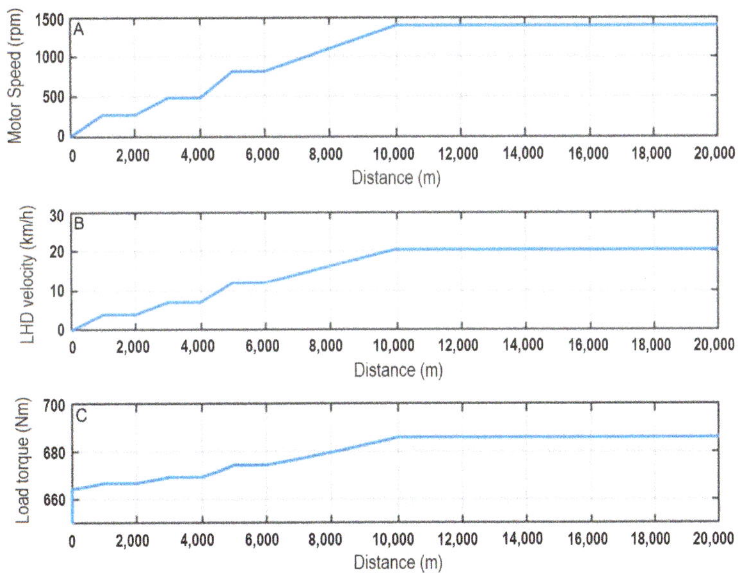

Figure 5. (**A**) engine rotation speed; (**B**) LHD velocity; and (**C**) load torque response.

3.2. Electric LHD Model

In this section, the inertia and load torque expressions obtained from the LHD load model are complemented with a dynamic model of the LHD electric drive to create and evaluate the full-electric LHD model.

As shown in Figure 2, the electric drive is composed of an induction motor, which provides the torque and speed for the LHD powertrain. Table 4 shows the technical parameters used to model the induction motor obtained by solving nonlinear equations derived from the motor equivalent circuit using regression algorithms as proposed in [27]. The power and speed for the LHD motor are assumed to be 132 kW and 1500, respectively, considering the LHD nameplate shown in Table 2.

Table 4. Summary of parameters determined for the induction-motor model.

Parameter	Value	Unit
Power	132	kW
Voltage	1	kV
Rotation speed	1470	rpm
Stator resistance	0.229	Ω
Stator reactance	0.4885	Ω
Rotor resistance	0.1255	Ω
Rotor reactance	1.1399	Ω
Magnetizing reactance	25.8843	Ω
Motor inertia	1.23	kgm^2
Number of poles	4	-

As shown in Figure 2, the induction motor is fed by a variable frequency drive (VFD) operating with indirect vector control considering decoupled speed and flux commands. Before implementing the vector control, it is necessary to perform a correct controller setting based on the frequency and phase-margin method. In this case, the crossover frequency of the flux and torque current controllers was set to 300 rad/s with an 86° of the phase margin. The speed controller was tuned considering a crossover frequency of 2 rad/s with a phase margin of 84°, while the flux controller considered a crossover frequency of 15 rad/s with 86° of phase margin.

Figure 6 shows the results of the speed command (ω_{ref}) and speed feedback (ω_m), torque current command ($i_{q,ref}$), and torque current feedback (i_q), load torque and electric torque, and electric power and output power generated by the electric drive, considering a load torque given by the parameters of Table 3. In this evaluation, a commanded speed ramp of 15 s accelerated the motor from 0 rpm up to 95% of the rated speed, which is equivalent to an LHD velocity ramp from 0 km/h to 20.5 km/h considering the tire dimensions and a gear ratio of 23.08.

Figure 6 shows that, at the end of the acceleration ramp (t = 16 s), the motor speed reaches 1396 rpm. As a result, the torque current command is increased above the rated value to increase the motor speed. It should be noted that the feedback current tracks the commanded signal with nearly zero tracking errors, showing a proper controller setting. Looking at Figure 6c, the electric torque has the same shape of i_{qs}, demonstrating an effective decoupled torque control during both dynamic and steady operation. During the acceleration stage (until t = 16 s), the electric torque increases above its rated value due to the significant inertia of the equipment that includes a loaded bucket. In Figure 6d, the electrical power of the LHD is significantly higher than the mechanical (output) power during the acceleration period. The mechanical power is obtained from the product between the load torque and speed while the electrical power is obtained by adding the internal losses of the induction motor to the mechanical power. Then, it can be inferred that the efficiency of the machine would tend to be lower during periods of acceleration; however, once the motor reaches steady operation this value would be close to 90%.

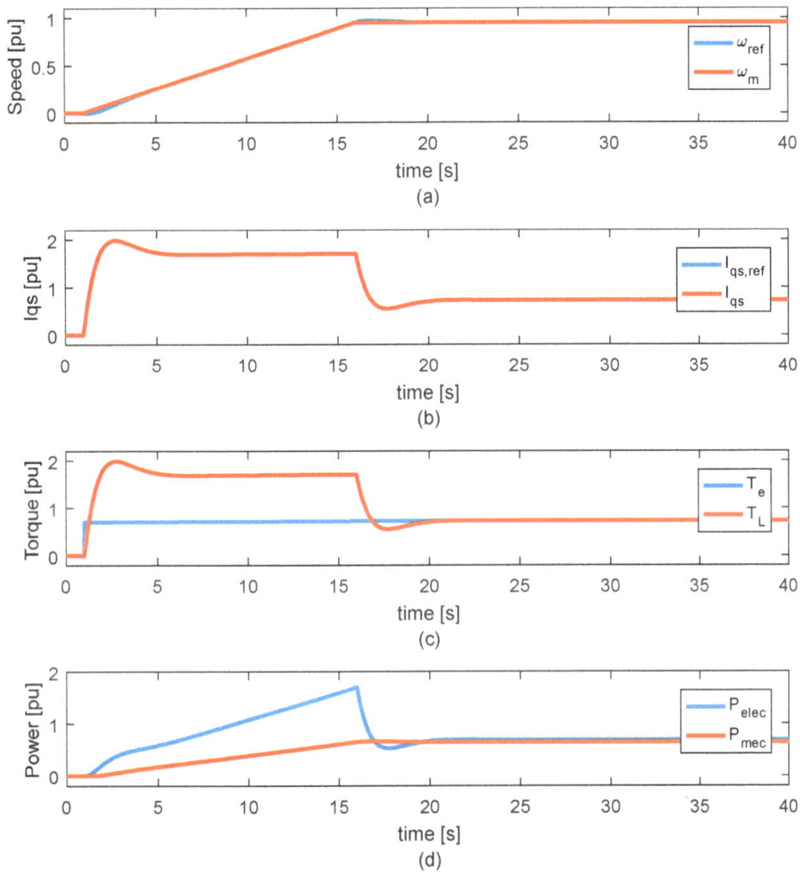

Figure 6. Numerical evaluation of the electric LHD model: (**a**) speed command and motor speed; (**b**) torque current command and motor torque current; (**c**) electric and load torque; and (**d**) electrical and mechanical power.

4. Electric LHD Model Based on Field Records

In this section, the Electric LHD model is evaluated considering the field record collected during the operation of a diesel LHD LH517 in an underground copper mine located in the central zone of Chile. Figure 7 shows the throttle, engine speed, engine torque and power, LHD velocity, and traveled distance during one hour of operation. The figure shows the evolution of these operational signals with clearly identifiable stops (at least six) of variable duration related to detentions where the operators wait to access the dumping zone or wait to return to the route after scooping or dumping. Here, the total traveled distance reaches 6.72 km. Despite the operational behavior, these field records can be used to assess the operation of the electric LHD model considering common work cycles developed in underground mines. The following section describes the field-signal selection and conditioning process required to evaluate the electric LHD model performance.

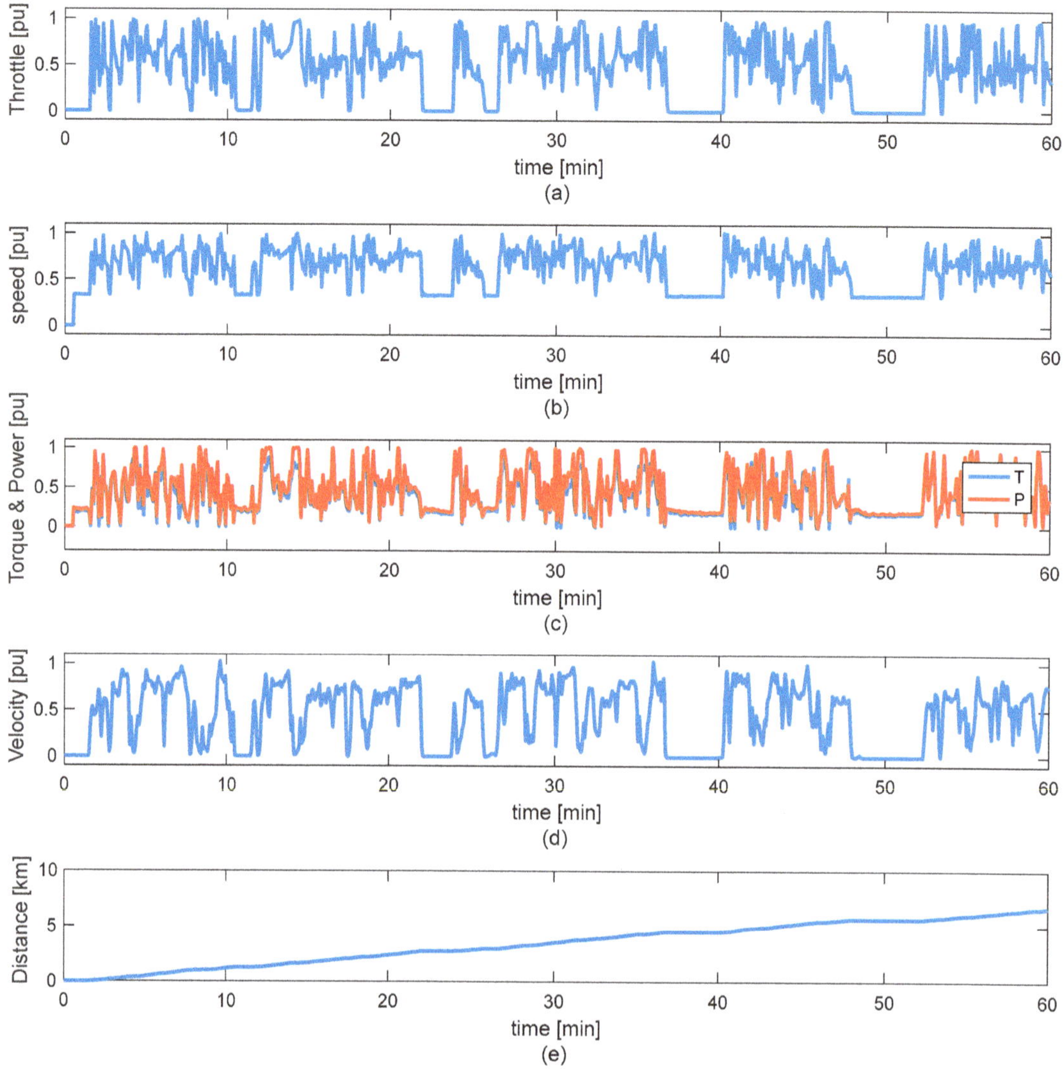

Figure 7. Operational variables of LHD LH517 during one hour of work: (**a**) throttle position; (**b**) engine speed; (**c**) engine torque and power; (**d**) LHD velocity; and (**e**) traveled distance.

4.1. Signal Conditioning of Field Records

To evaluate the electric LHD model presented in Section 3, the engine speed and engine torque of the diesel LHD were conditioned to be used as speed reference ($\omega_{ref,e}$) and load torque ($T_{L,e}$) inputs to the electric LHD model, respectively. Figure 8 shows the scheme of the field-record conditioning process. To obtain $\omega_{ref,e}$, the idle speed should be identified and subtracted from the engine speed. The idle speed is easily identified since it appears when the engine runs on, but the throttle position is zero representing the decoupling of the engine from the drivetrain. The disturbance input $T_{L,e}$, is directly obtained from the engine torque signal of the diesel LHD and is integrated into the electric LHD model along with the total inertia.

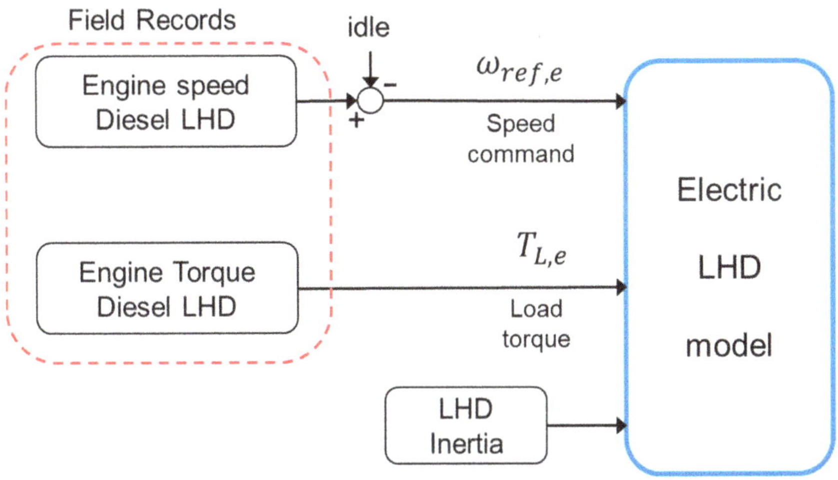

Figure 8. Field-record conditioning for electric LHD model evaluation.

4.2. Evaluation of Electric LHD Model

Once the field records of the engine speed and torque were properly conditioned, as described in the previous section, the electric LHD model was evaluated for the same time window (one hour) considered in Figure 7. Figure 9 shows the resulting signals of the speed command ($\omega_{ref,e}$) and speed feedback (ω_m), torque current command ($i_{qs,ref}$), and torque current feedback (i_{qs}), load torque and electric torque, and electric power demanded by the electric drive. To compute the total energy consumed during the period under evaluation, the motor input power was also included in Figure 9e. Figure 9a shows that the commanded speed is properly conditioned by eliminating the idle component. The speed and torque current commands are tracked by the motor speed and motor current with nearly zero tracking error, confirming the proper tuning of the electric drive controllers. As expected, the electrical torque developed by the induction motor is higher than the load torque during the acceleration periods while during the deceleration periods, the electrical torque is lower than the load torque, even reaching negative values that represent the regenerative operation of the electric LHD (Figure 9c). As a result, the power demanded by the electric drive shown in Figure 9d reaches peaks close to the rated power (1 p.u.) as well as negative values during regenerative operation. Figure 9e shows the energy demand of the electric LHD model during the evaluated period of one hour. This energy profile was obtained assuming that the operation of the hydraulic pump is contained in the load torque signal used to evaluate the electric LHD model. Additionally, to include the energy demand of the ventilation system, the energy computation considered a constant demand of 3.0 kW added to the motor power shown in Figure 9d during the period evaluated. With these considerations, the base value used to compute the total energy consumed by the electric LHD is 180 kW, which represents the sum of the induction motor, hydraulic pump, and ventilation-system-rated power. Based on these considerations, the total energy consumed by the electric drive at the end of the evaluation period reached 86.8 kWh.

The last evaluation compares the energy consumption between the electric LHD model to an equivalent diesel LHD. The energy consumption of the diesel LHD was obtained from the field records of diesel demand during one hour of operation of the LHD model LH517, with a load capacity of17 tons. The equivalent diesel consumption for a 14-ton capacity LHD was obtained using a ballpark figure by direct scaling of the diesel-demand signal with the load-capacity ratio (14/17). Figure 10a shows the estimated diesel demand for the 14-ton LHD, while Figure 10b shows the total diesel consumption obtained by direct integration of the diesel demand, considering the same one-hour operation period previously evaluated. Figure 10c compares the energy demand for the electric LHD and the diesel LHD. For the computation of diesel LHD energy demand, it was considered that diesel fuel has a specific energy of around 38 MJ per liter, equivalent to 10.6 kWh/L.

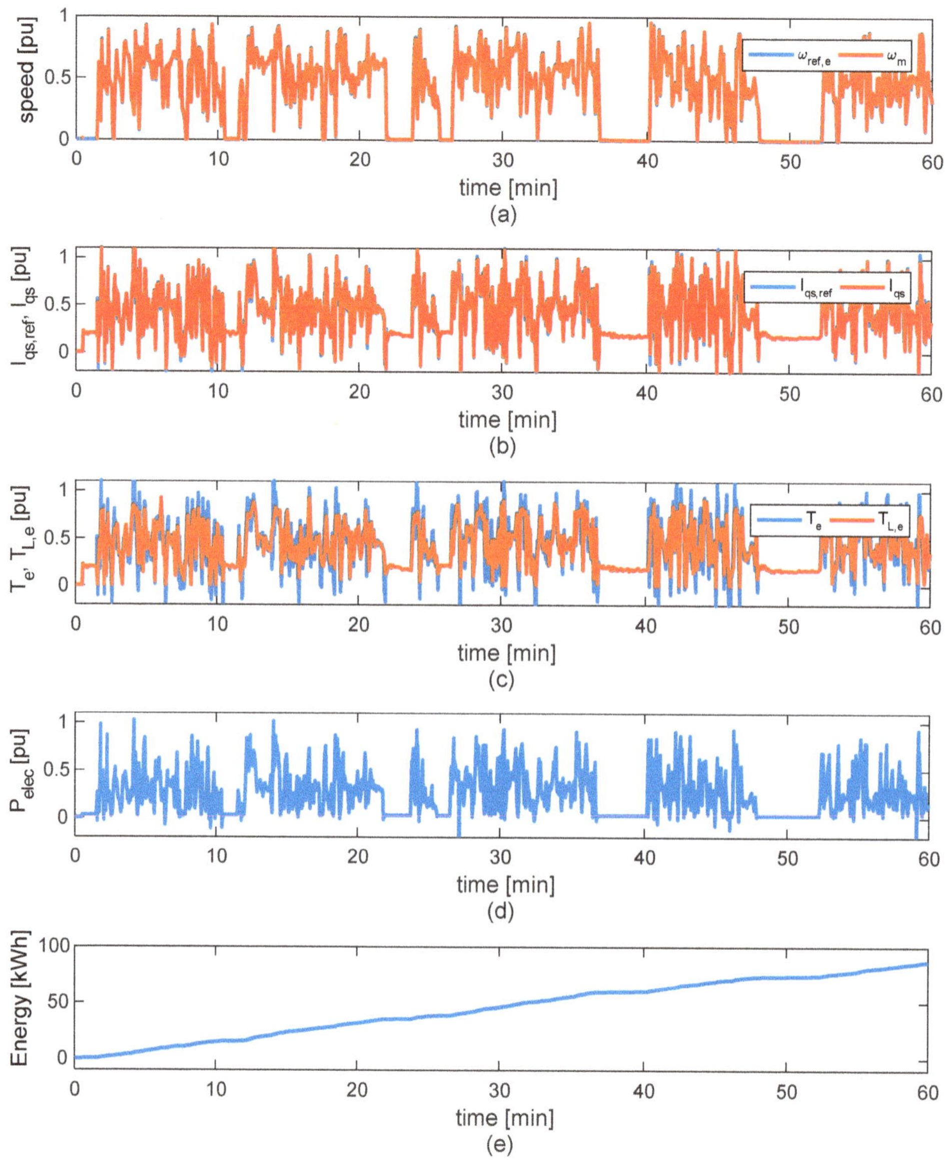

Figure 9. Evaluation of the electric LHD model based on the field records: (**a**) speed command and motor speed; (**b**) torque current command and motor torque current; (**c**) electric and load torque; (**d**) electric power; and (**e**) consumed energy.

Figure 10c shows that the diesel LHD has a much higher power consumption than the electric LHD. At the end of the evaluated operating hour, the electric LHD presents an energy consumption of 86.8 kWh, while the diesel LHD presents an energy consumption of 219.7 kWh. Therefore, the electric LHD saved 60.5% of energy compared with the diesel LHD for the hour of operation. This difference is mainly due to the low efficiency of diesel technology, which has significant heat losses that should be dissipated to the environment by radiators. Although in Chapter 4 the dynamic model of the gearbox was not considered due to its complexity, the gearbox efficiency for electric drivetrains ranges between 90% to 98% [28]. On the other hand, the electric LHD presents much greater energy efficiency, higher than 80%, if the power converter and motor efficiencies are combined.

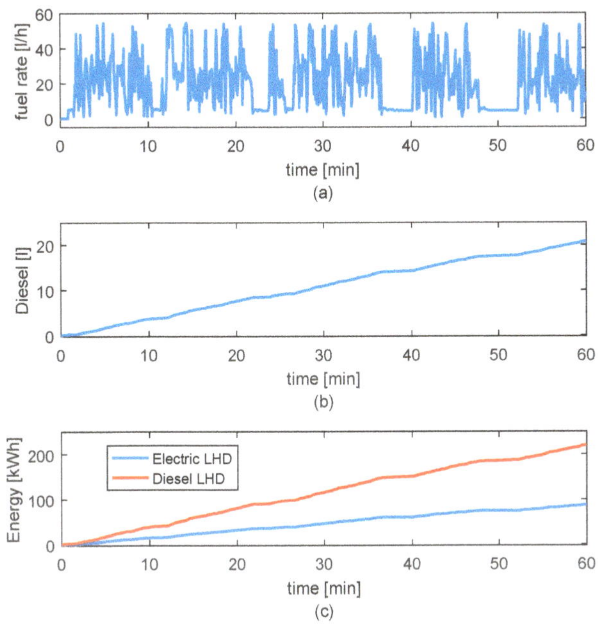

Figure 10. Analysis of energy consumption of electric and diesel LHD: (**a**) diesel-demand rate; (**b**) diesel demand; and (**c**) energy consumed by the electric and diesel LHD.

5. Conclusions

Although there is a current growing tendency towards incorporating electrical equipment in different industrial processes, hybrid and/or electric LHD technologies are not yet in high demand in underground mining, despite the variety of options available today. In the case of electric LHDs, this is probably mainly due to the limitations imposed by the trailing cable, or the limited-range battery packs. However, it is expected that the replacement of diesel-powered machines with their electrical counterparts will enable benefits such as exhaust gas reduction and a reduction of operational costs.

In this study, it was possible to characterize the work cycles of a diesel LHD in an underground mine in central Chile. This valuable information was used to propose a methodology that allows for the evaluation of the performance of an electric LHD based on field records already available for diesel equipment. The proposed methodology begins with the definition of the dynamic model of the electric LHD, which is the result of the integration of a detailed load model that provides the operational inertia and load torque with a vector-control electric drive.

Unlike the diesel LHD, the evaluation shows that the electric LHD can regenerate energy that contributes to the energy efficiency of the process, which results in power requirements 60.5% lower than the diesel LHD during one hour of continuous operation, representing significant energy savings. Thus, the proposed methodology represents a valuable tool for operators and process engineers to project the energy requirements for the adoption of battery or trailing-cable electric LHDs, based on the current and historical operational data of diesel LHDs.

Author Contributions: Conceptualization, G.R. and G.F.; methodology, G.R. and G.F.; validation, G.R., R.G. and K.S.; resources, G.R., K.S. and K.Z.; writing—original draft preparation, G.R. and R.G.; writing—review and editing, K.S. and K.Z.; visualization, G.R., R.G. and G.F. All authors have read and agreed to the published version of the manuscript.

Funding: This research was funded by AGH University of Science and Technology in Poland, scientific subsidy under number: 16.16.100.215, and Universidad Católica de la Santísima Concepción in Chile, project CIBAS-511317.

Data Availability Statement: Not applicable.

Conflicts of Interest: The authors declare no conflict of interest.

References

1. Flores, G. Future challenges and why cave mining must change. In *3rd International Symposium on Block and Sublevel Caving*; Castro, R., Ed.; Universidad de Chile: Santiago, Chile, 2014; pp. 23–52.
2. Salama, A.; Greberg, J.; Skawina, B.; Gustafson, A. Analyzing energy consumption and gas emissions of loading equipment in underground mining. *CIM J.* **2015**, *6*, 179–188. [CrossRef]
3. Halim, A.; Lööw, J.; Johansson, J.; Gustafsson, J.; van Wageningen, A.; Kocsis, K. Improvement of Working Conditions and Opinions of Mine Workers When Battery Electric Vehicles (BEVs) Are Used Instead of Diesel Machines—Results of Field Trial at the Kittilä Mine, Finland. *Min. Metall. Explor.* **2022**, *39*, 203–219. [CrossRef]
4. Varaschin, J.; De Souza, E. Economics of diesel fleet replacement by electric mining equipment. In Proceedings of the 15th North American Mine Ventilation Symposium, Blacksburg, VA, USA, 20–25 June 2015; Volume 8.
5. Tuck, M. Mine Ventilation. In *SME Mining Engineering Handbook*, 3rd ed.; Darling, P., Ed.; SME: Plymouth, MA, USA, 2011; pp. 1577–1594.
6. Michael, D.A.; Patricia, L.S.; Jay, H.L.; Aaron, B.; Patricia, A.S.; Roel, V.; Joseph, B.C.; Debra, T.S. The Diesel Exhaust in Miners Study: A Cohort Mortality Study with Emphasis on Lung Cancer. *J. Natl. Cancer Inst.* **2012**, *104*, 869–883. [CrossRef]
7. Jacobs, W.; Hodkiewicz, M.R.; Bräunl, T. A cost–benefit analysis of electric loaders to reduce diesel emissions in underground hard rock mines. *IEEE Trans. Ind. Appl.* **2014**, *51*, 2565–2573. [CrossRef]
8. Gleeson, D. Komatsu brings hybrid drive, narrow vein LHDs to Australian hard-rock market. *Int. Min.* **2021**, *129*, 120–134. [CrossRef]
9. Epiroc. Scooptram ST7 Battery. 8. 2022. Available online: https://www.epiroc.com/es-es/products/loaders-and-trucks/electric-loaders/scooptram-st7-battery (accessed on 15 March 2023).
10. Fiscor, S.; Jensen, J. The Best of MINExpo 2021. *Eng. Min.* **2021**, *222*, 42–53. Available online: https://www.proquest.com/docview/2604092726?pq-origsite=gscholar&fromopenview=true (accessed on 15 March 2023).
11. Sandvik. Sandvik LH518B—Battery Electric Loader. 8. 2021. Available online: https://www.rocktechnology.sandvik/en/products/underground-loaders-and-trucks/battery-electric-loaders-and-trucks/lh518b-battery-electric-loader/ (accessed on 15 March 2023).
12. Nieto, A.; Schatz, R.; Dogruoz, C. Performance analysis of electric and diesel equipment for battery replacement of tethered LHD vehicles in underground mining. *Min. Technol.* **2020**, *129*, 22–29. [CrossRef]
13. Grycan, W. Electric Vehicles in Mining for the Aspect of Operational Safety. *Prz. Elektrotechniczny* **2022**, 110–113. [CrossRef]
14. Katta, A.K.; Davis, M.; Kumar, A. Assessment of greenhouse gas mitigation options for the iron, gold, and potash mining sectors. *J. Clean. Prod.* **2020**, *245*, 118718. [CrossRef]
15. Rojas, G. Introducción de un LHD Híbrido a la Industria Minera y sus Posibilidades en el Mercado Chileno. Bachelor's Thesis, Universidad de Chile, Santiago, Chile, 2017.
16. Wolff, J.; Gómez, G. The Switched Reluctance Motor—An electric motor with high torque and small volume. *Energía* **1997**, *4*, 113–115. (In Spanish)
17. Schultz, J.W.; Huard, S. *Comparing AC Induction with Permanent Magnet Motors in Hybrid Vehicles and the Impact on the Value Proposition*; Parker Hannifin Corporation: Cleveland, OH, USA, 2013.
18. Hamouda, M.; Al-Amyal, F.; Odinaev, I.; Ibrahim, M.; Számel, L. A Novel Universal Torque Control of Switched Reluctance Motors for Electric Vehicles. *Mathematics* **2022**, *10*, 3833. [CrossRef]
19. Krause, P.C.; Wasynczuk, O.S.; Sudhoff, D. *Analysis of Electric Machinery and Drive Systems*; John Wiley & Sons: Hoboken, NJ, USA, 2002.
20. Ramirez, G. *Control of Electric Drives*; Universidad Católica de la Santísima Concepción: Concepción, Chile, 2017. (In Spanish)
21. Valenzuela-Cruzat, J.; Valenzuela, M.A. Integrated Modeling and Evaluation of Electric Mining Trucks During Propel and Retarding Modes. *IEEE Trans. Ind. Appl.* **2018**, *54*, 6586–6597. [CrossRef]
22. Meng, Y.; Fang, H.; Liang, G.; Gu, Q.; Liu, L. Bucket Trajectory Optimization under the Automatic Scooping of LHD. *Energies* **2019**, *12*, 3919. [CrossRef]
23. Ramírez, G.; Valenzuela, M.A.; Pittman, S.; Lorenz, R.D. Modeling and Evaluation of Paper Machine Coater Sections Part 1: 1-Coater Section and Tension Setpoints. *IEEE Trans. Ind. Appl.* **2019**, *55*, 2144–2154. [CrossRef]

24. Gillespie, T. *Fundamentals of Vehicle Dynamics*; SAE International: Warrendale, PA, USA, 2021.
25. Sanvik. Optime Monitoring. 2022. Available online: https://www.rocktechnology.sandvik/optimine (accessed on 15 March 2023).
26. Wicaksana, Y.; Widodo, N.P.; Kramadibrata, S.; Kresna, R. Determining rolling resistance coefficient on hauling road using dump-truck in open pit coal mine. In Proceedings of the International Symposium on Earth Science and Technology, Fukuoka, Japan, 6–7 December 2011.
27. Lee, K.; Frank, S.; Sen, P.K.; Polese, L.G.; Alahmad, M.; Waters, C. Estimation of induction motor equivalent circuit parameters from nameplate data. In Proceedings of the 2012 North American Power Symposium (NAPS), Champaign, IL, USA, 9–11 September 2012; pp. 1–6.
28. Břoušek, J.; Zvolský, T. Experimental study of electric vehicle gearbox efficiency. In *MATEC Web of Conferences*; EDP Sciences: Les Ulis, France, 2018; Volume 234, p. 02004.

energies

MDPI

Article

Geomechanical and Technical Aspects of Torpedo Blasting under Seismic and Rockburst Hazard Conditions in Legnica–Glogow Copper District Mines

Zbigniew Burtan [1,*], Jerzy Cieślik [1], Dariusz Chlebowski [1], Paweł Piasecki [2] and Krzysztof Gzik [2]

1 Faculty of Civil Engineering and Resource Management, AGH University of Krakow, Mickiewicza 30 av, 30-059 Cracow, Poland; jerz@agh.edu.pl (J.C.); chlebo@agh.edu.pl (D.C.)
2 KGHM Polska Miedź S.A., O/ZG Rudna, Henryka Dąbrowskiego 50 av, 59-100 Polkowice, Poland; pawel.piasecki@kghm.com (P.P.); krzysztof.gzik@kghm.com (K.G.)
* Correspondence: burtan@agh.edu.pl

Abstract: The dominant hazard in the Polish copper ore mining industry (LGCD mines—Legnica–Glogow Copper District) is the occurrence of mining tremors and rockbursts. One of the effective active methods of preventing this threat is torpedo blasting, which results in disturbing the structure of roof rocks. A change in the integrity of a roof, especially in the tremor-generating layer or in the contact between stiff rock layers, reduces the possibility of an elastic energy concentration and may also be a kind of stress concentrator, provoking the destruction of rocks. This article presents original solutions for determining the areas of rock mass within a mining area where it is advisable to conduct torpedo blasting, and the experience of blasting in the Rudna copper mine in the LGCD is described. The first part of this article presents the results and experiences of using torpedo blasting prevention in the LGCD mine conditions in the Rudna mine. Assuming that due to the stress of the rocks, torpedo blasting brings the greatest benefits in the areas of elastic energy concentration, the second part of this article presents the results of numerical modeling, on the basis of which the zones of elastic energy concentration in the form of total, shear and volume deformation were determined in the vicinity of mining activities for typical geological and mining conditions in the LGCD. The importance of methods for the verification of the proposed solutions based on the analysis of seismic activity and geophysical mechanisms of events' foci and seismo-acoustic emissions was also emphasized. The numerical simulations performed and conclusions from in situ observations allowed the formulation of general principles for the selection of torpedo blasting parameters.

Keywords: underground mining; rock mechanics; mining tremors; rockburst hazard; torpedo blasting; mining experience

Citation: Burtan, Z.; Cieślik, J.; Chlebowski, D.; Piasecki, P.; Gzik, K. Geomechanical and Technical Aspects of Torpedo Blasting under Seismic and Rockburst Hazard Conditions in Legnica–Glogow Copper District Mines. *Energies* **2024**, *17*, 1174. https://doi.org/10.3390/en17051174

Academic Editor: Manoj Khandelwal

Received: 30 September 2023
Revised: 12 February 2024
Accepted: 16 February 2024
Published: 1 March 2024

1. Introduction

Major natural hazards in copper mining in Poland include rockbursts and fault movements. Most rockbursts and rock mass destressing events registered in the Legnica–Glogow Copper District (LGCD) (Figure 1) have been the consequence of high-energy events, whilst the current rockbursting hazard level is associated with mining-induced seismicity. The potential for high-energy tremors is an inherent feature of copper fields within the LGCD region, mostly due to the occurrence of burst-prone roof rock strata over the worked-out areas and periodic faulting movements; these impacts are further exacerbated due to current mining activities [1]. The analysis of seismic activity recorded in recent years in the LGCD mine region shows that the most seismically active is the Rudna mine (Figure 1) [1,2].

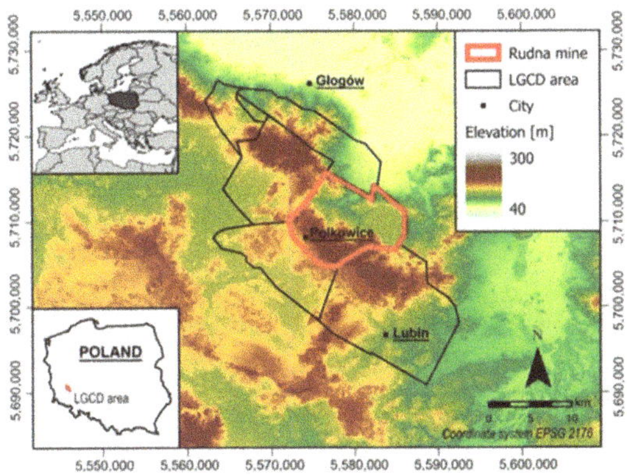

Figure 1. Location of the Rudna underground copper ore mine [3].

One of the effective active methods of preventing the threat of tremors and rockbursts is torpedo blasting performed in thick and highly stiff layers of the rock mass, which are the so-called tremor-prone layers. Torpedo blasting's aim is to provoke tremors or the local destruction of these layers (Figure 2) [4].

Figure 2. Scheme of torpedo blasting in a roof.

Considering the effect of torpedo blasting in terms of rockburst prevention, the effect of these blasts may be twofold, as follows [4]:

- Provoking a tremor, including a rockburst, as a result of disturbing the unstable stress balance of the rock mass.
- The creation of a local zone of destruction and weakening of the rock mass in the areas of expected stress concentration, without an immediate effect in the form of a tremor. Such a zone in the absorbing layer is a local place of stress concentration, which will result in the lower strength of the entire layer and the possibility of its destruction with the lower accumulation of elastic energy.

In the first case, the blasting should be characterized by a high intensity of paraseismic vibrations and should be adapted to the local mining and geological conditions due to possible difficulties in maintaining the working roof. It can then be combined with blasting in a larger number of exploitation workings to progress exploitation. In the second case, torpedo blasting may be an element of the so-called pre-emptive action for tremor prevention, planned and performed well in advance.

Torpedo blasting was used for the first time in Poland to reduce the risk of tremors and to ensure appropriate conditions for the collapse of roof rocks during the exploitation of coal seams in the Upper Silesian Coal Basin (USCB) [5]. Over time, these blastings became one of the elements of the active rockburst prevention system in the Polish coal mining industry (USCB mines) [5–10].

Torpedo blasting was also used in Polish copper mines in the LGCD [11,12], and as a result it was possible to exploit the deposit in rockburst-prone areas, in zones of geological anomalies [13], tectonic disturbances [14] and when operating in constrained conditions [12].

Blasting as a method of reducing the risk of rockbursts has been the subject of numerous international scientific works. Most of these works, however, are focused on stress relief blasts carried out at the working front or in the immediate roof in hard coal mines. Brauner [15–17] discussed the use of this method to limit rockbursts used in Germany.

The purpose of the destress blasting in seams or the roof and floor around the workings is to transfer the excessive induced stresses to the interior rock strata and to provide a protective barrier surrounding the excavation [16]. Kexin [18] and Xia et al. [19] demonstrated the application of the destress blasting to control of the floor heave in deep coal mines in China and reported a decrease in rock mass deformation after its application. Numerous publications have reported on stress relief blasting operations in Czech mines (part of USCB) [20–22]. Konicek et al. [23] highlighted the main benefits of using stress relief blasts as an effective method to overcome the high stresses causing rockbursts in these mines. In another article [24] by Konicek et al., the effectiveness of torpedo blasts in preventing rockbursts was estimated using the seismic effect method. Stress relief blasting has also been used in Indian coal mines [25], in South African ore mines [26] and in a Canadian copper mine [27,28]. Andrieux et al. [27] assessed the effectiveness of blasting using seismic tomography in one of the mines in Canada. Andrieux and Hadjigeorgiou [29] presented an empirical method of assessing the effectiveness of stress relief, based on various geotechnical and operational parameters. Fulawka et al. [30] proposed methods for assessing the effectiveness of stress relief blasts and also suggested innovative quantitative methods based on in situ seismic measurements. Modifications of stress relief blasting patterns aimed to improve their effectiveness in LGCD mines were presented by Mertuszka et al. [31]. The occurrence of high-stress zones determined by numerical calculations can be verified by a geophysical analysis of the mechanisms of foci events induced in these zones of tremors [32–34]. The widely used acoustic emission technology can also be used to monitor the destruction of roof rocks as a result of torpedo blasting [35,36], where the identification of seismo-acoustic signals allows for revealing the rock cracking mechanism.

The processes occurring within the rock strata as a result of stress relief blasting were also analyzed on the basis of numerical modeling. This type of analysis allows us to estimate the impacts of the blast energy depending on the sequence of blasts and the total amount of used explosives [30]. Numerical calculation techniques and their applications for analyzing the response of the rock strata to dynamic explosive load were proposed by Taylor et al. [37], who developed failure models capturing the characteristic dynamic fracture of rock under tension. Later, Maxwell and Young [38] developed numerical methods for examining objects under dynamic loads and simulated the extent of the failure zone under such loads.

Recently, numerical simulations have been used to describe the propagation of seismic waves in the rock mass after the detonation of explosives [39,40], and crack propagation around the blast hole [40–45]. Advanced simulations were also carried out to describe the initial destress of the rock mass after blasting [46–48]. Numerical methods were used to analyze the impact of blasting on underground infrastructure [49–52].

Geomechanical effects associated with large-scale rock mass destress were modeled by Vennes et al. [28,53]. The work of Miao et al. [54] reviewed contemporary methods of numerical modeling related to stress relief blasts.

The first part reported examples and case studies of torpedo blasting in LGCD conditions. The use of the rockburst prevention techniques was discussed recalling the blasting operations conducted in two sites within the Rudna mine, G-23/1 and G-11/8. In the first case, at the G-23/1 site, the torpedo blasting took place through long holes reaching high in the roof right to the contact zone between dolomite and anhydrite, in the region of the working face. In the second case, the torpedo holes were shorter and drilled near the edges of the mining site to be liquidated. These holes, blasted simultaneously with the blasting of excavation faces, resulted in the division of a thick layer of strong dolomites into smaller ones, taking advantage of their natural cleavability. Interestingly, the torpedo blasting holes were located in the corner of the G-11/8 panel, where numerical calculations showed the highest concentration of elastic energy. In both cases, good blasting efficiency was achieved, confirmed by the occurrence of provoked tremors after an appropriate waiting period.

In consideration of the effectiveness of torpedo blasting in LGCD, it is extremely important to determine the region and zones within the rock mass that correspond to the highest elastic energy densities. These zones are determined by the parameters of the exploitation system and the geological structure of the deposit, i.e., the arrangement of geological layers (the occurrence of so-called tremor-prone layers) and the discontinuity of rock strata determining the conditions of contact between the layers, possible separation and slippage at the interface. These factors determining the effectiveness of torpedo blasting were the subject of numerical analyses undertaken in the second part of this study. The proposed methodology for determining highly stressed areas in the tremor-generating layer, with respect to torpedo blasting, is a simplified example of the use of geomechanical numerical modeling aimed at identifying tremor and rockburst hazard zones. It should be added that the first part of the study does not involve an in situ test following the numerical modeling procedure. Therefore, the continuation of the conducted research should be aimed at determining the areas, the scope and timing of stress relief blasting in order to improve the effectiveness of stress relief in the rock strata.

2. Case Study of Rudna Mine and the Use of Torpedo Blasting as Active Rockburst Prevention

Polish copper mines have already gained a great deal of experience in conducting active rockburst prevention. In the geological and mining conditions of the LGCD, a number of single or cyclic torpedo blasts would be carried out in the roof in order to disintegrate it, thus limiting the capability of the rock strata to accumulate elastic energy. Taking into account different lengths of boreholes, their location in relation to the tremor-prone layers and the parameters of the exploitation system, blasting tests were planned and carried out, among others, in the sites G-23/1, G-1/7, G-7/5, X/1 and G-11/8 in the Rudna mine [12]. The underlying assumptions were similar and focused on the use of explosive charges in the contact zone between the layers of limestone and anhydrite. This article provides a detailed report on two blasting operations (G-23/1, G-11/8) illustrating technical solutions for the local mining and geological conditions.

At the G-23/1 site, the deposit was formed by dolomites, copper-bearing shales and sandstones, and was located at a depth of ~1070 m, its thickness was 3.9–5.6 m and the average uniaxial compressive strength was approximately 60 MPa. The roof was made of compact dolomites, divided into slabs up to 1 m thick, the surfaces of which were covered with clay, gypsum and, less often, anhydrite. The thickness of the dolomite layer was approximately 60 m and its uniaxial compressive strength was approximately 102 MPa. On top of the dolomite, there was 143 m of anhydrite layers separated by dolomitic inclusions (9.7 m) and clay shales (4 m) (Figure 3).

After the rockburst that occurred in the final phase of exploitation of the G-23/1 panel, a single torpedoing of the roof was carried out as part of additional rockburst prevention, with four blast holes approx. 60 m (ϕ96 mm) in length, inclined at an angle of approx. 75 degrees to the horizontal. The holes were drilled from workings located along the front such that their ends reached the roof part of compact dolomites (Figure 4) and at the same time were found in a working area between 45 and 55 m of the working face (Figure 5). The total amount of emulsion explosives charge (emulgit 82 GP, cartridged) was ~1.0 Mg, which gave rise to a tremor with the energy rating of 8.6×10^3 J. However, the actual effect of the torpedo blasts is difficult to estimate because the blasts were carried out at the end of the exploitation, when it was practically coming to an end.

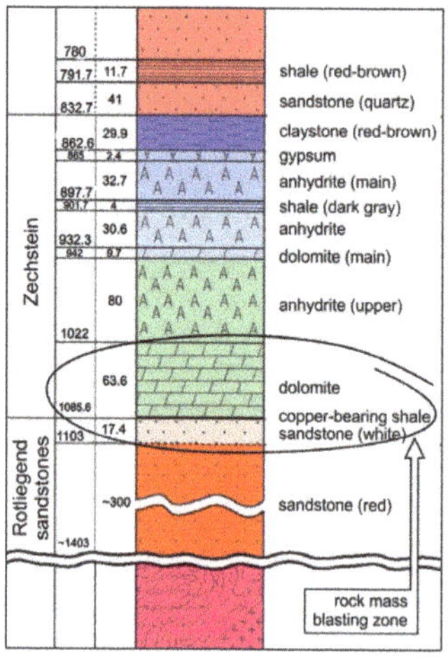

Figure 3. Lithological profile (part) of the deposit in the G-23/1 area.

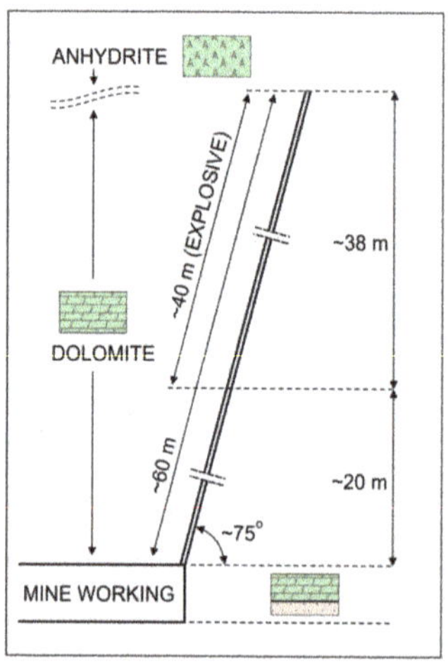

Figure 4. General scheme of torpedo blasting in the roof (G-23/1).

Figure 5. Sketch and location of blasts in the roof (G-23/1).

In the G-11/8 field, the deposit consisted of all lithological series, including dolomite, Zechstein shale and sandstone. The deposit was located at a depth of ~1125 m and its thickness was 2.5–3.1 m. The uniaxial compressive strength of the rock strata in the working face zone was around 78 MPa. In the roof of the workings, there was a dolomite layer approximately 15 m in thickness. These were cryptocrystalline dolomites, massive and compact, with a clear plate division (0.1–0.8 m), with the compressive strength of the order of 137 MPa. Directly overlying the dolomite formations, there was a 64 m layer of anhydrites, topped by rock salt (194 m), clay shales (1.5 m) and anhydrite again (48 m) (Figure 6).

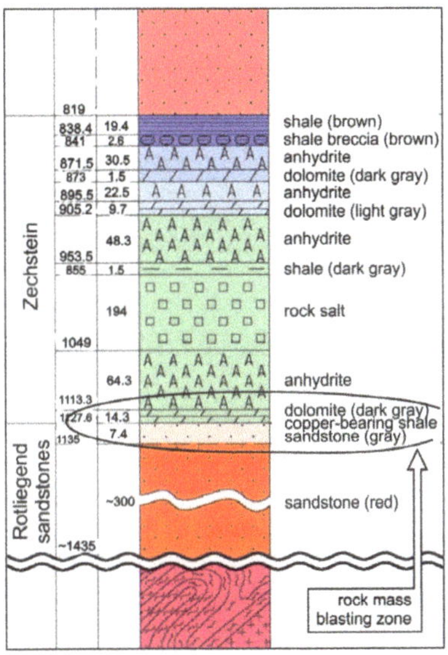

Figure 6. Lithological profile (part) of the deposit in the G-11/8 area.

After the rockburst, which destroyed a part of the workings, further mining activity in the area was carried out using an additional (non-routine) active prevention strategy, consisting of cyclical torpedo blasting in the roof (every 30 m of the face advance), simultaneously with group blasting in the face front. Blast holes with a length of ~15 m, ϕ45 mm reaching the roof section composed of carbonates (Figure 7), were drilled from workings located in the corner of the rear part of the working area (indicated with yellow color on the map in Figure 8) at an angle of approximately 70 degrees to the horizontal. In the period of 4 years from the resumption of mining activity, 20 torpedo blasting operations were carried out, consuming an average of 1.7 Mg explosive materials in each series, detonated only in the roof (not taking into account group blasting at the face front). Each blasting operation was performed using 60–80 blast holes, charged with emulsion explosives (emulgit RPT, in bulk) mechanically, in an amount of approximately 24 kg per hole. During the waiting period after the blasting works, a total of 42 rock mass tremors with energies above 10^3 J were recorded, including 17 events with energy ratings of the order of 10^5 J, 14 of the order of 10^4 J and 11 of the order of 10^3 J, respectively.

Based on the expertise gained during torpedo blasting at the Rudna mine, it should be stated [12] that drilling, loading and blasting ~60 m long holes in the roof (G-23/1, G-1/7) is feasible and well within the technical and organizational capabilities. Due to the relatively long period of implementation of blasting works, the continuous presence of miners in the area of the face front is required. This is a major security issue.

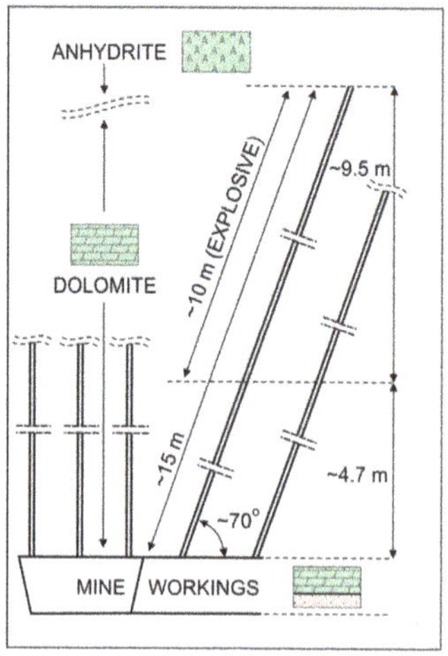

Figure 7. General scheme of torpedo blasting in the roof (G-11/8).

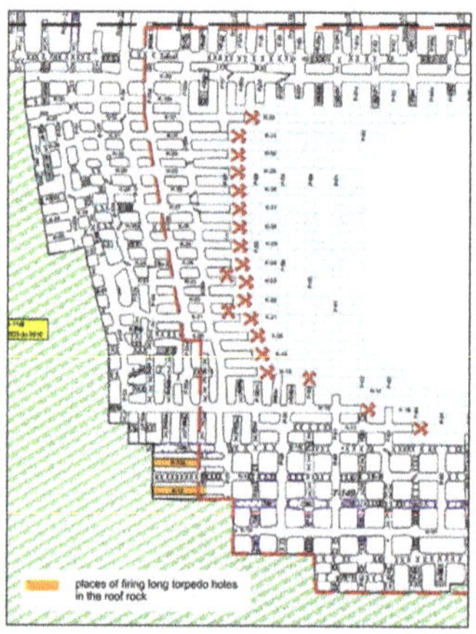

Figure 8. Sketch and location of blasts in the roof (G-11/8).

As a result of arranging the holes along the face front and mine working (e.g., in fields G-7/5, X/1) [12], portions of roof layers in the vicinity of the workings would be damaged, thus prompting the need to shut down these workings and disturbing the mining operations. In turn, blasting in the vicinity of a geological disturbance in the form of a syncline (section X/1) [12] allowed for stabilization of the convergence rate of mine workings in the working zone. Apparently, active prevention based on torpedo blasting is generally effective, yet further research and analyses (including the use of numerical tools) aimed at further improving its effectiveness are still fully merited.

3. Materials and Methods

Geomechanical Conditions for Torpedo Blasting—FEM Analysis

In the light of general assumptions regarding torpedo blasting, it is vital that the state of effort in the rock strata in the area to be mined should be known as well as the elastic strain energy accumulated in the rock strata, especially in the tremor-generating layers. The purpose of torpedo blasting is to induce a tremor by disturbing the state of stress equilibrium in the rock strata, where the failure is imminent. This is possible by supplying elastic energy that will provoke a tremor under given conditions. Assuming the continuity of the rock strata and simplifying the site geometry, it is possible to determine the distribution of elastic strain energy and effort to identify regions where the use of torpedo blasting should be most effective.

Accordingly, 3D numerical FEM models were built emulating the general geological conditions and the room-pillar mining system of the LGCD and the G-23/1 and G-11/8 sites in the Rudna mine (Figure 1). Tremor-prone layers in these areas responsible for seismic activity are represented by dolomitic limestone and anhydrites of variable thicknesses (Table 1 and Figure 9).

Table 1. Parameters of rock layers in the Rudna mine used in the numerical analyses.

Rock Layer	Thickness [m]	E_m [GPa]	ν [-]
Anhydrite	100	17.9	0.22
Dolomite	40	13.3	0.22
Dolomite deposit	4.5	8.8	0.24
Sandstone	100	5.7	0.35
Liquidation	4.5	0.2	0.4

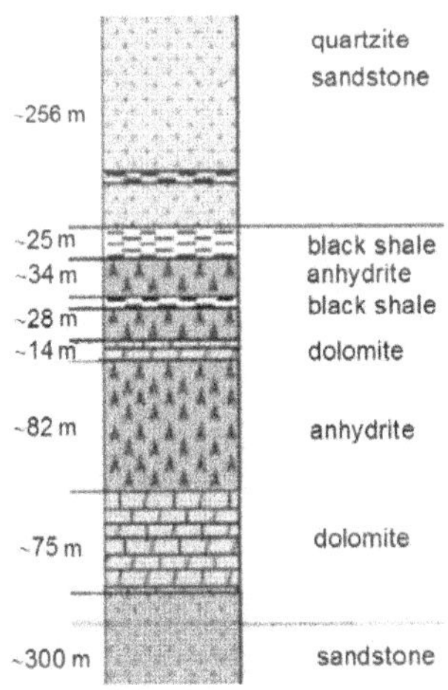

Figure 9. Example of the geological profile of the Rudna mine.

The geometry of the numerical model, the FEM mesh and the boundary conditions are shown in Figure 4. The calculations were performed on a linear-elastic constitutive model of the rock strata, whose parameters were determined on the basis of geological data from Rudna mine and recalling the Hoeck–Brown relationship [55]. Underlying the model are the general assumptions of the room and pillar exploitation system used in the LGCD, i.e., the front section, working area (Figure 10b—red), the mining liquidation zone (Figure 10b—green) and the abandoned liquidation zone (Figure 10b—grey) These aspects were incorporated by adopting equivalent deformation parameters of the corresponding spaces. In this approach, small operational pillars were not modeled. The conditions assumed in the adopted FEM calculations in terms of site geometry correspond to the mining operations in the G-23/1 and G-11/8 sections of the Rudna mine (Figure 10a).

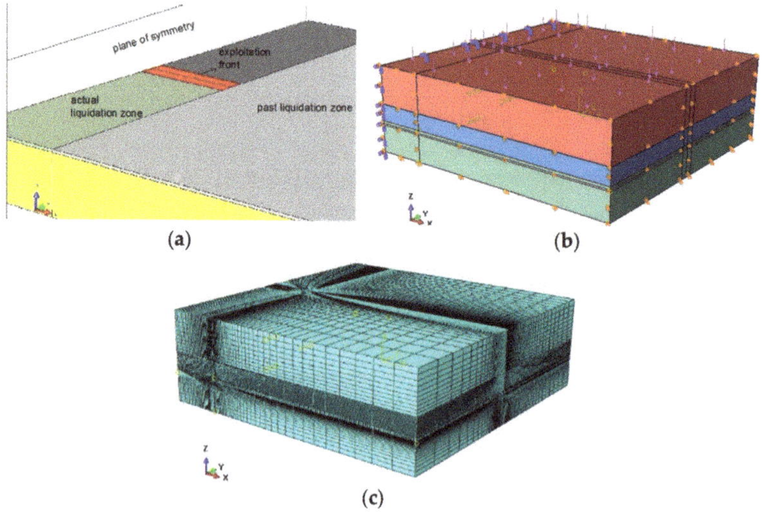

Figure 10. The computational model assumptions, (**a**) geometry, (**b**) boundary conditions, (**c**) FEM mesh.

The proposed method of analysis is a simplified one and the numerical procedure was applied without handling each and every variant of the room-pillar working system used in the LGCD. This simplification strategy yields only the results of analyses that are adequate to typical mining situations encountered in LGCD conditions. A solution was adopted whereby separate pillars should be replaced by zones with equivalent elastic parameters, whose impacts on the high roof layers are similar to those exerted by separate pillars. These parameters were adopted on the basis of previous calculations and recalling the authors' expertise in modeling typical convergences measured in the mine working area. Obviously, advanced calculations could and should take into account the modeling of separate pillars, as discussed in [56,57].

The calculation procedure relies on the displacement boundary conditions of the model corresponding to zero displacements on the respective edges of the model (Figure 10b). The acting load was the vertical normal stress exerted on the upper horizontal edge of the model corresponding to the mining depth of 1100 m. The problem was solved in a multi-step computational procedure reflecting the conditions of the initial state of stress and the progress of mining operations. The problem was treated as geometrically non-linear; therefore, the FEM equations in the ABAQUS v.6.12 computing system were solved using the Newton–Raphson incremental iterative procedure.

4. Results and Discussion—Rock Mass Effort and Elastic Strain Energy as Factors Determining the Effectiveness of Torpedo Blasting

4.1. *The Effort of the Tremor-Prone Layer in the Working Area*

In the context of effectiveness of torpedo blasting, two aspects are of primary importance, i.e., the direction of the blast holes and the size of the explosive. These two factors are related to the geological and mining conditions, but they can be determined using the analysis of effort and distribution of elastic energy accumulated in the rock strata.

The dynamic stresses necessary to trigger a tremor can be determined by Expression (1) resulting from the Mohr–Coulomb failure criterion (Figure 11).

$$\sigma_d(x,y,z) \geq W_{M-C} - W_S(x,y,z) \quad (1)$$

where

W_{M-C}—a measure of critical effort according to the Mohr–Coulomb M-C criterion;
W_{M-C} = $2c\cos\varphi + (\sigma_1 + \sigma_3)\sin\varphi$;

$W_S(x,y,z)$—a measure of effort due to static stress tensor (before blasting);
$W_S(x,y,z) = \sigma_1 - \sigma_3$;
$\sigma_d(x,y,z)$—a effort required to trigger a tremor under the dynamic load caused by torpedo blasting;
σ_1, σ_3—greatest and least principal stress.

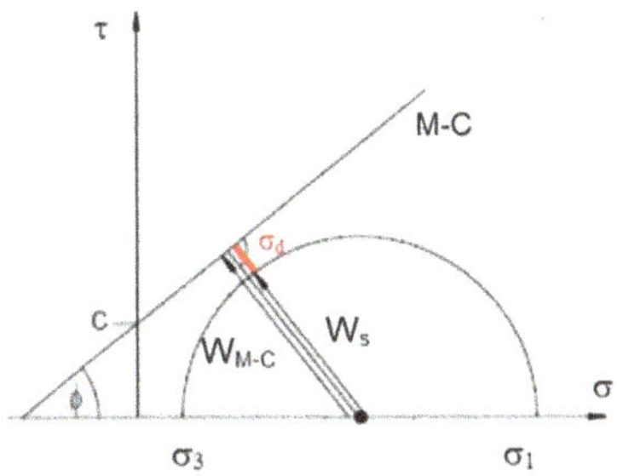

Figure 11. Mohr–Coulomb failure envelope. The effort needed to trigger tremors is marked in red.

Expression (1) applies only to regions within the rock strata where the effort W is defined as:

$$W = \frac{W_S}{W_{M-C}} \tag{2}$$

taking the value $W < 1$. According to the authors, the regions of the rock strata where the effort value falls in the range of $0.8 < W < 1$ can be considered the most favorable in the context of locating blast holes, because the value of dynamic stresses σ_d required to fracture the rock strata there is relatively small. The map of the strain factor W (2) covering the rock strata in the dolomitic limestone layer treated as a tremor-prone layer in LGCD conditions is shown in Figure 12a.

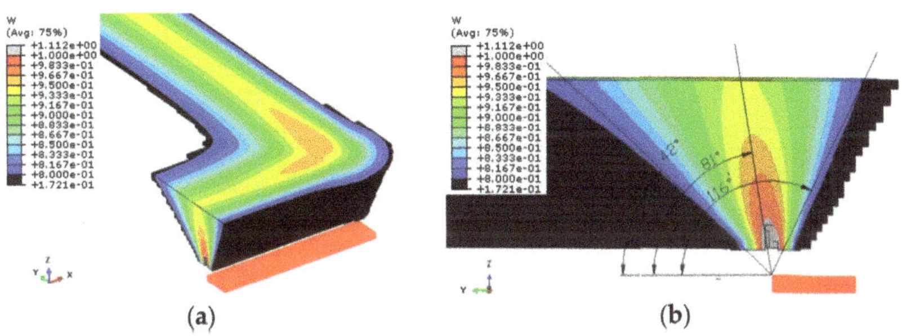

Figure 12. Effort map in the dolomitic limestone layer for the Rudna mine conditions. (**a**) Perspective view, (**b**) cross section view.

In the tremor-prone layer, the area with the effort value of $0.8 < W < 1$ is found in front of the excavation and is wedge- shaped with a wall inclination of 42–116 degrees. The highest effort value in the analyzed layer is registered at an angle of 81 degrees to the horizontal, towards the working zone (Figure 12b). In the corner of the working zone (Figure 12a), the high effort section is locally wider. Drilling holes for torpedo blasting in the indicated range should prove most effective in the context of triggering tremors. In addition, torpedo holes in this region are a good choice as they enhance the effect of additional discontinuity between the layers. Damaging the natural contact between the tremor-prone layer and the surrounding strata in this area should be the most effective.

4.2. Analysis of the Distribution of Elastic Strain Energy in the Vicinity of the Working Zone

Underlying the analyses of the elastic energy distribution in the vicinity of the working zone was the model in which limestone, anhydrite and dolomite layers were able to slip against each other, in accordance with Coulomb's friction law with a friction coefficient of $\mu = 0.6$.

The calculation results were presented in the form of maps of the distribution of elastic strain energy density coefficients: k_{Ac}—total, k_{Af}—shear strain and k_{Av}—volumetric strain elastic energy, defined as follows:

$$A_c = \frac{1}{2E}\left[\sigma_x^2 + \sigma_y^2 + \sigma_z^2 - 2\nu(\sigma_x\sigma_y + \sigma_y\sigma_z + \sigma_z\sigma_x) + 2(1+\nu)\left(\tau_{xy}^2 + \tau_{zx}^2 + \tau_{yz}^2\right)\right] \quad (3)$$

$$A_f = \frac{1+\nu}{6E}\left[(\sigma_x - \sigma_y)^2 + (\sigma_y - \sigma_z)^2 + (\sigma_z - \sigma_x)^2 + 6\left(\tau_{xy}^2 + \tau_{zx}^2 + \tau_{yz}^2\right)\right] \quad (4)$$

$$A_v = \frac{1-2\nu}{6E}(\sigma_x + \sigma_y + \sigma_z)^2 \quad (5)$$

$$k_{Ac} = \frac{A_c}{A_c^p};\ k_{Af} = \frac{A_f}{A_f^p};\ k_{Av} = \frac{A_v}{A_v^p} \quad (6)$$

where

A_c—density of total elastic energy; A_f—density of shear strain elastic energy; A_v—density of volumetric strain elastic energy; the index p indicates the values of initial energy corresponding to the initial state of stress.

The spatial distributions of the elastic strain energy density coefficients around the working zone are illustrated in Figure 13.

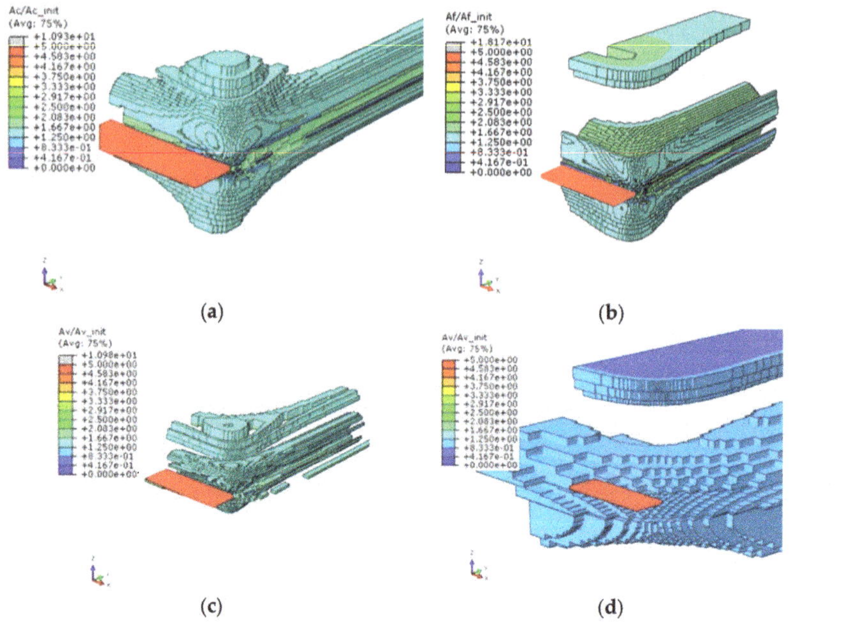

Figure 13. Volumes of rock around the working zone corresponding to the values of the elastic energy density coefficients in the roof layers: (**a**) Total $1.5 < k_{Ac} < 1.7$; (**b**) shear $1.5 < k_{Af} < 1.7$; (**c**) volumetric $1.5 < k_{Av} < 1.7$ and (**d**) volumetric $k_{Av} < 1$.

Maps of the total elastic energy coefficient in the cross-sections perpendicular to the face front are shown in Figure 14a, for various distances from the corner of the front section and in cross-sections parallel to the front section in Figure 14b.

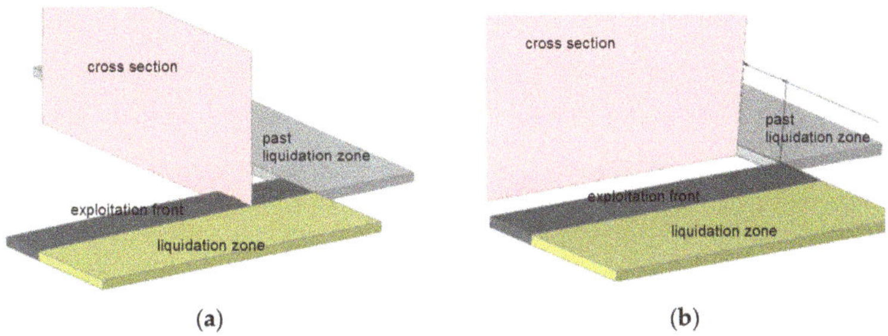

Figure 14. Cross-sections perpendicular (**a**) and parallel (**b**) to the working face, showing elastic energy density maps.

First, Figure 9 shows maps of three types of energy density in the cross-section perpendicular to the working front (Figure 14a), in the middle of the working front section.

It should be noted that extreme values of individual energy concentration coefficients occur in the front of working zone and in the roof in the dolomitic limestone layer (tremor-prone layer) ahead of the front section. Ahead of the working front, there is an increase in the concentration of strain energy k_{Af} (Figure 15b) in the floor and roof of the limestone layer, the maximum of which is located approximately 40 m in front of the working zone. Above the working space, the load acting on the roof layers is relieved, which is accompanied by a decrease in the value of the volume strain energy density, which goes down below $k_{Av} < 1$ (Figure 15c). This is important in the context of the stress analysis in the roof layers in this region, because with a small value of the stress axiator, a smaller value of the stress deviator is also required to trigger the rock failure. A good indicator of energy concentration in the rock strata is the total energy elastic density coefficient k_{Ac} (Figure 15a), which comprehensively illustrates the accumulation of individual types of strain elastic energy. Therefore, further analysis of elastic energy distributions around the entire working zone, in individual sections perpendicular and parallel to the working front (Figure 13a,b), was conducted basing on the total elastic energy density coefficient k_{Ac} (Figure 16).

The distributions of the total strain energy density coefficient values k_{Ac} for cross-sections parallel to the working front are shown in Figure 17.

Additionally, calculations were performed for this variant of the model in which slippage between the tremor-prone layer and the surrounding rocks would be precluded. The results of the numerical simulations performed for both variants: that precluding slippage between the layers (Figure 18a) and that admitting slippage (Figure 18b) yield the spatial distribution of the total energy density coefficient k_{Ac}. In regard to the symmetry of the working zone, the drawing covers half of the area of the modeled exploitation site.

Obviously, the extreme values of total elastic energy coefficient k_{Ac} are registered in the roof strata in the vicinity of the working front, in the direction inclined to the horizontal at an angle of 63–69°. In the corner of the site, the value of total elastic strain energy is the highest and decreases over a distance of approximately 90 m, as we move away towards the runway and the center part of the working front.

These analyses reveal that concentrations of the strain energy density coefficient k_{Af} occur in regions located in front of the working front (Figure 15b). The total energy concentration k_{Ac} also corresponds to this situation (Figures 15 and 16). This is consistent with observations and analyses of the occurrence of tremor epicenters, the largest number of which occur in front of the working front at a distance of less than 50 m [58,59]. As regards the potential of high-energy tremor occurrence, the most unfavorable conditions prevail when there is a large accumulation of shear strain energy in a given region. In such situations, seismicity analyses should allow for the possibility of correlating the mechanism of rock failure through shearing with the geophysical mechanism of the tremor focus, in which the shear component plays a dominant role [33,34].

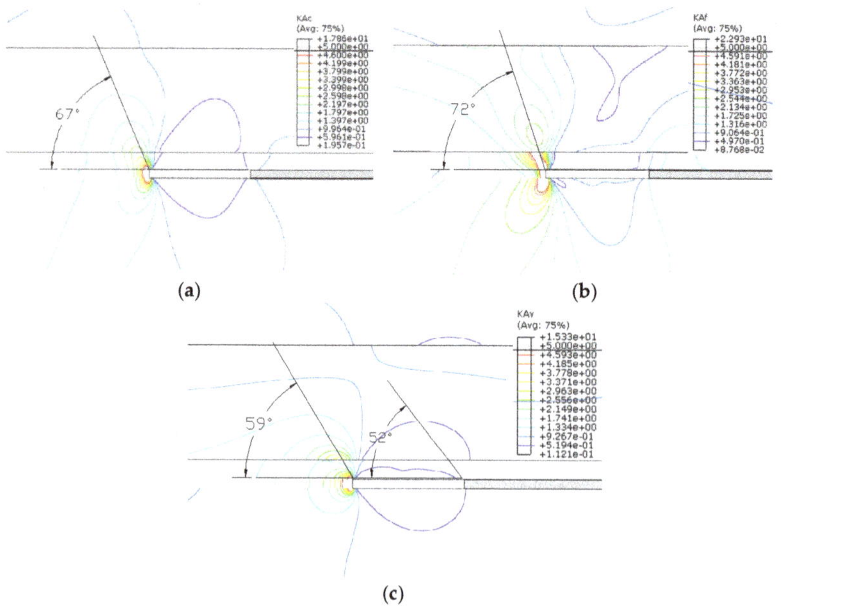

Figure 15. Map of strain energy coefficients in roof layers. (**a**) k_{Ac} total, (**b**) k_{Af} shear and (**c**) k_{Av} volumetric. The location of extreme values is marked with straight lines.

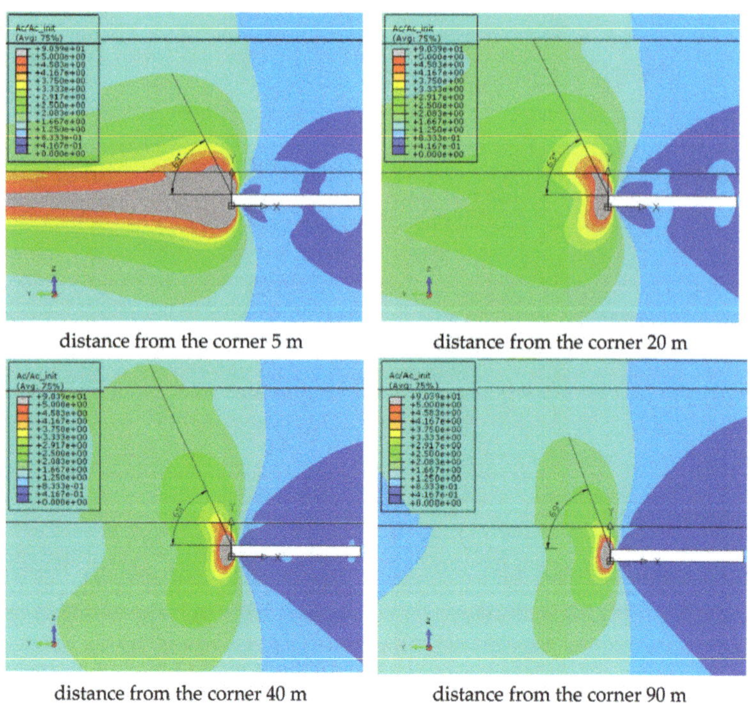

Figure 16. Maps of the total energy elastic density coefficient k_{Ac} in the cross-section perpendicular to the working front (Figure 14a).

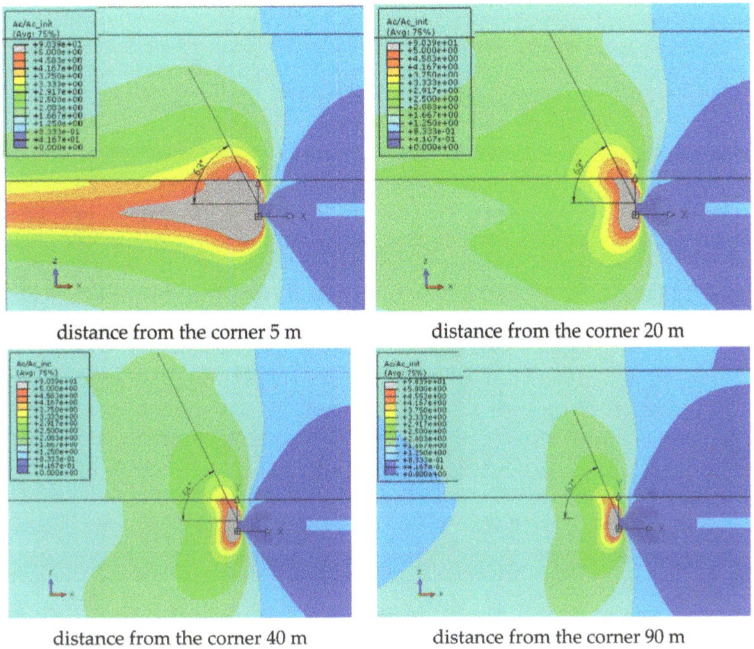

Figure 17. Maps of the total energy elastic density coefficient in the roof layers k_{Ac}, parallel to the working front (Figure 14b).

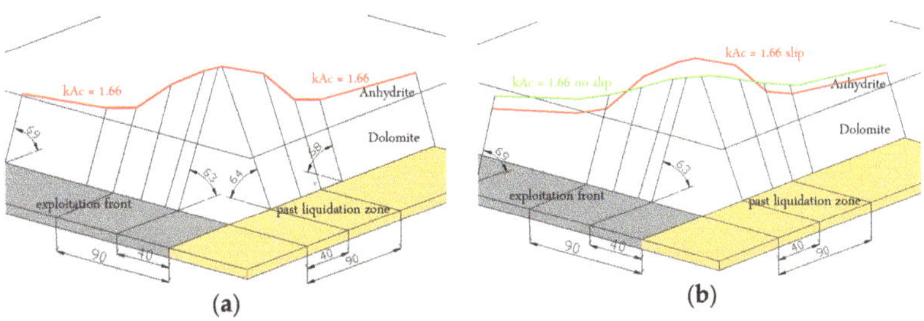

Figure 18. Spatial visualization of the distribution of the range of coefficient k_{Ac} with a value of 1.66 for models (**a**) with slippage between layers (**b**) without slippage.

Low values of the volume strain energy coefficient k_{Av} are registered in regions where low-energy tremors may potentially occur, such as the roof rock above the working zone and in the goafs (Figure 15c). These analyses lead us to the conclusion that coefficients of shear strain energy density k_{Af} and the density of total elastic energy k_{Ac} can be treated as measures of the potential risk of high-energy tremors of the rock strata, and the coefficients of volumetric strain energy density k_{Av} is associated with the regions of low-energy tremor occurrence.

5. Conclusions

The dominant hazard in the Polish copper mining industry (LGCD mines—Legnica-Glogow Copper District) is the occurrence of mining tremors and rockbursts. One of the effective active methods of preventing this hazard is torpedo blasting, which results in disturbing the structure of the roof rocks. These blastings should be considered with two effects in mind: the "triggering" a tremor and "weakening" of the rock strata.

Currently, torpedo blasting in the roof rock is still continued in the Rudna mine as a part of the rockburst prevention strategy. These are primarily cyclical blasts aimed at locally weakening the roof layers in the mining liquidation zones. They are performed to reduce the capability of roof rocks to accumulate large energies and, consequently, to provoke low-energy tremors.

Extensive research as well as practical experience acquired in the Rudna mine (LGCD) during blasting in compact roof layers lead us to the following conclusions:

- There are possibilities of active rockburst prevention in the torpedo blasting form based on long blast holes up to approx. 60 m long and it is not a problem from a technical point of view, although it is complicated in implementation,
- Sections of torpedo holes filled with explosive material should be located in zones with the maximum effort and accumulation of elastic energy (at present or in the future); in relation to the geomechanical conditions in the LGCD area, the blast holes should be inclined at an angle of approximately 65 ± 5 degrees,
- Blast holes should reach the contact zone between limestone and anhydrite; by changing the type of contact a weakening effect will be achieved and, consequently, a reduction in the value of the accumulated energy of elastic deformations; events caused by the cracking of the tremor-producing layer will be more frequent, but their seismic energies will be lower (which will result in a reduced risk of rockbursts),
- Torpedoing the roof has positive effects in the form of induced tremors, especially when used in the area of the edges of old goafs and in tectonic disturbance zones,
- The implementation of torpedo blasting may, however, lead to local destruction of roof layers in the immediate vicinity of active workings, thus disturbing the technological process of mining works.

Based on the numerical calculations performed, it can be concluded that appropriate geomechanical computer simulations, compared with analyses of geophysical mechanisms of tremor foci and with the results of monitoring of the level of roof failure by seismo-acoustic emission methods, may help identify the potential hazard areas. From the standpoint of active rockburst prevention in the form of torpedo blasting, these regions are considered to be areas where this type of prevention should be used.

This research work should be continued to determine the areas, scope and timing of stress relief blasting operations in order to improve the effectiveness of rock mass relaxation in conditions of high rockburst risk.

Author Contributions: Research/conceptualization of the paper, data analysis and verification, stress/energy state modelling, results comparison, conclusions, final editing, data collecting and preliminary analysis, results visualization, draft text editing, Z.B., J.C., D.C., P.P. and K.G. All authors have read and agreed to the published version of the manuscript.

Funding: The research was funded by AGH University of Krakow; subsidy numbers 16.16.100.215, 501.696.7996.

Data Availability Statement: Data is contained within the article.

Conflicts of Interest: Authors Paweł Piasecki and Krzysztof Gzik were employed by the company KGHM Polska Miedź S.A. The remaining authors declare that the research was conducted in the absence of any commercial or financial relationships that could be construed as a potential conflict of interest.

References

1. Burtan, Z. The influence of regional geological settings on the seismic hazard level in copper mines in the Legnica-Głogów Copper Belt Area (Poland). In Proceedings of the AG 2017-3rd International Conference on Applied Geophysics. E3S Web of Conferences, Gniew, Poland, 21–23 June 2017; Volume 24, p. 01004. [CrossRef]
2. Cieślik, J.; Burtan, Z.; Chlebowski, D.; Zorychta, A. Geomechanical analysis of location and conditions for mining-induced tremors in LGOM copper mines. *J. Sustain. Min.* **2017**, *16*, 94–103. [CrossRef]
3. Owczarz, K.; Blachowski, J. Application of DInSAR and Spatial Statistics Methods in Analysis of Surface Displacements Caused by Induced Tremors. *Appl. Sci.* **2020**, *10*, 7660. [CrossRef]

4. Zorychta, A.; Cieślik, J.; Burtan, Z.; Chlebowski, D. Geomechaniczne warunki poprawy efektywności strzelań torpedujących w kopalniach LGOM. *CUPRUM—Czas. Nauk.-Tech. Górnictwa Rud Miedzi* **2015**, *3*, 83–93. (In Polish)
5. Pawłowicz, K. *Strzelania Torpedujące Jako Metoda Zapobiegania Tąpaniom*; Prace Naukowe GiG; Główny Instytut Górnictwa: Katowice, Poland, 1996; Volume 803. (In Polish)
6. Krzyżowski, A. Rockburst prevention in hard coal mines (in Poland). In Proceedings of the 13th International Scientific-Technical Symposium ROCKBURSTS 2002, Ustron, Poland, 12–15 November 2002; Central Mining Institute in Katowice: Katowice, Poland, 2002; pp. 193–236.
7. Drzewiecki, J.; Kabiesz, J. Dynamic events in roof strata—Occurrence and prevention. *Coal Sci. Technol. Mag.* **2008**, *235*, 55–57.
8. Wojtecki, Ł.; Mendecki, M.J.; Zuberek, W.M. Determination of Destress Blasting Effectiveness Using Seismic Source Parameters. *Rock Mech. Rock Eng.* **2017**, *50*, 3233–3244. [CrossRef]
9. Wojtecki, Ł.; Konicek, P.; Schreiber, J. Effects of torpedo blasting on rockburst prevention during deep coal seam mining in the Upper Silesian Coal Basin. *J. Rock Mech. Geotech. Eng.* **2017**, *9*, 694–701. [CrossRef]
10. Olechowski, S.; Ćwiękała, M. Rock bursts prevention in conditions of increased seismic activity during the exploitation of seam 713/1-2 with wall II-E1 in KWK ROW Ruch Rydułtowy coal mine. *Przegląd Górniczy* **2021**, *10*, 26–33. (In Polish)
11. Butra, J. *Eksploatacja Złoża rud Miedzi w Warunkach Zagrożenia Tąpaniami i Zawałami*; KGHM Cuprum CBR: Wrocław, Poland, 2010. (In Polish)
12. Anderko, A.; Laskowski, M.; Mirek, A.; Osman, R.; Wróbel, J. Additional prevention against tremors with the use of blasting in roof layers—Tests of KGHM Polska Miedź S.A. Rudna copper ore mine. *Przegląd Górniczy* **2015**, *2*, 1–9. (In Polish)
13. Gzik, K.; Laskowski, M.; Wróbel, J. *Wpływ Strzelania Krawędziowego na Zachowanie Górotworu w Polu X/1 O/ZG Rudna, Prowadzonym w Szczególnie Trudnych Warunkach Geologiczno-Górniczych*; Górnicze Zagrożenia Naturalne—Praca Zbiorowa; Wydawnictwo GIG: Katowice, Poland, 2010. (In Polish)
14. Gzik, K.; Laskowski, M.; Świder, M.; Mirek, A. *Wpływ Uwarunkowań Tektonicznych na Prowadzenie Robót Eksploatacyjnych w Polu G-7/5 O/ZG Rudna*; Kabiesza, J., Ed.; Zagrożenia i Technologie—Praca Zbiorowa; Wydawnictwo GIG: Katowice, Poland, 2012. (In Polish)
15. Brauner, G. *Experiences with Destress Blasting Technique in Radbod Colliery*; Verlag Gluckauf GmbH: Essen, Germany, 1976; Volume 112, pp. 951–957. (In German)
16. Bräuner, G. *Rockbursts in Coal Mines and Their Prevention*; Balkema: Rotterdam, The Netherlands, 1994; 144p.
17. Velsen-Zerweck, R. *Destress Coal Blasting in Ruhr Hardcoal Mining in Rockburst Prevention*; Nobel Hefte; Sprengtechnischer Dienst der Orica Germany: Troisdorf, Germany, 1984; Volume 50, pp. 85–92. (In German)
18. Kexin, D. Maintenance of roadways in soft rock by roadway-rib destress blasting. *China Coal Soc.* **1995**, *20*, 311–316. (In Chinese)
19. Xia, H.; Xu, Y.; Zong, Q.; Fu, J. Research on the technology of unloading blast for a deep soft rock tunnel and its engineering application. *J. Anhui Univ. Sci. Technol. (Nat. Sci.)* **2007**, *27*, 13–16. (In Chinese)
20. Holecko, J.; Ptacek, J.; Takla, G.; Konecny, P. Rock bursts in the Czech part of the Upper Silesian Coal Basin—Features, theoretical models and conclusions for practice. In Proceedings of the 9th ISRM Congress, International Society for Rock Mechanics, Paris, France, 25–28 August 1999; pp. 1101–1104.
21. Dvorsky, P.; Konicek, P. Systems of rock blasting as a rockburst measure in the Czech part of Upper Silesian Coal Basin. In Proceedings of the 6th International Symposium on Rockburst and Seismicity in Mines, Perth, Australia, 9–11 March 2005; pp. 493–496.
22. Konicek, P.; Konecny, P.; Ptacek, J. Destress rock blasting as a rockburst control technique. In Proceedings of the 12th ISRM Congress, International Society for Rock Mechanics, Beijing, China, 18–21 October 2011; pp. 1221–1226.
23. Konicek, P.; Saharan, M.R.; Mitri, H. Destress Blasting in Coal Mining—State-of-the-Art Review. In Proceedings of the First International Symposium on Mine Safety Science and Engineering. Procedia Engineering, Beijing, China, 26–29 October 2011; Volume 26, pp. 179–194.
24. Konicek, P.; Soucek, K.; Stas, L.; Singh, R. Long-hole destress blasting for rockburst control during deep underground coal mining. *Int. J. Rock Mech. Min. Sci.* **2013**, *61*, 141–153. [CrossRef]
25. Chouhn, R.K.S. Induced seismicity of Indian coal mines. *Phys. Earth Planet. Inter.* **1986**, *44*, 82–86. [CrossRef]
26. Toper, A.Z.; Adams, D.J.; Janse Van Rensburg, A.L. The effects of preconditioning blast in confined rock. In Proceedings of the ISRM International Symposium on Integral Approaches to Applied Rock Mechanics, Santiago, Chile, 10–14 May 1994.
27. Andrieux, P.P.; Brummer, R.K.; Liu, Q.; Simser, B.P.; Mortazavi, A. Large-scale panel destress blast at Brunswick Mine. *CIM Bull.* **2003**, *96*, 78–87.
28. Vennes, I.; Mitri, H.; Chinnasane, D.R.; Yao, M. Large-scale destress blasting for seismicity control in hard rock mines: A case study. *Int. J. Min. Sci. Technol.* **2020**, *30*, 141–149. [CrossRef]
29. Andrieux, P.; Hadjigeorgiou, J. The destressability index methodology for the assessment of the likelihood of success of a large-scale confined destress blast in an underground mine pillar. *Int. J. Rock Mech. Min. Sci.* **2007**, *45*, 407–421. [CrossRef]
30. Fuławka, K.; Mertuszka, P.; Pytel, W.; Szumny, M.; Jones, T. Seismic evaluation of the destress blasting efficiency. *J. Rock Mech. Geotech. Eng.* **2022**, *14*, 1501–1513. [CrossRef]
31. Mertuszka, P.; Szumny, M.; Fuławka, K.; Kondoł, P. Novel approach for the destress blasting in hard rock underground copper mines. *J. Sustain. Min.* **2022**, *21*, 141–154. [CrossRef]
32. Lurka, A.; Stec, K. *Charakterystyka Radiacji Fal Sejsmicznych w Obszarze Epicentralnym dla Wstrząsów w LGOM, Warsztaty—Zagrożenia Naturalne w Górnictwie*; Wyższy Urząd Górniczy: Katowice, Poland, 2005. (In Polish)

33. Orlecka-Sikora, B.; Papadimitriou, E.E.; Kwiatek, G. A study of the interaction among mining induced seismic events in the Legnica-Glogow Copper District, Poland. *Acta Geophys.* **2009**, *57*, 2. [CrossRef]
34. Urban, P.; Wiejacz, P.; Domański, B. *Analiza Mechanizmów Wstrząsów Górniczych Występujących na Obszarze Legnicko-Głogowskiego Okręgu Miedziowego*; Geofizyka w Geologii i Górnictwie, wyd.; Wydział Nauk o Ziemi UŚ: Katowice, Poland, 2010. (In Polish)
35. Ren, F.Q.; Zhu, C.; He, M.; Shang, J.; Feng, G.; Bai, J. Characteristics and Precursor of Static and Dynamic Triggered Rockburst: Insight from Multifractal. *Rock Mech. Rock Eng.* **2023**, *56*, 1945–1967. [CrossRef]
36. Ren, F.; Zhu, C.; Yuan, Z.; Karakus, M.; Tang, S.; He, M. Recognition of shear and tension signals based on acoustic emission parameters and waveform using machine learning methods. *Int. J. Rock Mech. Min. Sci.* **2023**, *171*, 105578. [CrossRef]
37. Taylor, L.M.; Chen, E.P.; Kuszmaul, J.S. Microcrack-induced damage accumulation in brittle rock under dynamic loading. *Comput. Methods Appl. Mech. Eng.* **1986**, *55*, 301–320. [CrossRef]
38. Maxwell, S.C.; Young, R.P. Propagation effects of an underground excavation. *Tectonophysics* **1998**, *289*, 17–30. [CrossRef]
39. Chen, S.G.; Zhao, J. A study of UDEC modelling for blast wave propagation in jointed rock masses. *Int. J. Rock Mech. Min. Sci.* **1998**, *35*, 93–99. [CrossRef]
40. Ma, G.W.; Hao, H.; Zhou, Y.X. Modelling of wave propagation induced by underground explosion. *Comput. Geotech.* **1998**, *22*, 283–303. [CrossRef]
41. Zhu, Z.; Mohanty, B.; Xie, H. Numerical investigation of blasting-induced crack initiation and propagation in rocks. *Int. J. Rock Mech. Min. Sci.* **2007**, *44*, 412–424. [CrossRef]
42. Zhu, W.C.; Wei, C.H.; Li, S.; Wei, J.; Zhang, M.S. Numerical modeling on destress blasting in coal seam for enhancing gas drainage. *Int. J. Rock Mech. Min. Sci.* **2013**, *59*, 179–190. [CrossRef]
43. Ma, G.W.; An, X.M. Numerical simulation of blasting-induced rock fractures. *Int. J. Rock Mech. Min. Sci.* **2008**, *45*, 966–975. [CrossRef]
44. An, H.M.; Liu, H.Y.; Han, H.; Zheng, X.; Wang, X.G. Hybrid finite-discrete element modelling of dynamic fracture and resultant fragment casting and muck-piling by rock blast. *Comput. Geotech.* **2017**, *81*, 322–345. [CrossRef]
45. Baranowski, P.; Mazurkiewicz, Ł.; Małachowski, J.; Pytlik, M. Experimental testing and numerical simulations of blast-induced fracture of dolomite rock. *Meccanica* **2020**, *55*, 2337–2352. [CrossRef]
46. Tang, B.; Mitri, H. Numerical modelling of rock preconditioning by destress blasting. *Proc. Inst. Civ. Eng.-Ground Improv.* **2001**, *5*, 57–67. [CrossRef]
47. Baranowski, P.; Damaziak, K.; Mazurkiewicz, Ł.; Mertuszka, P.; Pytel, W.; Małachowski, J.; Pałac-Walko, B.; Jones, T. Destress Blasting of Rock Mass: Multiscale Modelling and Simulation. *Shock Vib.* **2019**, *2019*, 2878969. [CrossRef]
48. Xu, J.; Kang, Y.; Wang, X.; Feng, G.; Wang, Z. Dynamic characteristics and safety criterion of deep rock mine opening under blast loading. *Int. J. Rock Mech. Min. Sci.* **2019**, *119*, 156–167. [CrossRef]
49. Zheng, X.; Hua, J.; Zhang, N.; Feng, X.; Zhang, L. Simulation of the Load Evolution of an Anchoring System under a Blasting Impulse Load Using FLAC3D. *Shock Vib.* **2015**, *2015*, 972720. [CrossRef]
50. Sainoki, A.; Emad, M.Z.; Mitri, H.S. Study on the efficiency of destress blasting in deep mine drift development. *Can. Geotech. J.* **2017**, *54*, 518–528. [CrossRef]
51. Pytel, W.M.; Mertuszka, P.P.; Jones, T.; Paprocki, H. Numerical simulations of geomechanical state of rock mass prior to seismic events occurrence—Case study from a Polish copper mine aided by FEM 3D. In *Proceedings of the 27th International Symposium on Mine Planning and Equipment Selection*; Springer: Cham, Switzerland, 2019; pp. 417–427.
52. Yi, C.; Nordlund, E.; Zhang, P.; Warema, S.; Shirzadegan, S. Numerical modeling for a simulated rockburst experiment using LS-DYNA. *Undergr. Space* **2021**, *6*, 153–162. [CrossRef]
53. Vennes, I.; Mitri, H. Geomechanical effects of stress shadow created by large-scale destress blasting. *J. Rock Mech. Geotech. Eng.* **2017**, *9*, 1085–1093. [CrossRef]
54. Miao, S.; Konicek, P.; Peng-Zhi Pan, P.Z.; Mitri, M. Numerical modelling of destress blasting—A state-of-the-art review. *J. Sustain. Min.* **2022**, *21*, 278–297. [CrossRef]
55. Hoek, E.; Carranza-Torres, C.T.; Corkum, B. Hoeck-Brown failure criterion—2002 edition. In Proceedings of the NARMS-TAC Conference, Toronto, ON, Canada, 7–10 July 2002; Volume 1, pp. 267–273.
56. Pytel, W.; Mertuszka, P.; Fuławka, K.; Lurka, A.; Pałac-Walko, B. Resultant axial stresses in instrumented rockbolts induced by dynamic effects occurred due to multi-face blasting in the working areas. *Tunn. Undergr. Space Technol.* **2021**, *116*, 104088. [CrossRef]
57. Pytel, W.; Fuławka, K.; Pałac-Walko, B.; Mertuszka, P. Numerical and Analytical Determination of Rockburst Characteristics: Case Study from Polish Deep Copper Mine. *Appl. Sci.* **2023**, *13*, 11881. [CrossRef]
58. Kozłowska, M. Ilościowa analiza odległości epicentralnych wstrząsów względem frontu eksploatacji w O/ZG Rudna (LGOM). *Przegląd Górniczy* **2012**, *68*, 75–85. (In Polish)
59. Burtan, Z.; Zorychta, A.; Chlebowski, D.; Cieślik, J. Analiza sejsmiczności indukowanej w O/ZG „Rudna" w aspekcie rozmieszczenia ognisk wstrząsów względem frontu eksploatacyjnego. In Proceedings of the XXIII Międzynarodowa Konferencja Naukowo-Techniczna z Cyklu Górnicze Zagrożenia Naturalne, Szczyrk, Poland, 8–10 November 2016. (In Polish).

energies

MDPI

Article

Numerical Investigation of the Influence of Roof-Cutting Parameters on the Stability of Top Coal Gob-Side Entry Retaining by Roof Pre-Fracturing in Ultra-Thick Coal Seam

Yongkang Yang [1,*], Peipeng Gao [1], Chao Zhang [1] and Chenlong Wang [2]

1 Key Laboratory of In-Situ Property-Improving Mining of Ministry of Education, Taiyuan University of Technology, Taiyuan 030024, China
2 College of Mechanical and Vehicle Engineering, Taiyuan University of Technology, Taiyuan 030024, China
* Correspondence: yangyongkang@tyut.edu.cn; Tel.: +86-132-0341-3056

Abstract: Gob-side entry retaining by roof cutting, a pillarless mining technique, plays a critical role in maintaining continuous production, rapid connection, and enhancing the coal recovery rate in fully mechanized top coal caving working faces. This technique stands as a sustainable development method in coal mining. The present research, set against the backdrop of the Yitang Coal Mine 100602 top coal gob-side entry retaining by roof cutting, investigates the influence of roof-cutting borehole depth, borehole dip angle, mining height, and coal seam thickness on stability in an ultra-thick coal seam under 12 distinct mining conditions. A typical model of overburden structure post-roof pre-splitting was established to study the failure mechanism of the top coal roof. The results reveal that the dip angle and depth of the roof pre-fracturing borehole significantly impact the movement characteristics of the overlying strata. Optimal conditions are found when the dip angle and depth of the roof pre-fracturing borehole, the mining height, and the top coal thickness are 10°, 16 m, 4 m, and 4 m, respectively. Under these circumstances, the load transfer from the goaf to the gob-side entry can be effectively intercepted, mitigating the influence of roof fracture activities on the top coal gob-side entry. Field measurements confirm that suitable anchor-net support can stabilize the roof's rock structure. This research underpins the significance of roof pre-fracturing for the promotion and application of top coal gob-side entry retaining by roof cutting in ultra-thick coal seams.

Keywords: top coal gob-side entry; roof pre-fracturing; roof-cutting parameters; ultra-thick coal seam

Citation: Yang, Y.; Gao, P.; Zhang, C.; Wang, C. Numerical Investigation of the Influence of Roof-Cutting Parameters on the Stability of Top Coal Gob-Side Entry Retaining by Roof Pre-Fracturing in Ultra-Thick Coal Seam. *Energies* **2023**, *16*, 4788. https://doi.org/10.3390/en16124788

Academic Editor: Krzysztof Skrzypkowski

Received: 20 May 2023
Revised: 8 June 2023
Accepted: 14 June 2023
Published: 18 June 2023

1. Introduction

In thick coal seam mining, the roadway is either arranged along the bottom of the coal seam to form a top coal roadway, which is difficult to support, or arranged along the top of the coal seam to form the bottom maintaining coal roadway [1]. For the bottom maintaining coal roadway, the method of slicing mining is adopted to avoid the problem of difficult support of the roadway surrounding rock [2,3], which has a long recovery period, high mining cost, and serious floor heave problems [4,5]. For the top coal roadway, the coal was initially mined using the common fully mechanized caving mining method, and a pillar of more than ten meters or even dozens of meters was left between the working faces to resist the impact of mining [6,7], which caused a large waste of coal resources [8–10]. After continuous efforts and attempts by scholars, the width of the pillar was continuously improved under the premise of ensuring mining safety and was supplemented with corresponding support means [11–13], and small pillar mining was successfully tested and gradually promoted [14,15]. Under this type of roadway protection method, the coal recovery rate is improved and the stress concentration of the coal pillar is reduced [16]; however, the effect of the support for deep or soft roadway surrounding rock is still unsatisfactory [17], and there are problems such as slow mining speed and gas accumulation, such as those identified in the works of Bosikov et al. (2022 and 2023) [18,19] and Zhijun Tian et al. [20].

After the introduction of gob-side entry retaining technology, initially due to the low degree of mechanization, the support materials were not advanced, and the surrounding rock is easy to fail under roof pressure, which will strengthens the labor intensity of workers and reduces mining efficiency. Gob-side entry retaining technology can only be tested and applied in thin coal seams [20]. With the introduction of mechanized mining equipment and support technology, high-strength support materials such as scalable hydraulic supports, anchors, and cables were used for supporting, and high-water materials and flexible membrane filling bodies were used to replace coal pillars [21–24]. At the same time, the gob-side entry retaining system suitable for use in China was proposed in combination with the domestic coal seam mining conditions, and its continuous application and development [25–27], application trend in the transition from simple geological conditions to complex geological conditions, and the retaining of the gob-side entry using the roof cutting coal pillarless mining technique has achieved certain results in the comprehensive mining of medium-thick coal seams and steep conditions [28–30]. This roadway protection method significantly improved the coal recovery rate and solved the problems of mining relay and gas accumulation [31,32], and became the main stream for thick coal seam mining, but the solidification of the filling material and the mining speed need to be accurately coordinated and it is difficult to determine the strength of the surrounding rock in time. However, with the continuous deterioration of the mining environment, the problems of the large deformation of the filling body and stress concentration have been exposed one after another [33], and the accidents caused by the large deformation of the roadway surrounding rock have become more and more serious; thus, the conventional gob-side entry retaining approach has been challenged.

For this reason, scholars have explored other roadway-retaining methods and support materials, among which the most representative is gob-side entry retaining using the roof cutting coal pillarless mining technique. The researchers in [34] cut off the stress transmission path of the goaf area while cutting the roof using directional rupture blasting technology, and made use of the rock expansion to fill the goaf with collapsed rocks and reinforced them into the roadway wall with support equipment such as an NPR anchor/cable, giving full play to its self-bearing capacity. At present, gob-side entry retaining technology, as an emerging roof cutting coal pillarless mining technique, has been widely promoted and used in thin and medium-thick coal seam conditions [35–37]. In top coal gob-side entry retaining, the top coal is softer and more fragile, and under the thick coal seam conditions, the gangue caving is more violent and takes longer to fill the goaf, so the process of gob-side entry retaining has stricter requirements for the roof and goaf side road wall support compared with the passive bearing of the gob-side entry. The active pressure relief of gob-side entry retaining has obvious superiority [38] and provides a new way of thinking to solve the problem of the difficult-to-support top coal roadway. However, there is almost no literature or field practice related to top coal gob-side entry retaining at home or abroad, and the theory has delayed engineering practice, which has led to a delay in starting field engineering tests. Taking Yitang coal mine as the engineering background, the author uses a combination of theoretical analysis, numerical simulation, and field tests to investigate the evolution law of the displacement field of the surrounding rock and the feasibility of top coal gob-side entry retaining, and develops a reasonable support strategy to provide a theoretical reference and a technical reference for the gob-side entry retaining test under similar mining conditions.

2. Field Practice of 100602 Top Coal Gob-Side Entry Retaining

2.1. Production Geology

The sixth mining area of Yitang coal mine has an east–west strike length of 1340 m, a north–south dip length of 1400 m, and a mining elevation of +175 m~+335 m. The geological and structural conditions of the working face are simple, the dip angle of the coal seam is 2~5°, the average coal thickness is 7.6 m, the return roadway height is 4 m, and the width is 6 m; the lithology of the roof and floor is shown in Figure 1. The open-off cut length is 150 m and the retain roadway length is 750 m; the roadway layout is shown in Figure 2. The current mining method of Yitang coal mine is slicing mining, with a top coal thickness of 1.2 m, a bottom coal thickness of 2.4 m, and an average mining height of the working face of 5.0 m. After mining, the mining height is large, the mine pressure is strong, the top coal is soft and broken, there is serious floor heave, and it is difficult to control the end face. In order to solve the above problems, it was decided that an engineering test would be carried out on the top coal gob-side entry retaining for the tailgate of the 100602 working face.

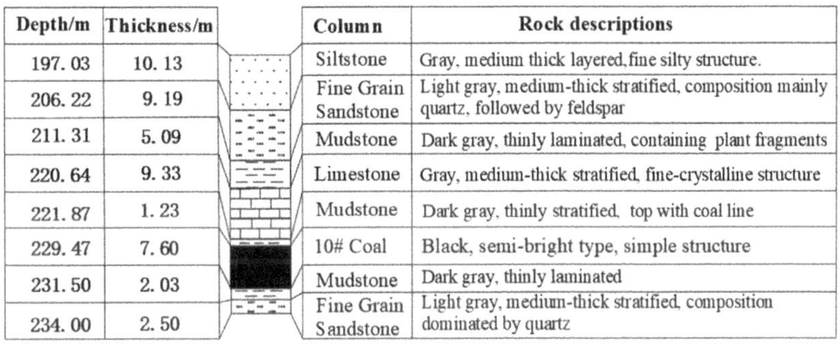

Depth/m	Thickness/m	Column	Rock descriptions
197. 03	10. 13	Siltstone	Gray, medium thick layered, fine silty structure.
206. 22	9. 19	Fine Grain Sandstone	Light gray, medium-thick stratified, composition mainly quartz, followed by feldspar
211. 31	5. 09	Mudstone	Dark gray, thinly laminated, containing plant fragments
220. 64	9. 33	Limestone	Gray, medium-thick stratified, fine-crystalline structure
221. 87	1. 23	Mudstone	Dark gray, thinly stratified, top with coal line
229. 47	7. 60	10# Coal	Black, semi-bright type, simple structure
231. 50	2. 03	Mudstone	Dark gray, thinly laminated
234. 00	2. 50	Fine Grain Sandstone	Light gray, medium-thick stratified, composition dominated by quartz

Figure 1. Integrated rock layer column diagram.

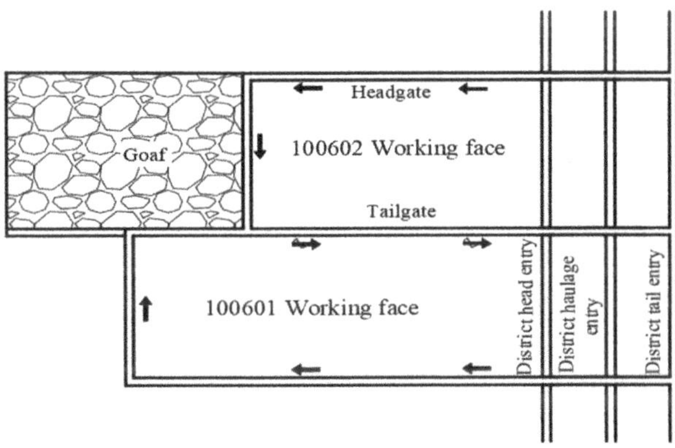

Figure 2. Roadway layout plan.

2.2. Gob-Side Entry Retaining Technology

The process of gob-side entry retaining is summarized as four steps; the process flow is shown in Figure 3.

(1) Within 80 m of the advance working face, the designed roof support parameters are used to make up the constant resistance large deformation anchor cable and reinforcement anchor cable to reinforce the roof and improve its pressure-bearing capacity, so as to lay the foundation for resisting the stress disturbance in the process of advance presplitting blasting and coal mining. The anchor cable should be long enough to reach the solid rock layer and produce a stable anchoring effect.

(2) Within 60 m of the advance working face, the equipment is used to set up the presplitting drill holes on the roof according to the design angle, ensuring that all holes are in the same plane during the drilling process. Then, the explosives are placed in the cavity pipe and into the drill holes, and are finally detonated and form joint-cutting.

The roof of the goaf side is more likely to collapse under the influence of mining pressure after slitting while cutting off the stress transmission path and weakening the stress of the roadway surrounding rock.

(3) During mining, retractable U-shaped steel and metal mesh for gangue support are settled up within 6 meters behind the working face, so that the gangue fills the goaf and form a wall with a certain strength. In addition, a hydraulic pillar with a π beam is set up to strengthen the roof support, inhibit the roof on the side of the goaf sinking, reduce the bending moment on the roof, and enhance the stability of the roadway.

(4) Under the support of the blocking gangue structure, the collapsed gangue in the goaf is gradually compacted, the main roof subsidence is contained, and the intense quarry stress caused by it is weakened. The surrounding rock of the lagging working face enters the stable state; then, some temporary support equipment is withdrawn and concrete is sprayed on the gangue wall surface to prevent ventilation leakage in the goaf, and the final gob-side entry retaining is complete.

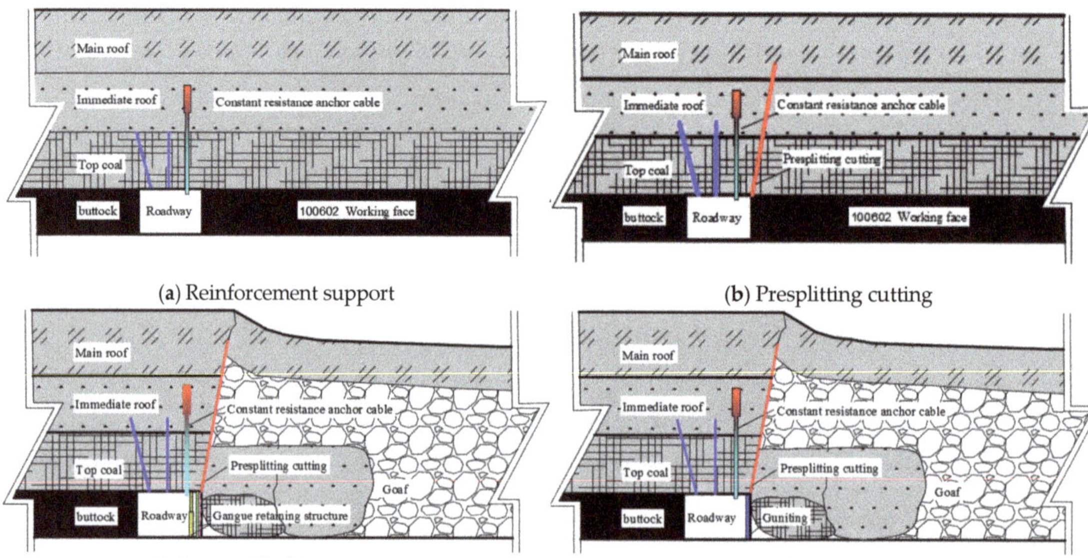

(**a**) Reinforcement support (**b**) Presplitting cutting

(**c**) Gangue-blocking support (**d**) Guniting

Figure 3. Process flow for gob-side entry retaining.

2.3. *Roof-Cutting Technology*

(1) Shaped charge technology

This blasting technology is a new type of concentrated energy blasting technology developed on the basis of the comparative study of many kinds of concentrated energy blasting and directional blasting methods. The construction process is simple: the application only requires the construction of gun holes in the pre-fracture line and the use of a two-way concentrated energy device charge, and the direction of the concentrated energy corresponds to the pre-fracture direction of the rock body. The blast product will form a concentrated energy flow in both set directions and generate concentrated tension stresses, causing the blast hole to penetrate along the concentrated energy direction and form a pre-cracked surface. Since the rock between the holes is fractured, the unit consumption of the blasting explosives will be greatly reduced, and at the same time, the damage to the rock around the holes will be greatly reduced due to the protection of the surrounding rock by the shaped charge device, which can achieve presplitting and protect the roof of the roadway at the same time. The principle diagram of the shaped charge is shown in Figure 4.

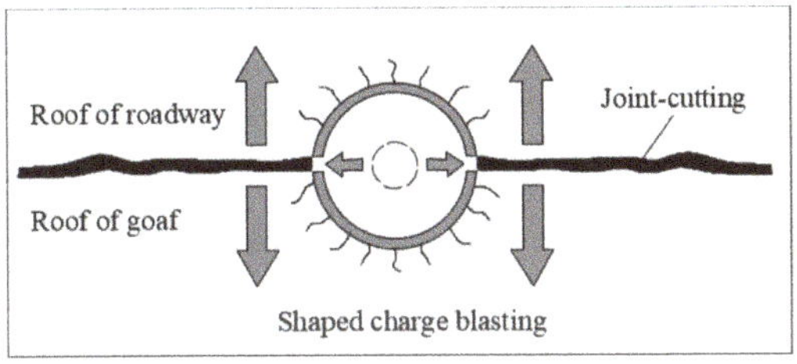

Figure 4. Schematic diagram of the mechanism of shaped charge blasting.

This study considers the geological conditions of the 100602 working face, applying actual engineering expertise to derive the parameters for pre-splitting joints on the roof of the roadway along the trough:

(1) After reinforcing with the anchor cable, a two-way energy-concentrating tensile blasting hole is constructed 60 m ahead of the working face using design parameters. This forms a roof-cutting and pressure-relief presplit cut suture over a 30 m range.
(2) The two-way energy-gathering tube, constructed from special energy-gathering PVC, measures 42 mm in outer diameter, 36 mm in inner diameter, and 1500 mm in length. A detonator is placed in each blast hole, charged forwards, and connected in series. Blast holes are initially designed to be 10 m deep, with the final depth determined on-site.
(3) Concentrated energy blasting employs grade 3 emulsion explosives, using 2~4 rolls of 32 mm × 300 mm powder per tube. The sealing hole length is at least 2.5 m, using bentonite gun mud for sealing.
(4) Considering on-site construction convenience, the distance between the drilling rigs is 3.8 m, and between the blast holes is 760 mm (500 mm if the hole depth is less than 8 m).
(5) Accurate directional presplitting blasting on the roof employs energy-concentrating tube uncoupling and charging, tested using five methods. The charging structure is 4 + 4 + 4 + 2 + 2, with a mud-sealing depth of 2500 mm.
(6) Before mining, energy-focused blasting kerfs are implemented on the gob-side roof, as shown in Figure 5.

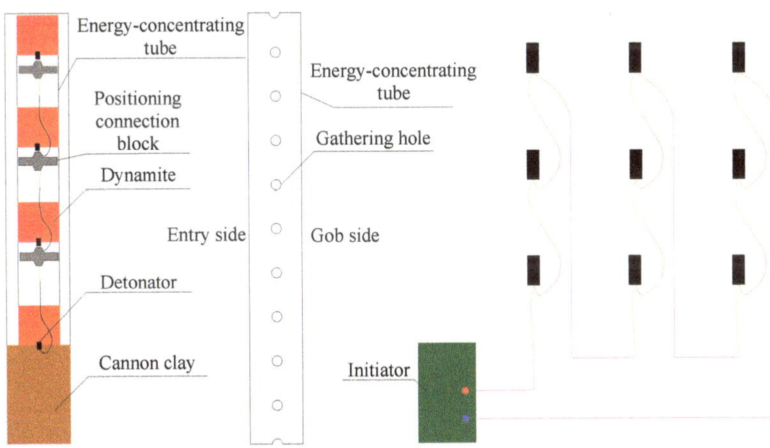

Figure 5. Schematic diagram of the mechanism of shaped charge blasting.

(2) Parameter design

The vertical distance from the top of the roadway to the farthest end of the cut line is the roof-cutting height. The collapsed form of the rock, the strength of the roadway by the mining impact, and whether the gangue can fill the goaf depend on the roof-cutting height. Based on the theory of rock fragmentation and expansion, taking into account the amount of roof subsidence, the amount of floor heaving, and the characteristics of the top coal gob-side entry retaining, the formula for calculating the roof-cutting height is as follows:

$$H_q = \frac{H_c + H_f - \Delta H_1 - \Delta H_2}{K - 1} \tag{1}$$

where H_q is the roof-cutting height, m; H_c is the coal seam mining height, m; H_f is the thickness of the top coal, m; ΔH_1 is the roof subsidence, m; ΔH_2 is the floor heaving, m; and K is the coefficient of rock fragmentation and expansion, generally taken as 1.3~1.5.

The borehole dip angle is an important process parameter that affects the effect of gob-side entry retaining. If the cutting angle is too large, the difficulty of the construction increases and the load on the cantilever beam itself increases. If the cutting angle is too small, it will cause the roof on the goaf side to compress and damage the gob-side entry roof during the falling process, causing the roof strata of the goaf to not collapse along the joint-cutting. When designing the cutting angle, we need to fully consider the rock engineering environment and the influence of mining activities, which, according to the engineering conditions, can be calculated using the following formula:

$$\alpha = \frac{1}{2} arcsin \sqrt{\frac{3rH_q}{2[\sigma_t](K-1)}} \tag{2}$$

where H_q is the roof-cutting height, m; σ_t is the tensile strength of the roof, MPa; α is the borehole dip angle between the joint-cutting and the vertical direction, °; and K is the coefficient of rock fragmentation and expansion, generally taken as 1.3~1.5. Engineering experience shows that the borehole dip angle should not be greater than 20°.

In order to form interpenetrated cracks between two boreholes so that the roof of the gob-side entry eventually forms presplitting joint-cutting along the set direction, it is necessary to select suitable borehole spacing. Based on cohesive explosive C-J theory and combined with the conservation of momentum theorem and rock blasting damage theory, the range of rock damage caused by the detonation wave generated by the explosion of the explosives in the borehole on both sides can be expressed as:

$$R = r_0 \left[\frac{K_s T_m}{(1 - D_0)\sigma_0 + P} \right]^{\frac{1}{c}} \tag{3}$$

where r_0 is the radius of the borehole, mm; D_0 is the initial damage to the roof rock; σ_0 is the tensile strength of the roof, MPa; P is the original rock stress, MPa; c is the explosive blasting decay index, $c = \frac{2-\mu}{1-\mu}$; T_m is the maximum shockwave pressure of the borehole wall, MPa; and K_s is the side pressure coefficient.

After the explosive blasting in the borehole, the rock damage area disturbed by the blasting impact should form a superimposed area between the two boreholes. The borehole spacing d should be less than or equal to 2R in order to achieve a better presplitting effect.

3. Numerical Simulation Analysis

3.1. Model Establishment and Simulation Scheme

Combined with the engineering geological conditions of Yitang coal mine, a numerical calculation model with a width of 240 m and a height of 60 m was established using PFC^{2D} 5.0 numerical simulation software. Taking into account the influence of the boundary, the lower left and right ends of the model were set as fixed boundaries, and a vertical stress of 6 MPa was applied to the upper boundary to equate the overburden load. The length of the working face was 150 m, the width of the roadway was 6 m, and the height was 4 m. The numerical simulation process is as follows: the original model is balanced, the roadway is excavated and the roof cut, the face is mined gradually, and the roadway deformation law is analyzed after the collapse of the goaf is complete. The numerical model is shown in Figure 6. The physical and mechanical parameters of coal and rock are shown in Table 1.

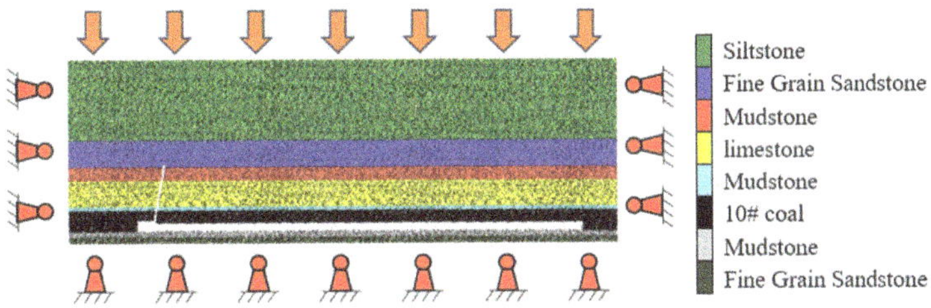

Figure 6. Numerical model of cut top formation.

Table 1. Mechanical parameters of each rock formation.

Lithology	Density/g·cm^{-3}	Compressive Strength/MPa	Tensile Strength/MPa	Cohesion/MPa	Elastic Modulus/MPa	Poisson's Ratio	Friction Angle/°
Siltstone	2.57	31.58	2.88	9.32	4238	0.31	30.33
Fine sandstone	2.60	51.79	5.65	13.00	5836	0.24	27.14
Mudstone	2.52	30.10	3.82	5.01	3813	0.26	28.20
Limestone	2.54	40.83	4.45	8.02	3023	0.36	34.53
Mudstone	2.53	34.77	3.78	4.78	3865	0.30	27.56
No. 10 coal	1.45	11.00	0.90	3.10	1100	0.28	28.00
Mudstone	2.53	34.77	3.78	4.78	3865	0.30	27.56
Fine sandstone	2.60	51.79	5.65	13.00	5836	0.24	27.14

PFC^{2D} is a discrete element numerical simulation software package. In order to ensure the simulation results are true and reliable, it is necessary to calibrate the microscopic parameters and bind the simulation parameters with the parameters of real rock mechanics. Based on the on-site retrieval of rock cores and the completion of indoor mechanical tests, the mechanical parameters of each rock layer were obtained. Based on this, the coal and rock were calibrated. The parameter verification stress–strain curve is shown in Figure 7, and the microscopic parameters of the verified model are shown in Table 2.

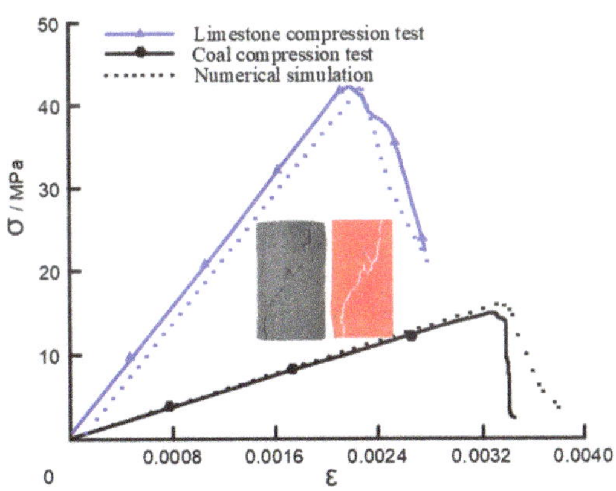

Figure 7. Stress–strain curve for parameter verification of numerical model.

Table 2. Numerical model microscopic parameters.

Lithology	Pb_emod/GPa	Pb_kratio	Pb_ten/MPa	Pb_coh/MPa	Pb_fa/°
Siltstone	0.47	1.35	10	30	20
Fine sandstone	0.44	1.20	23	54	28
Mudstone	0.50	1.25	18	14	30
Limestone	0.49	1.00	25	36	22
Mudstone	0.48	1.40	14	12	30
No. 10 coal	0.28	2.00	2	8	16
Mudstone	0.48	1.43	16	13	32
Fine sandstone	0.43	1.22	24	55	27

According to the technical characteristics of top coal gob-side entry retaining and the engineering mining conditions, four factors were selected: roof-cutting height, borehole dip angle, mining height, and coal seam thickness. Three levels were set for each factor to investigate their influence on the effect of top coal gob-side entry retaining, and a total of 12 groups of numerical simulation schemes were designed according to the principle of the single variable, as shown in Table 3.

Table 3. Numerical simulation scheme for cutting the top into a lane.

Factor Scheme Number	Height of Roof Cutting/m	Borehole Dip Angle/°	Mining Height/m	Coal-Seam Thickness/m
1–3	12, 16, 20	10	4	8
4–6	16	0, 10, 20	4	8
7–9	16	10	3, 4, 5	8
10–12	16	10	4	6, 8, 10

3.2. Analysis of Displacement Field of Surrounding Rock

From Figure 8a–c, it can be seen that the roof subsidence displacement is 142 cm, 62 cm, and 106 cm under different top cutting heights. When the roof-cutting height is 12 m, the cutting seam only partially cuts off the limestone roof. The extrusion pressure generated by the main roof bending and sinking is transferred from the rock roof to the roadway, the coal beam is fractured, and the coal seam roof is completely dislodged. When the roof-cutting height is 16 m, the joint-cutting completely cuts off the limestone roof, the main roof bending and sinking process has less influence on the cantilever beam, the integrity of the coal seam roof is better, and the roof subsidence is more moderate. When the roof-cutting height is 20 m, the joint-cutting completely cuts off the mudstone layer below the main roof, the main roof subsidence space becomes larger, the cantilever beam structure self-weight increases, and the displacement of the roof on the joint-cutting side increases obviously.

From Figure 8d–f, it can be seen that the roof subsidence displacement is 44 cm, 62 cm, and 122 cm under different borehole dip angles. When the borehole dip angle is 0°, the collapse of the rock in the goaf will be obstructed by the rock on the other side of the joint-cutting, which means the collapse of the goaf is not complete under the same calculation conditions. The bending and sinking of the main roof is smaller, and the roof subsidence displacement is also smaller. When the borehole dip angle is 10°, the collapse of the rock layer in the goaf is not obstructed by the rock body on the other side of joint-cutting. The collapse of each rock layer is complete, the structure of the cantilever beam is intact, and the roof subsidence displacement increases slowly. When the borehole dip angle is 20°, the cantilever beam structure volume becomes bigger and the weight of the cantilever beam itself increases because the angle is too inclined. Under the extrusion of main roof deformation, the roof near to the buttock structural body is torn, the cantilever beam structure as a whole is destabilized, and the roadway produces large deformations.

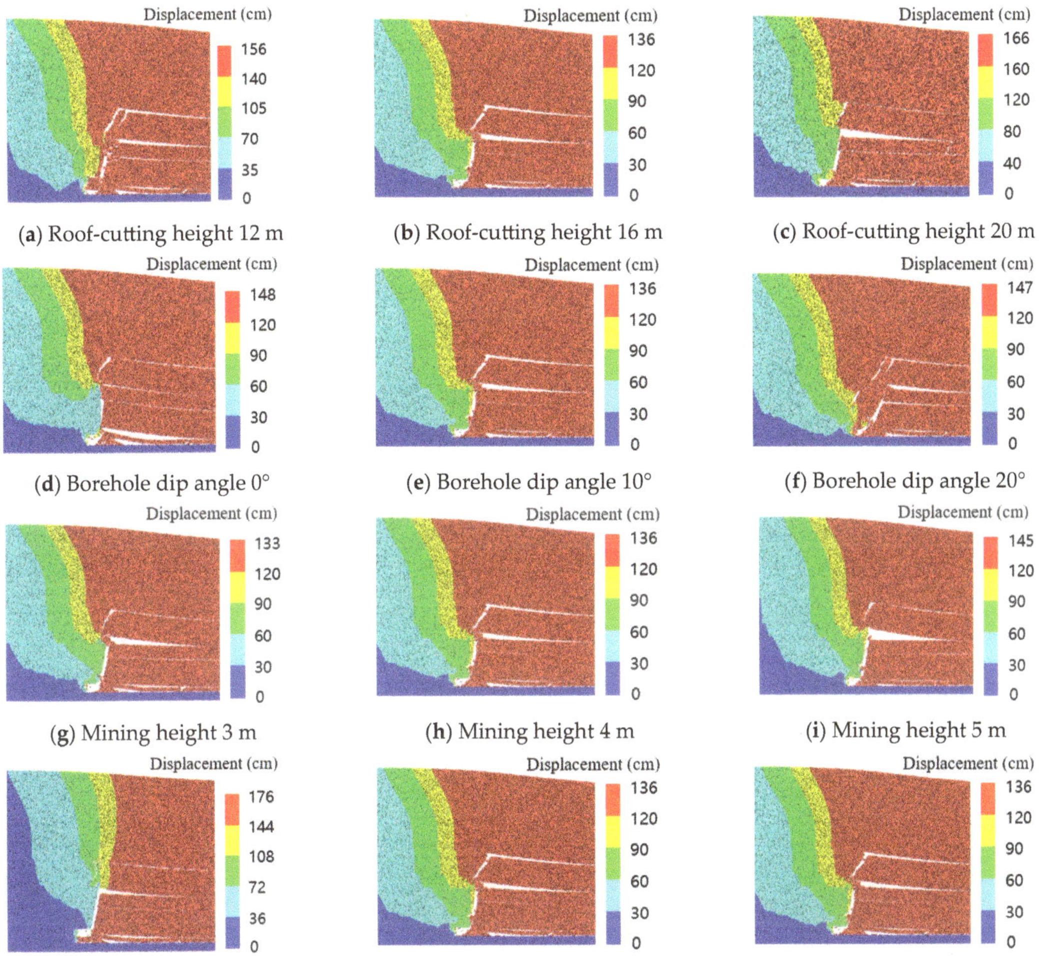

(**a**) Roof-cutting height 12 m (**b**) Roof-cutting height 16 m (**c**) Roof-cutting height 20 m
(**d**) Borehole dip angle 0° (**e**) Borehole dip angle 10° (**f**) Borehole dip angle 20°
(**g**) Mining height 3 m (**h**) Mining height 4 m (**i**) Mining height 5 m
(**j**) Coal-seam thickness 6 m (**k**) Coal-seam thickness 8 m (**l**) Coal-seam thickness 10 m

Figure 8. Comparison of roadway displacement.

From Figure 8g–i, it can be seen that under different mining heights, the roof subsidence displacement is 59 cm, 62 cm, and 66 cm, respectively. After the roof is cut off, the influence of mining height on the surrounding rock of the roadway is greatly weakened, and with the increase in the mining height, the roof subsidence displacement only increases a little; however, the impact of the collapsed rock on the gangue-retaining structure obviously increases, and the gap between the main roof and the cut-off rock layer will also increase. The coal seam roof and cantilever beam structure under each mining height are in a stable state, and considering that the height difference between the crossheading and the working face will increase the difficulty of the end support, consistency should be maintained between the two as much as possible when mining.

From Figure 8j–l, it can be seen that the roof subsidence displacement is 106 cm, 62 cm, and 96 cm, respectively, under different coal seam thicknesses. When the thickness of the coal seam is 6 m, the top coal is thin and the shear resistance is weak, so it breaks at the shoulder corner of the roadway under the influence of mining and the whole coal seam roof falls. When the thickness of the coal seam is 8 m, the top coal thickness increases and the shear resistance is stronger. The roof of the coal seam and the roof of the rock seam are not separated from each other, and the deformation of the compound roof follows the same trend. When the thickness of the coal seam is 10 m, the increase in the coal thickness means that the roof-cutting height can only reach the upper part of the limestone roof, which cannot be completely cut off. The percentage of fragile coal body in the cantilever beam structure increases and the percentage of hard rock body decreases, so that its bearing capacity decreases greatly, and, finally, vertical tearing occurs on the side of the cantilever beam near the buttock, and the whole structure collapses downward.

The comprehensive analysis of the above indicates that the roof-cutting height calculations should try to ensure that the immediate roof or main roof can be completely cut off. It is not meaningful to increase the roof-cutting height; under the premise of ensuring that the collapse of the cut rock is not hindered, the borehole dip angle should be reduced to improve the stability of the cantilever beam structure; when the thickness of the coal seam roof is thin, it delamination should be prevented, and when the thickness of the coal seam roof is thick, the effect of roof cutting is poor. After the increase in the mining height, the gangue-retaining structure should be strengthened. The process parameters of Yitang top coal gob-side entry retaining are determined as follows: roof-cutting height, 16 m; borehole dip angle, 10°; mining height, 4 m; coal seam thickness, 8 m. In addition, the roof subsidence displacement is larger in the simulation results, which is because no support structure such as anchors and cables are applied in the simulation, and the roof subsidence displacement is expected to be smaller under a reasonable support system when field tests are conducted.

3.3. Analysis of Stress Field in Surrounding Rock

From the analysis of Figures 9–12, it can be seen that the variation pattern of roof stress under different factors is basically the same. After cutting the roof, the roof structure changes, and the stress rapidly increases. The early mining face is far from the roadway, and the overburden load and mining stress are borne by the gravel in the distant goaf. During the process of gradually stabilizing the surrounding rock structure, the roof stress decreases, and the overburden load on the roadway decreases with the increase in mining progress. When the mining face approaches the roadway, it is included in the mining affected area; the stress it is subjected to rapidly increases, and after collapsing and stabilizing, the stress rapidly decreases.

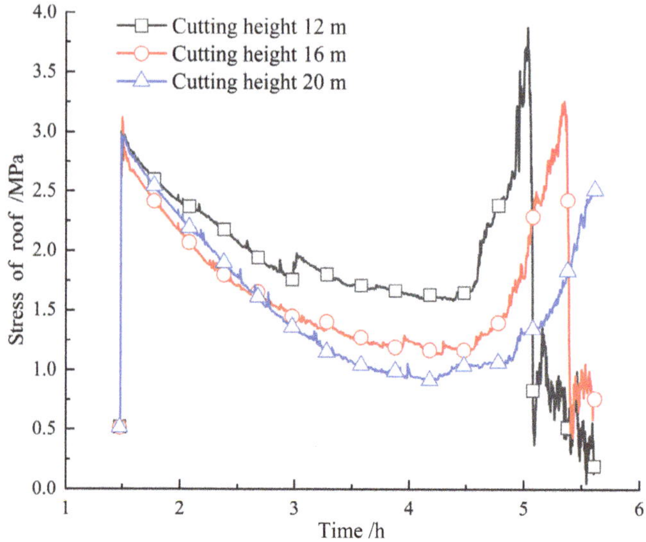

Figure 9. Roof stress variation curve under different roof-cutting heights.

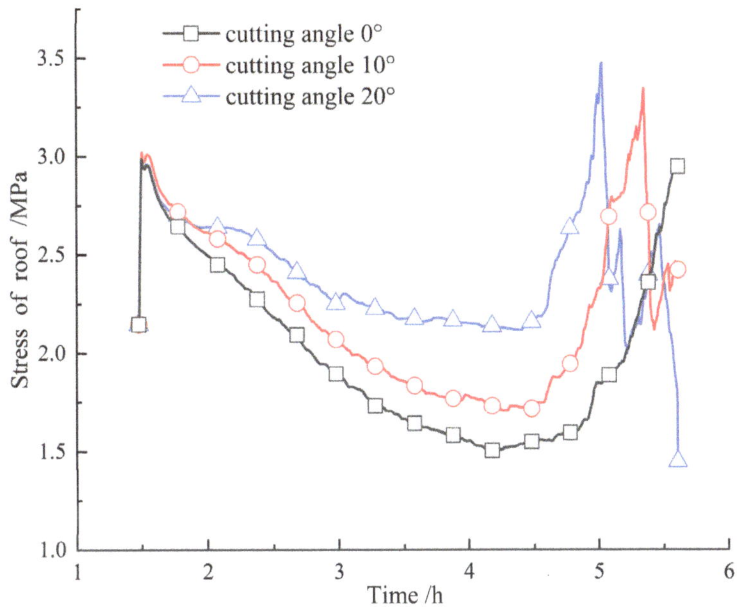

Figure 10. Roof stress variation curve under different borehole dip angles.

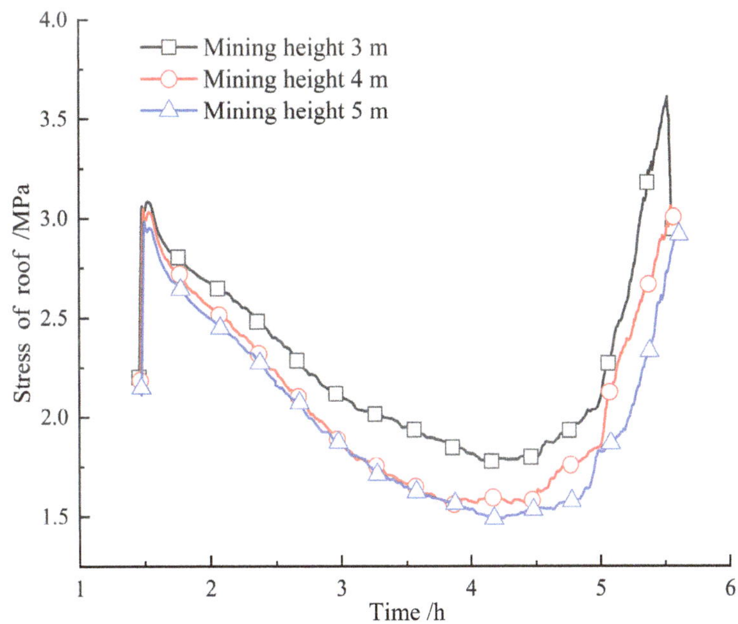

Figure 11. Roof stress variation curve under different mining heights.

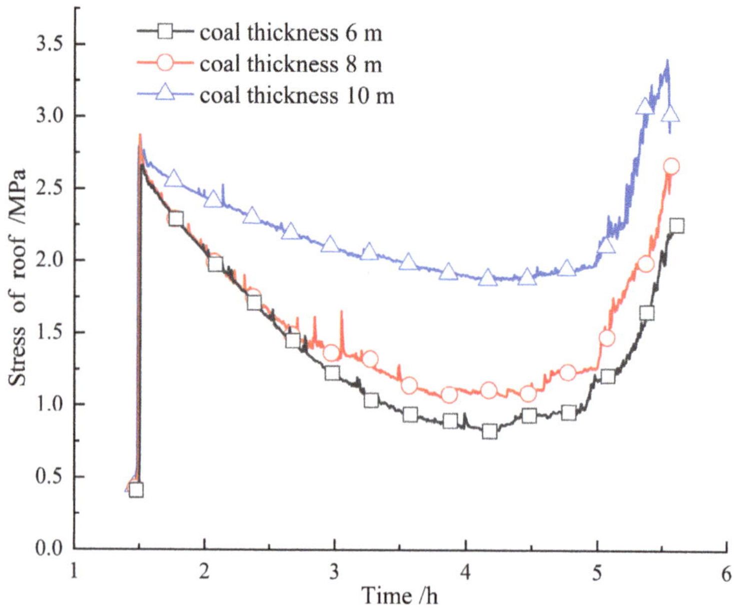

Figure 12. Roof stress variation curve under different coal thicknesses.

Under different roof-cutting heights, the peak stress of roof is 3.87 MPa, 3.27 MPa, and 2.63 MPa, respectively. As the roof-cutting height increases, the stress on the roof gradually decreases. Under different borehole dip angles, the peak stress on the roof is 2.88 MPa, 3.27 MPa, and 3.51 MPa, respectively. As the borehole dip angle increases, the stress on the roof gradually increases. Under different mining heights, the peak stress values of the roof are 3.63 MPa, 3.27 MPa, and 2.85 MPa, respectively. As the mining height increases, the stress on the roof gradually decreases. Under different coal seam thicknesses, the peak stress values of the roof are 2.26 MPa, 3.27 MPa, and 3.47 MPa, respectively. As the coal seam thickness increases, the stress on the roof gradually increases.

4. Top Coal Roadway Support Strategy and Stay Roadway Effect

4.1. Roadway Damage Form and Mechanical Model Analysis

The rock structure of a traditional top coal roadway can be regarded as a coal–rock composite roof. The roof of the coal seam is soft and fragile, while the roof of the rock seam is hard. The weak roof of the coal seam bends and sinks under the action of stress; the gap between the coal seam and the rock seam is generated and further developed into delamination, and the shoulder corner of the roadway and the middle of the coal beam first reach the limit of tensile–shear damage and damage occurs. Under the stress, the bending moment of the coal beam increases while the weak surface inside the coal seam slips, and the cracks between adjacent weak surfaces penetrate each other to cut the coal beam into several parts laterally. The roof of the coal beam is severely broken, and its mechanical structure model is shown in Figure 13.

Based on the equivalent buttock-type solid support beam structure, it can be seen that the bending moment of the structure is mainly the effect of horizontal stress. The vertical stress has a small effect on the bending moment. Ignoring the effect of vertical stress, the mechanical formula can be calculated to produce the bending distance of the structure:

$$M = \frac{q_1 a^2}{12} \quad (4)$$

where q_1 is the horizontal direction uniform load, kN/m; and a is the height of the roadway, m.

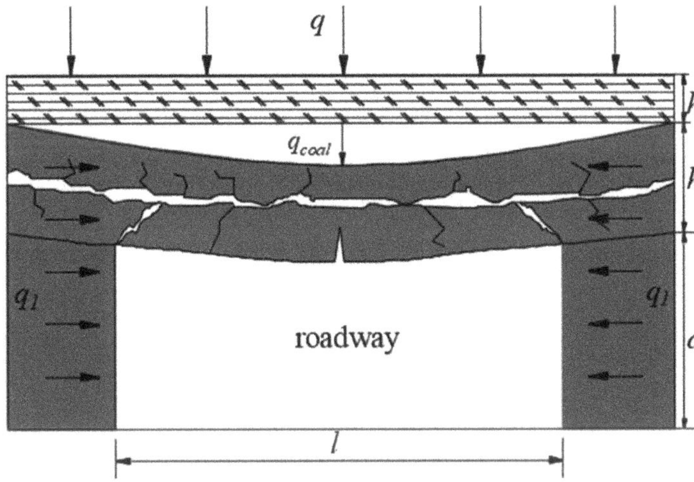

Figure 13. Roadway damage mechanics model of the top coal.

After the separation of the coal roof and the rock roof, the two can be regarded as two independent beam structures. At this time, the vertical stress can be ignored, and the horizontal stress is the main cause of the coal beam damage. Using the mechanical formula, the maximum tensile stress inside the single-layer beam structure can be found:

$$\sigma_{max} = \frac{6M}{bh^2} - \sigma_x \tag{5}$$

where σ_{max} is the maximum tensile stress inside the beam structure, MPa; σ_x is the horizontal structural stress of the beam structure, MPa; h is the thickness of the beam structure, m; b is the cross-sectional width of the beam structure, m; and M is the maximum bending moment to which the beam structure is subjected, kN·m.

Substituting Equation (4) into Equation (5), when b = 1, the maximum tensile stress inside the roof of coal beam can be obtained:

$$\sigma_{coalmax} = \frac{3q_{coal}l^2 - q_1a^2}{2h_2^2} - \sigma_x \tag{6}$$

where q_{coal} is the self-weight of the coal beam and overburden load transferred to the coal beam, MPa; and σ_x is the horizontal structural stress of the beam structure, MPa.

After the roof is cut off, the mining stress transmission path is cut off, the surrounding rock structure is changed, the horizontal stress effect is weakened, and the vertical stress effect on the roof is enhanced. As can be seen from Figure 14, the bending moment of the composite roof under stress gradually increases, and the internal tensile stress of the roof accumulates. Damage occurs first at the corner of the roadway, and gradually develops into a large area of roof collapse. In this state, the coal–rock composite roof is sinking jointly, and the coal–rock composite roof can be regarded as a stacked beam. The maximum tensile stress inside this structure is:

$$\sigma_{max} = \frac{6M}{bk_1^2} - \sigma_x \tag{7}$$

where k_1 is the moment of inertia discount factor of the laminated beam, which is 0.75.

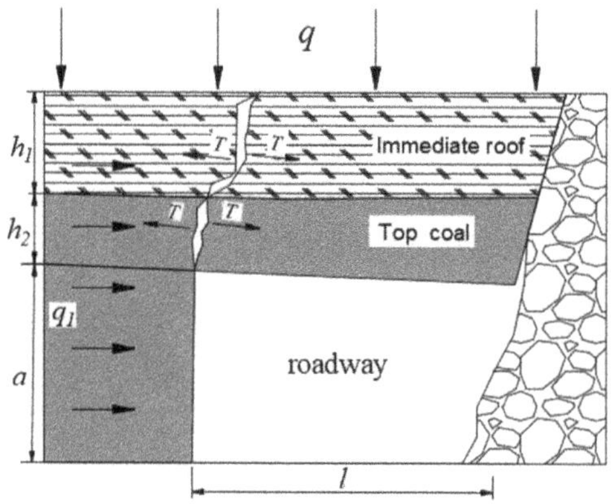

Figure 14. Mechanical model of roof damage after roof cutting.

The material mechanics stacked beam equation is:

$$\begin{cases} M = \frac{ql^2}{24} \\ I = \frac{bh^2}{12} \\ \rho = \frac{EI}{M} \end{cases} \quad (8)$$

where M is the maximum bending moment of the laminated beam structure, kN m; q is the upper load of the laminated beam structure, kN/m; l is the span of the laminated beam structure, m; I is the moment of inertia of the laminated beam structure; b is the cross-sectional width of the laminated beam structure, taken as 1 m; h is the thickness of the laminated beam structure, m; ρ is the radius of curvature of the laminated beam structure, m; and E is the modulus of elasticity of the laminated beam structure, MPa.

If the coal beam of the roof is affected by vertical tectonic stress, which is M_1, the bending moment of the rock beam is M_4, and the bending moment of the coal beam due to the influence of the buttock is M_2, the superposition moment M_3 of the coal beam by the buttock and vertical stress is $M_1 - M_2$, and the sum of the bending moment M of the coal beam and rock beam under the action of vertical tectonic stress is $M_1 + M_4$. The maximum tensile stress that the coal rock beam undergoes when the supporting top coal roadway undergoes coordinated subsidence of the coal–rock beam is:

$$\sigma_{coalmax} = \frac{E_1 h_1 (ql^2 - 2q_1 a^2)}{4k_1 (E_1 h_1^3 + E_2 h_1^3)} - \sigma_x \quad (9)$$

Since the bottom of the composite roof is the coal body, the stability of the coal beam determines the stability of the whole roadway. When the maximum tensile stress $\sigma_{coalmax}$ inside the coal beam is greater than the ultimate tensile strength σ_{coal} of the coal beam, the coal beam will be damaged and the whole structure will be destabilized.

4.2. Roadway Support Strategy

Based on the above discussion, it can be seen that the main control factor for the destabilization of the roadway when the coal seam roof is damaged in the process of top coal gob-side entry retaining needs to focus on the support, and its support strategy is as follows: increase the density and preload of the anchor cable to prevent the coal beam from leaving the layer and the weak surface from slipping; increase the length of the anchor cable to anchor into the deep stable rock layer, and suspend the top coal or rock beam that cannot be self-stabilized below to control its deformation; increase the density of the anchor cable to enhance the strength of the anchor beam; increase the length of the anchor cable to form a ladder-bearing structure together with the longer reinforcement anchor cable and the slightly shorter common anchor cable to give full play to the self-supporting capacity of the surrounding rock. It is also possible to consider laying metal mesh and erecting a metal beam to form a joint support structure to provide a synergistic support effect; surface spraying and shallow grouting can also be applied to the coal seam roof to improve its overall structural strength. In addition, additional single support equipment is required for mining disturbance and periodic pressure in order to maintain the stability of the roadway.

Detailed support parameters are shown in Figure 15, and the support scheme is as follows.

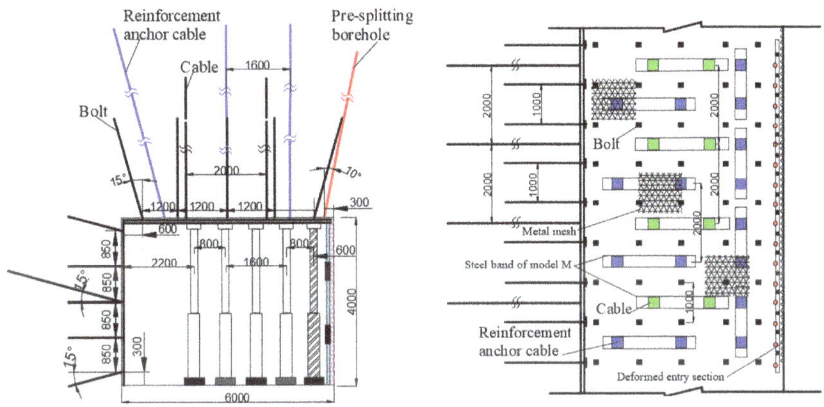

(a) (b)

Figure 15. Roadway support layout diagram. (**a**) Cross-sectional drawing of the roadway support; (**b**) roadway top plate support plan.

For roof support, the roof of the roadway is protected by an anchor mesh cable and M steel belt, with anchor rods of the size ϕ 18 mm × 4000 mm and an inter-row distance of 1200 mm × 1000 mm. The edge anchor rods are arranged at an inclination of 15°, and the rest of the anchor rods are arranged perpendicular to the roof. The anchor ropes are ϕ 21.6 mm × 10000 mm with inter-row spacing of 2000 mm × 2000 mm, and the reinforcement anchor ropes are ϕ 21.6 mm × 18,000 mm with a spacing of 1.6 m, with a row spacing of 1 m on the near-cut side and 2 m on the remaining two rows.

The anchor ropes are ϕ 18 mm × 2400 mm with an inter-row spacing of 850 mm × 1000 mm. The edge anchor ropes are arranged at 15° deviation, and the rest of the anchor ropes are arranged perpendicular to the roadway gang. The anchor ropes are ϕ 17.8 mm × 4000 mm, with a row spacing of 2 m, and are arranged at 15° from the horizontal.

For the temporary reinforcement of support, a π beam with a mono-hydraulic pillar form a "one beam and five columns" support structure. After the impact of mining, two mono-columns can be withdrawn so that it is transformed into a "one beam and three columns" form. After the roadway is stabilized, only the monolithic pillar on the side of the mining area is retained.

The gob-side entry retaining support is a combination of steel mesh, individual pillars, and expandable U-shaped steel used for gangue-retaining support within 6 m of the lagging working face. Scalable U-shaped steel with a row spacing of 600 mm and a steel mesh size of 2500 mm × 4000 mm, with an overlap of 100 mm at the joint of the steel mesh and tied with iron wire for fixation is used. If there is serious leakage on site, the steel mesh or iron wire mesh can be laid in multiple layers to strengthen the support of the blocking gangue.

4.3. Effect of the Retention Lane

The purpose of mine pressure observation is to grasp the spatiotemporal distribution patterns of deformation and stress in the surrounding rock of the roadway by analyzing the data collected during mining, providing a scientific basis for the subsequent design of gob-side entry retaining and roadway support and ensuring safe construction. During the process of gob-side entry retaining, monitoring points should be set up to monitor the subsidence displacement of the roof, the separation of the roof layer, and the force and deformation of the anchor rod and cable. A roof separation meter and anchor rod and cable dynamometer should be installed 30 m ahead of the working face, and the first roof separation meter, roof and floor displacement sensor, and anchor rod and cable dynamometer should be installed at a distance of 30 m behind the working face. Five sets of roof separators and anchor cable force gauges are arranged at intervals of 50 m, while one set of roof and floor displacement sensors is arranged at intervals of 20 m for a total of 11 sets, all distributed within 200 m of the lagging working face. As the working face continues to advance, measurement stations within the monitoring distance of the front and rear are also added. Any measurement equipment damaged during the advancing process of the working face should be replaced and supplemented in a timely manner. The specific layout of the measurement points is shown in Figure 16.

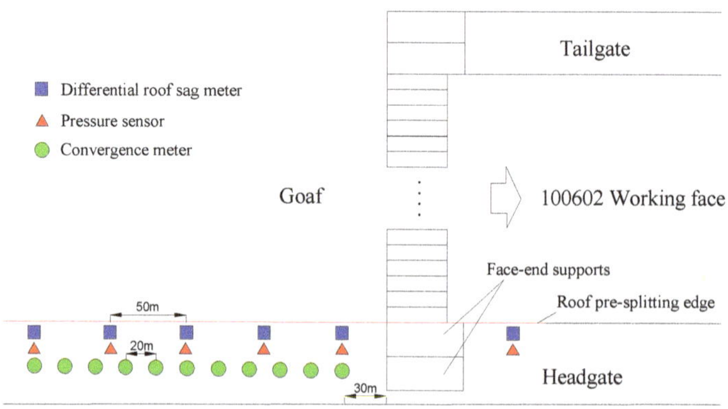

Figure 16. Field measurement configuration.

By comparing the data obtained from various measurement stations, it can be found that as the working face continues to advance, the rock pressure patterns reflected by each measurement point are basically the same. Therefore, only some typical data in the deformation of the roof and roadway sides are selected for analysis. The monitoring results are shown in Figure 17.

From the site monitoring results shown in Figure 17, the roof subsidence displacement near the joint-cutting side is larger than near the buttock side. The roof subsidence displacement near the joint-cutting side is about 308 mm, while the final roof subsidence displacement near the buttock side is about 196 mm. The final gangue wall convergence is about 164 mm, while the buttock convergence is about 116 mm. The roof subsidence is affected by the incoming pressure, showing a periodic rapid growth, while the buttock convergence is slower. Finally, at about 140 m lagging the working face, the surrounding rock of the roadway entered a stable state.

After gob-side entry retaining, the surrounding rock control effect is good. The ventilation system of the working face is improved, the speed of mining is improved, a large number of coal pillars are recovered, and good technical and economic benefits are achieved. The effect of gob-side entry retaining is shown in Figure 18.

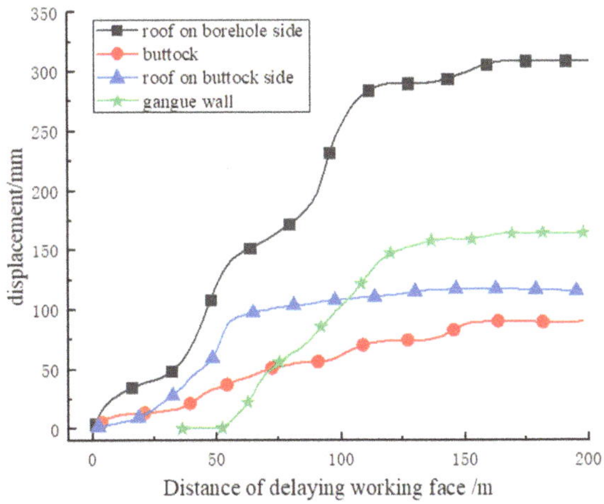

Figure 17. Field measurement data.

Figure 18. Effect of gob-side entry retaining.

5. Discussion

This research significantly contributes to the understanding of gob-side entry maintenance using roof-cutting techniques in fully mechanized longwall mining. By investigating the key parameters of roof pre-fracturing boreholes, we identified the optimal settings that enhance the stability of gob-side entries in ultra-thick coal seams. Moreover, field measurements verified the efficacy of these settings and the benefits of appropriate anchor-net support in forming a stable rock structure.

Nonetheless, our study has certain limitations. It focused on a specific engineering background, which might not represent all mining conditions. Therefore, the universality of our findings could be a subject for future research. Moreover, other potentially impactful factors, such as geological conditions and mining techniques, were not deeply examined. It is also important to consider alternative methods, particularly hydraulic fracturing, which involves opening the reservoir with vertical or inclined wells, placing hydraulic sandblasting perforators at given formation intervals, and pumping the working fluid into the hydraulic perforator. The nozzles of the blaster create grooves (cavities) in the formation and then fracture the formation through the gaps (cavities) that form.

Moving forward, we recommend conducting comprehensive research considering various mining conditions and additional impacting factors. This could broaden the applicability of our findings and contribute more effectively to the sustainable development of coal mines.

6. Conclusions

(1) Based on the geological conditions of the Yitang coal mine, numerical simulation analysis was conducted on top coal gob-side entry retaining, and the influence of various factors on the effectiveness of gob-side entry retaining was explored. The simulation results indicate that under the conditions of a roof-cutting height of 16 m, borehole dip angle of 10°, mining height of 4 m, and an average thickness of 8 m in the coal seam, the roof of the coal seam did not detach and the bending settlement was small. The cantilever beam has good load-bearing capacity, and the overall structure is stable. The rock strata in the goaf collapse completely and the filling condition is good.

(2) During the process of gob-side entry retaining, it was found that the variation pattern of the roof stress under different factors is basically the same, with a peak stress of around 3 MPa. As the roof-cutting height and mining height increase, the stress on the roof gradually decreases; as the borehole dip angle and coal seam thickness increase, the stress on the roof gradually increases.

(3) The mechanical analysis found that the strength of the roof coal beam directly affects the stability of the roadway, and the roof stability is related to the changeable factors such as the thickness of the top coal, the span of the roadway, and the elastic modulus of the coal seam. When increasing the thickness of the top coal appropriately, the elastic modulus of the roof anchorage area is enhanced by using step support, and the span distance is reduced by using reduced span support to enhance the strength of the roof of the coal seam and maintain the stability of the surrounding rock.

(4) The on-site monitoring data of the mine pressure shows that under a reasonable support plan, the maximum subsidence of the roof is 308 mm, and the total convergence of the roadway movement is 280 mm. The total affected area is within 140 m of the lagging working face, and the deformation of surrounding rock is well controlled, achieving certain technical and economic benefits.

Author Contributions: Project administration, Funding acquisition, Supervision, Y.Y.; Data curation, Formal analysis, Investigation, P.G.; Formal analysis, Writing—original draft, C.Z.; Methodology, Writing—review & editing, Funding acquisition, C.W. All authors have read and agreed to the published version of the manuscript.

Funding: This work was supported by the National Natural Science Foundation of China (Grant No. 51404167); Natural Science Foundation of Shanxi Province (Grant No. 201901D211066); Scientific Research Grant Project for Returned Overseas Students in Shanxi Province (HGKY2019038); Teaching Reform Innovation Project for Higher Education Institutions in Shanxi Province (J2019055); China Postdoctoral Science Foundation funding project (Grant No. 2016M590151).

Data Availability Statement: The data used to support the findings of this study are included within the article.

Conflicts of Interest: The authors declare no conflict of interest.

Nomenclature

H_q	the roof-cutting height, m
H_c	the mining height, m
H_f	the thickness of the top coal, m
ΔH_1	the roof subsidence displacement, m
ΔH_2	the floor heaving displacement, m
K	the coefficient of rock fragmentation and expansion

σ_t	the tensile strength of the coal seam roof, MPa
α	the borehole dip angle, °
r_0	the radius of the borehole, mm
D_0	the initial damage to the roof rock
σ_0	the tensile strength of the roof rock, MPa
P	the original rock stress, MPa
c	the explosive blasting decay index
T_m	the maximum shockwave pressure of the borehole wall, MPa
K_s	the side pressure coefficient
q_1	the horizontal direction overburden load, kN/m
α	the height of the roadway, m
σ_{max}	the maximum tensile stress inside the beam structure, MPa
σ_x	the horizontal structural stress of the beam structure, MPa
h	the thickness of the beam structure, m
b	the cross-sectional width of the beam structure, m
M	the maximum bending moment to which the beam structure is subjected, kN·m
q_{coal}	the self-weight of coal beam and overburden load transferred to coal beam, MPa
k_1	the moment of inertia discount factor of the laminated beam
q	the upper load of the laminated beam structure, kN/m
l	the span of the laminated beam structure, m
I	the moment of inertia of the laminated beam structure
b	the cross-sectional width of the laminated beam structure, m
ρ	the radius of curvature of the laminated beam structure, m
E	the modulus of elasticity of the laminated beam structure, MPa

References

1. Wang, K.; Huang, Y.; Gao, H.; Zhai, W.; Qiao, Y.; Li, J.; Ouyang, S.; Li, W. Recovery Technology of Bottom Coal in the Gob-Side Entry of Thick Coal Seam Based on Floor Heave Induced by Narrow Coal Pillar. *Energies* **2020**, *13*, 3368. [CrossRef]
2. Lv, H.; Cheng, Z.; Liu, F. Study on the mechanism of a new fully mechanical mining method for extremely thick coal seam. *Int. J. Rock Mech. Min. Sci.* **2021**, *142*, 104788. [CrossRef]
3. Liu, S.; Yang, K.; Tang, C.; Chi, X. Rupture and Migration Law of Disturbed Overburden during Slicing Mining of Steeply Dipping Thick Coal Seam. *Adv. Civ. Eng.* **2020**, *2020*, 8863547. [CrossRef]
4. Liu, S.; Huang, J.; Huang, Q.; Li, T.; Tang, S.; Liu, X. Floor Pressure-Relief During Top Slice Mining of Extra-Thick Coal Seams and Its Implications for Gas Drainage Application. *Geotech. Geol. Eng.* **2019**, *37*, 3113–3125. [CrossRef]
5. Li, Z.-L.; He, X.-Q.; Dou, L.-M.; Song, D.-Z. Comparison of rockburst occurrence during extraction of thick coal seams using top-coal caving versus slicing mining methods. *Can. Geotech. J.* **2018**, *55*, 1433–1450. [CrossRef]
6. He, W.; He, F.; Zhao, Y. Field and simulation study of the rational coal pillar width in extra-thick coal seams. *Energy Sci. Eng.* **2019**, *8*, 627–646. [CrossRef]
7. Li, X.; Zhao, Y.; He, W.; Li, L.; He, F. Study on Coal Pillar Width and Surrounding Rock Control of Gob-Side Entry in Extra-thick Coal Seam. *Geotech. Geol. Eng.* **2020**, *38*, 6855–6868. [CrossRef]
8. Wang, Q.; Jiang, Z.; Jiang, B.; Gao, H.; Huang, Y.; Zhang, P. Research on an automatic roadway formation method in deep mining areas by roof cutting with high-strength bolt-grouting. *Int. J. Rock Mech. Min. Sci.* **2020**, *128*, 104264. [CrossRef]
9. Çelik, A.; Özçelik, Y. Investigation of the efficiency of longwall top coal caving method applied by forming a face in horizontal thickness of the seam in steeply inclined thick coal seams by using a physical model. *Int. J. Rock Mech. Min. Sci.* **2021**, *148*, 104917. [CrossRef]
10. Liang, Y.; Li, L.; Li, X.; Wang, K.; Chen, J.; Sun, Z.; Yang, X. Study on Roof-Coal Caving Characteristics with Complicated Structure by Fully Mechanized Caving Mining. *Shock. Vib.* **2019**, *2019*, 6519213. [CrossRef]
11. Wang, E.; Xie, S. Determination of coal pillar width for gob-side entry driving in isolated coal face and its control in deep soft-broken coal seam: A case study. *Energy Sci. Eng.* **2022**, *10*, 2305–2316. [CrossRef]
12. Ma, Z.; Chen, C.; Liang, X.; Chen, A.; Song, W. Field and numerical investigation on the stability of coal pillars of gob-side entry driving with top coal. *Arab. J. Geosci.* **2020**, *13*, 1193. [CrossRef]
13. Wang, B.; Dang, F.; Gu, S.; Huang, R.; Miao, Y.; Chao, W. Method for determining the width of protective coal pillar in the pre-driven longwall recovery room considering main roof failure form. *Int. J. Rock Mech. Min. Sci.* **2020**, *130*, 104340. [CrossRef]
14. Chen, A. Width Design of Small Coal Pillar of Gob-Side Entry Driving in Soft Rock Working Face and Its Application of Zaoquan Coal Mine. *Adv. Civ. Eng.* **2021**, *2021*, 9999957. [CrossRef]
15. Qi, F.; Zhou, Y.; Cao, Z.; Zhang, Q.; Li, N. Width optimization of narrow coal pillar of roadway driving along goaf in fully mechanized top coal caving face. *J. Min. Saf. Eng.* **2016**, *33*, 475.

16. Wang, Y.; Li, Y.; Yu, Z.; Zhang, H.; Ding, D. Research on the Surrounding Rock Control Technology of Gob-Side Entry with a Narrow Coal Pillar Reserved in a Fully Mechanized Caving Face with Large Mining Height. *Geotech. Geol. Eng.* **2021**, *40*, 285–300. [CrossRef]
17. Sakhno, I.; Sakhno, S. Numerical Studies of Floor Heave Control in Deep Mining Roadways with Soft Rocks by the Rock Bolts Reinforcement Technology. *Adv. Civ. Eng.* **2023**, *2023*, 2756105. [CrossRef]
18. Bosikov, I.I.; Martyushev, N.V.; Klyuev, R.V.; Savchenko, I.A.; Kukartsev, V.V.; Kukartsev, V.A.; Tynchenko, Y.A. Modeling and Complex Analysis of the Topology Parameters of Ventilation Networks When Ensuring Fire Safety While Developing Coal and Gas Deposits. *Fire* **2023**, *6*, 95. [CrossRef]
19. Balovtsev, S.; National University of Science and Technology; Skopintseva, O.; Kolikov, K. Misis Aerological risk management in preparation mining of coal mines. *Sustain. Dev. Mt. Territ.* **2022**, *14*, 107–116. [CrossRef]
20. Tian, Z.; Zhang, Z.; Deng, M.; Yan, S.; Bai, J. Gob-Side Entry Retained with Soft Roof, Floor, and Seam in Thin Coal Seams: A Case Study. *Sustainability* **2020**, *12*, 1197. [CrossRef]
21. Xie, S.; Wang, E.; Chen, D.; Li, H.; Jiang, Z.; Yang, H. Stability analysis and control technology of gob-side entry retaining with double roadways by filling with high-water material in gently inclined coal seam. *Int. J. Coal Sci. Technol.* **2022**, *9*, 52. [CrossRef]
22. Cheng, Y.; Jiang, F.; Lin, J.; Chen, Q.; Zhang, D.; Zhang, F. Experimental Study on Gob Side Entry Retaining by Roadside Flexible Packing Under Hard Roof. *J. Min. Saf. Eng.* **2012**, *29*, 757.
23. Zhang, Y.; Wang, X. Modification test of high-water packing material and application in the gob-side entry retained engineering. *Arab. J. Geosci.* **2022**, *15*, 579. [CrossRef]
24. Sun, X.; Li, G.; Song, P.; Miao, C.; Zhao, C.; Li, Q.; Xia, X. Application Research on Gob-Side Entry Retaining Methods in No. 1200 Working Face in Zhongxing Mine. *Geotech. Geol. Eng.* **2018**, *37*, 185–200. [CrossRef]
25. Zhang, X.; Chen, L.; Gao, Y.; Hu, J.; Yang, J.; He, M. Study of An Innovative Approach of Roof Presplitting for Gob-Side Entry Retaining in Longwall Coal Mining. *Energies* **2019**, *12*, 3316. [CrossRef]
26. Zhao, H. State-of-the-art of standing supports for gob-side entry retaining technology in China. *J. S. Afr. Inst. Min. Metall.* **2019**, *119*, 891–906. [CrossRef]
27. Han, C.; Yang, H.; Zhang, N.; Li, X.; Liu, Y.; Liu, W.; Yao, W. Study on Reasonable Coal Pillar Width of Gob-Side Roadway Excavating with Multi-Layer Hard Roof in Western Deep Mine. *Geotech. Geol. Eng.* **2021**, *40*, 1417–1428. [CrossRef]
28. Zhou, X.; Li, H.; Li, X.; Wang, J.; Meng, J.; Li, M.; Mei, C. Research on Gob-Side Entry Retaining Mining of Fully Mechanized Working Face in Steeply Inclined Coal Seam: A Case in Xinqiang Coal Mine. *Sustainability* **2022**, *14*, 10330. [CrossRef]
29. Yang, X.; Sun, Y.; Wang, E.; Mao, W. Study and Application of Gob-Side Entry Retaining Formed by Roof Cutting and Pressure Relief in Medium-Thickness Coal Seam with Hard Roof. *Geotech. Geol. Eng.* **2020**, *38*, 3709–3723. [CrossRef]
30. Wu, H.; Li, Q.; Zhu, C.; He, L. Study on the failure law of surrounding rock in inclined coal seam with gob side entry. *Sci. Rep.* **2023**, *13*, 973. [CrossRef]
31. Ma, X.; He, M.; Wang, J.; Gao, Y.; Zhu, D.; Liu, Y. Mine Strata Pressure Characteristics and Mechanisms in Gob-Side Entry Retention by Roof Cutting under Medium-Thick Coal Seam and Compound Roof Conditions. *Energies* **2018**, *11*, 2539. [CrossRef]
32. Fu, W. Study on Surrounding Rock Control of Gob-Side Entry Retaining in Inclined and Thick Longwall Face. *Geotech. Geol. Eng.* **2022**, *40*, 3477–3491. [CrossRef]
33. Kong, D.; Pu, S.; Cheng, Z.; Wu, G.; Liu, Y. Coordinated deformation mechanism of the top coal and filling body of gob-side entry retaining in a fully mechanized caving face. *Int. J. Geomech.* **2021**, *21*, 04021030. [CrossRef]
34. Wang, H.; Wang, J.; He, M.; Ma, Z.; Tian, X.; Liu, P. A novel non-pillar coal mining technology in longwall top coal caving: A case study. *Energy Sci. Eng.* **2023**, *11*, 1822–1841. [CrossRef]
35. Xie, S.; Wu, Y.; Guo, F.; Zou, H.; Chen, D.; Zhang, X.; Ma, X.; Liu, R.; Wu, C. Application of Pre-Splitting and Roof-Cutting Control Technology in Coal Mining: A Review of Technology. *Energies* **2022**, *15*, 6489. [CrossRef]
36. Wang, Q.; He, M.; Yang, J.; Gao, H.; Jiang, B.; Yu, H. Study of a no-pillar mining technique with automatically formed gob-side entry retaining for longwall mining in coal mines. *Int. J. Rock Mech. Min. Sci.* **2018**, *110*, 1–8. [CrossRef]
37. Ma, Z.; Wang, J.; He, M.; Gao, Y.; Hu, J. Key Technologies and Application Test of an Innovative Noncoal Pillar Mining Approach: A Case Study. *Energies* **2018**, *11*, 2853. [CrossRef]
38. Zhen, E.; Wang, Y.; Yang, J.; He, M. Comparison of the macroscopical stress field distribution characteristics between a novel non-pillar mining technique and two other current methods. *Adv. Mech. Eng.* **2019**, *11*, 1687814019850418. [CrossRef]

energies

Article

The Impact of Vibrating Screen Startup Time on Vibration Amplitude and Energy Consumption in Transient State

Jacek Feliks and Paweł Tomach *

AGH University of Krakow, Faculty of Mechanical Engineering and Robotics, Department of Machinery Engineering and Transport, 30-059 Kraków, Poland; feliks@agh.edu.pl
* Correspondence: tomach@agh.edu.pl; Tel.: +48-12-617-30-73

Abstract: The article presents the results of research on the trajectory and amplitude values during the startup of a vibrating screen, depending on the set startup time. The research was conducted using a suspended vibrating screen with a segmental vibration trajectory, where vibration excitation was achieved through two vibratory motors directly attached to the screen frame. The vibratory motors were powered with three-phase voltage and connected to the grid through frequency converters, allowing for current control in steady-state and startup conditions. The analysis focused on the vibration trajectory and the resulting amplitude values for nine selected startup times (ranging from 0 s to 8 s). Two measurement methods were employed: using piezoelectric acceleration sensors and optical recording. Additionally, the momentary power consumption by the screen drive was recorded, and the maximum current flowing in the power supply circuit was determined. These studies enabled the determination of the most favourable startup parameters that reduce the power demand of the drive without significantly affecting the kinematic parameters of the screen during its startup. The results of experimental research presented in the article regarding the startup time of the screening machine and its influence on vibration parameters have shown that the startup time significantly impacts the amplitude values that occur during startup. In the analysed case, the most favourable startup time was 3 s. Power consumption studies during startup revealed that startup time has a notable influence on momentary power consumption. The conducted research demonstrated that for startup times below 0.2 s, the maximum momentary power was the highest (reaching 5 kW, with a current consumption of 4.5 times the rated current). Regarding energy efficiency, the most favourable startup times exceed 2.0 s. Extending these startup times can significantly reduce the power demand of the drive, even up to threefold, thereby enabling the selection of motors with much lower power ratings.

Keywords: vibrating screen startup; screen power demand; vibration measurement

Citation: Feliks, J.; Tomach, P. The Impact of Vibrating Screen Startup Time on Vibration Amplitude and Energy Consumption in Transient State. *Energies* **2023**, *16*, 7129. https://doi.org/10.3390/en16207129

Academic Editors: Sergey Zhironkin, Krzysztof Skrzypkowski and Maxim Tyulenev

Received: 13 September 2023
Revised: 4 October 2023
Accepted: 13 October 2023
Published: 18 October 2023

1. Introduction

The fundamental operation in material enrichment processes is the screening process. This process enhances the material's utility, allowing for the attainment of a uniform grain composition [1–3]. In most industrial sectors, screens are used for this purpose. One of the basic types of screens is the super-resonance vibrating screen [4], which has undergone numerous modifications [5–7] to improve process efficiency, reduce noise levels, and minimize dynamic effects on the foundation or supporting structures [8]. Presently, the way to reduce energy consumption during the screening process is a significant concern. In the 21st century, vibrating screens with circular (WK) and linear (WP) vibrations are the most prevalent types of screens, whereas screens with elliptical vibrations (PZ, PWE) are less common [9]. More complex trajectories are also utilized to enhance screening efficiency, albeit at the expense of a more intricate drive system [10,11]. The product range of the manufacturers also encompasses vibrating flip-flow screens (VFFS), which, due to their slightly different design, are extensively employed for the separation of highly viscous and fine materials [12,13].

The primary component of any super-resonance screen is the sieve frame set in vibrating motion along a defined trajectory. Elastic support systems are employed to reduce dynamic effects on the foundation [14]. The vibrating motion of the sieve frame is driven by a drive system, typically consisting of an electric motor, flexible couplings, and an inertia vibrator or vibratory motors [15,16], occasionally electromagnetic ones [17]. In super resonant screens in which an inertial vibrator generates vibrations, the value of the vibration amplitude depends on the mass of elements (weights) attached to the vibrator shafts (or unbalanced masses in the vibration motor). However, the vibration frequency depends on the type of engine used (its rotational speed) or the method of power supply and control (e.g., use of a frequency converter). Most inertia vibrators offer the option to adjust the force applied to induce vibrations by adding or subtracting additional unbalanced masses or altering the configuration of unbalanced masses [9].

The operation of a screening machine can be divided into three stages: startup [16], steady-state operation [15], and shutdown. The startup phase refers to the period from the drive initiation until the rated vibration frequency is reached and the parameters of the sieve frame vibrations stabilize. This phase is particularly critical in selecting the screen's mechanical parameters and the electrical parameters of its drive, as the screen passes through the resonance frequency during this period, which can potentially have destructive effects on the equipment. The starting time is an important parameter that affects the values of the vibration amplitude achieved during this phase of operation and the energy demand of the engines—which can be demonstrated by conducting appropriate tests.

The steady-state period represents a phase during which the screen maintains a stable vibration trajectory defined for a given technological process, while the shutdown phase is the time interval between the drive's deactivation and the screen coming to a complete stop.

Screen drives consume power during the first two phases of operation. Such drives are designed to handle the startup torque, which is the most demanding in terms of power. During steady-state operation, the drive operates with a power level typically at 20–30% of the rated power. Using motors with excessive power reserves can adversely affect the power grid (significant reactive power) and increase the vibrating mass. In most cases, research on screens is limited to steady-state studies [15,18] and studies focused solely on the screening process using DEM simulations [19–24]. There are also attempts to develop vibration models that would consider the screened material's influence on the vibration parameters. [25,26]. In order to aim for a reduction in the rated power of the drive, conducting startup research is crucial.

This research aims to determine the most favourable startup parameters that reduce the power demand of the drive without significantly affecting the kinematic parameters of the screen during its startup.

2. The Startup of a Linear Vibration Screener—Theoretical Analysis

During steady-state operation, vibratory screens operate at vibration frequencies several times higher than their natural resonance frequency. Compared to the mass of the screened material, the substantial mass of the riddle plate ensures the attainment of stable operating parameters despite fluctuations in material flow rate through the screener. The main challenge associated with operating at such vibration frequencies is passing through the resonant frequency during system startup and shutdown. Consequently, there is a highly unfavourable multiple increase in vibration amplitude compared to the steady-state condition [9]. Numerous scientists have analysed and researched these phenomena, enabling the determination of fundamental relationships that facilitate the estimation of the maximum amplitude value. Determining the maximum amplitude value allows for selecting an appropriate support system (suspension) for the screen, ensuring its trouble-free operation. Current capabilities for numerical simulations enable increasingly precise determination of vibration amplitudes appearing in transient states and the verification of theoretical models. In order to conduct theoretical analyses of the screen's operation during startup, it was necessary to develop a physical model of the super-resonance screen with linear vibrations (WP)—this model is presented in Figure 1.

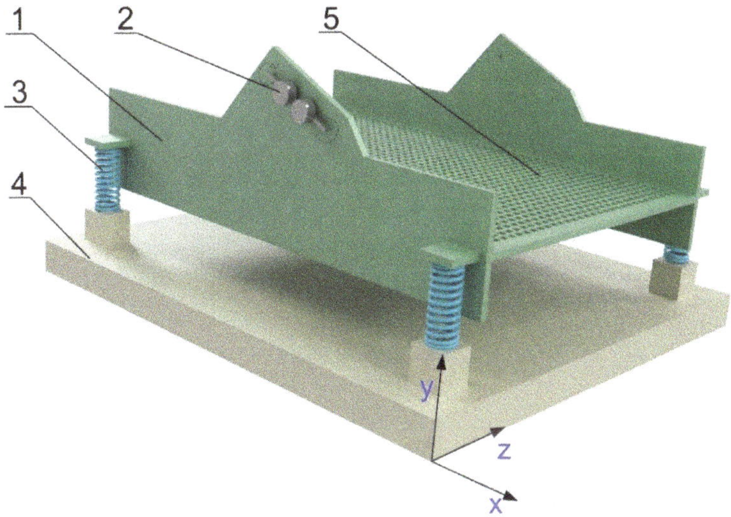

Figure 1. Physical representation of the linear vibrating screen model utilized in the analysis: 1—riddle plate; 2—inertial vibrators; 3—spring support; 4—supporting structure; 5—screen surface.

Figure 2 presents the structural diagram of a two-mass inertial vibrator-supported screen. It consists of a screen with a mass of m1, supported by four springs with a spring constant of k/4, along with two unbalanced masses $m_0/2$, acting as vibrators. The vibrator system is rigidly connected to the screen. It operates in a counter-directional motion, causing only inertial forces in the direction of I-I axis to act on the screen. The plane of the I-I axis passes through the centre of mass of the vibrating mass and the axis between the two vibrators. This direction is inclined to the horizontal by an angle α1. The analysis of such a system requires certain simplifying assumptions, where the direction of the acting driving force passes through the centre of mass of the screen, which is satisfied in the case of self-synchronization of the drive, and damping is neglected. Such a system performs linear motion, as no torsional moments causing rotary vibrations of the screen will occur. The value of the driving force acting in the I-I direction is determined by the formula:

$$P = m_0\left(\mathrm{r}_0\omega^2 sin\omega t - \ddot{S}\right) \tag{1}$$

where (Figure 2):
m_1—mass of the riddle;
m_0—unbalanced mass;
m—vibrating mass;
k—equivalent elastic modulus of the suspension along the x and y axes;
r_0—eccentric radius of unbalanced mass;
α_1—the angle of inclination of the I-I axis relative to the horizontal;
ω—vibration frequency;
t—time.
When decomposing this force into two components, we obtain:

$$P_x = m_0\left(\mathrm{r}_0\omega^2 cos\alpha_1 sin\omega t - \ddot{x}\right) \tag{2}$$

$$P_y = m_0\left(\mathrm{r}_0\omega^2 sin\alpha_1 sin\omega t - \ddot{y}\right) \tag{3}$$

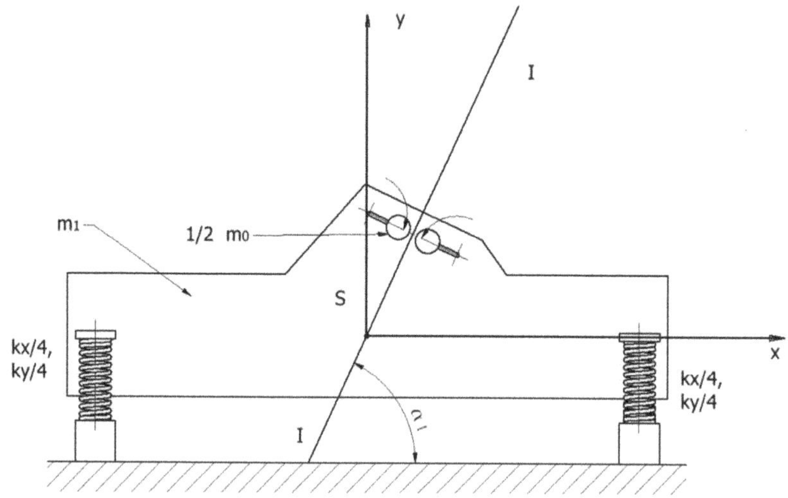

Figure 2. Two-dimensional physical representation of a linearly vibrating screen model.

After comparing with the value of the driving force and further transformation, we obtain the formula for the amplitude of vibrations in the x and y directions. The most significant, considering the interaction with the ground, is the formula for the y direction, which looks as follows:

$$A_y = \frac{m_0 r_0 \omega^2 sin\alpha_1}{k - m\omega^2} \tag{4}$$

After considering the formula for the natural frequency:

$$\omega_0 = \sqrt{\frac{k}{m}} \tag{5}$$

It is obtained:

$$A_y = \frac{\frac{m_0 r_0 sin\alpha_1}{m}}{\left(\frac{\omega_0}{\omega}\right) - 1} \tag{6}$$

On Figure 3, the amplitude of vibrations in the y-direction as a function of the frequency ratio ω/ω_0 is shown.

As seen from Figure 3, at the resonance frequency, the amplitude of the screen's vibrations tends toward infinity, and beyond this frequency, it rapidly decreases and approaches a constant value. Assuming an operating range of the screen at least four times above the resonance, we can conclude that in the steady state, the vibration amplitude will be constant and equal to:

$$A_y = \frac{m_0 r_0 sin\alpha_1}{m} \tag{7}$$

During startup, this amplitude will be higher and dependent on damping and primarily on the time and manner of passing through resonance. Amplitude studies during startup are presented in the article [16], although they did not account for the startup time and its impact on electrical power consumption.

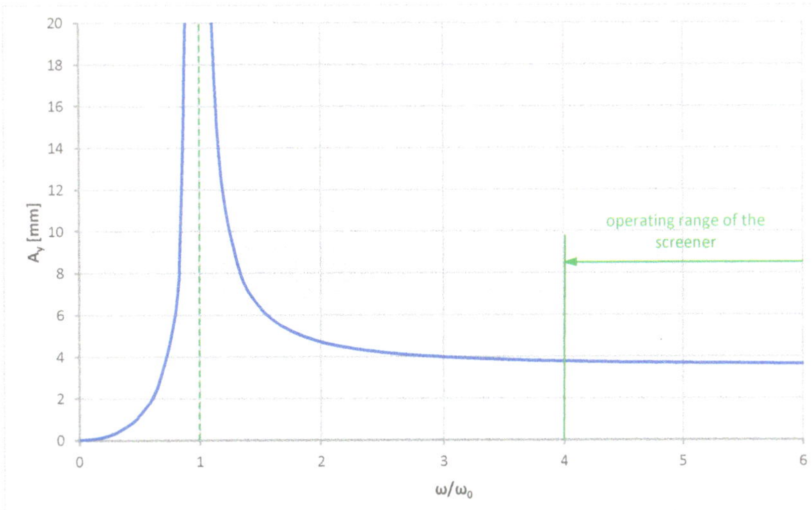

Figure 3. Theoretical amplitude response of vibrations in the y-axis (vertical) as a function of ω/ω_0.

3. Description of Test Stands and Applied Research Methods

The research was conducted at Department of Machinery Engineering and Transport. The object of the study was a WP screen with a linear trajectory of vibrations, with the screen suspended on a system of four steel cables attached to the screen on one side and to the elastic support system on the other. Vibrations with a linear trajectory were generated by a system of two vibratory motors, 3PH—6 poles type MVE 500/1E-50A0, with a power of 0.55 kW and a nominal speed of 1000 RPM (50 Hz). These motors are mounted on a frame attached to the screen above the screen deck. The engines have opposite rotational directions, and their close proximity allows for the occurrence of the "self-synchronization" effect of the vibrators. A view of this research station with the most important components marked is presented in Figure 4.

One of the research objectives was to demonstrate the impact of the preset startup time on the parameters of the vibrating screen's motion. A frequency converter (variable frequency drive) Lenze AC Tech SMV ESV552 (Lenze: Hameln, Germany) with a power of 5.5 kW was used for startup. This converter increased the supply frequency from 0 to 50 Hz within a specified time frame in a linear manner. The applied power supply and control system (position 3—Figure 4) also enabled the measurement and recording of power using a digital three-phase wattmeter Metrix PX0120 (Metrix: Asnières-sur-Seine, France) and current intensity with a digital meter DPM V14 (DPM: Kowanówkowo, Poland). The measurement results presented in the article consider the energy consumed by the entire system, including the 50 W power consumption by the converter itself.

The study of vibrational motion was conducted using the experimental setup depicted in Figure 5. Vibration measurements using this setup were carried out using a three-axis PCB (piezoelectric) accelerometer, which was mounted on the screen of the screener. Accelerometers have many advantages and are often used for precise vibration measurements and in machine diagnostics [27–30]. The signal from the sensor was recorded by a NI9232 data acquisition card housed in a CompactDAQ (National Instruments Corporation: Austin, TX, USA) enclosure and then processed using custom software developed in the LabVIEW environment. The sampling frequency for all measurements was set at 1 kHz. This setup, where two accelerometers were employed, was also presented in a prior work [16].

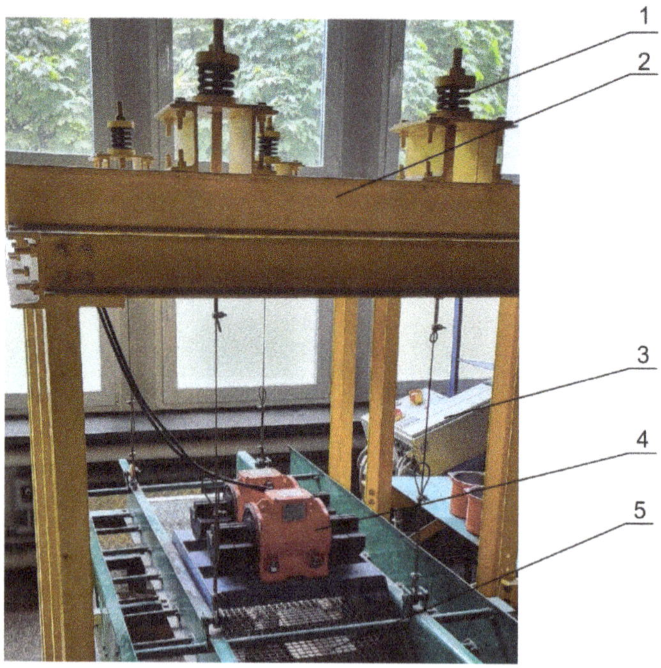

Figure 4. View of the WP rectilinear vibration screen test stand. 1—spring support; 2—frame; 3—power and control system; 4—vibrating motors; 5—screen surface.

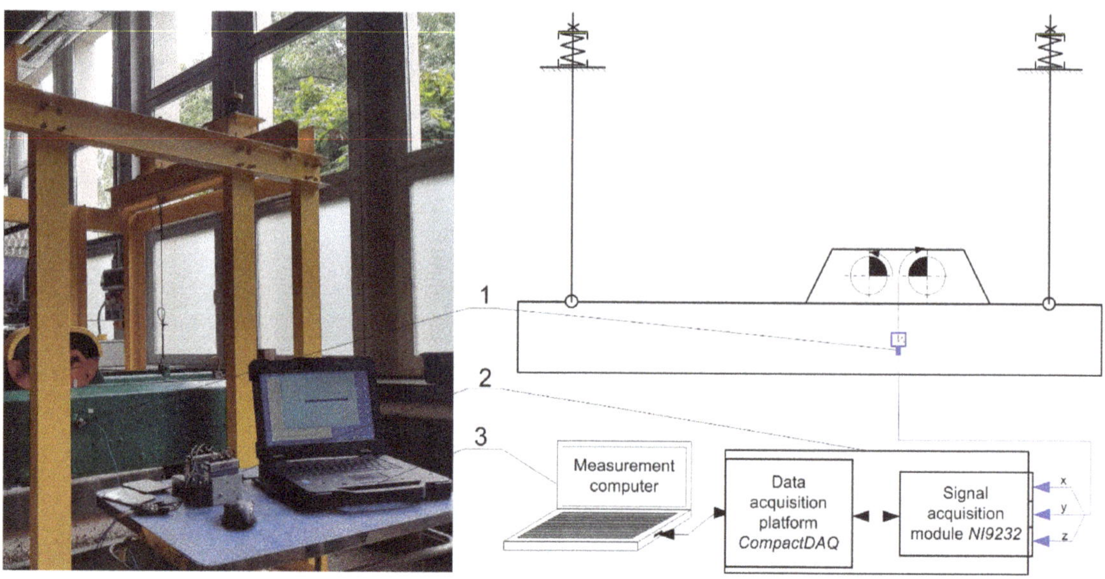

Figure 5. View of the research stand and schematic of the vibration measurement system using an acceleration sensor: 1—three-axis acceleration sensor PCB 356A02; 2—measurement card; 3—measurement computer.

The investigation into the vibrating screen's motion characteristics at the commencement of operation took place at a specific location on the screen deck (denoted as point 3 in

Figure 5). This location (point) was aligned with the screen deck's surface, coinciding with the plane perpendicular to the side of the sieve. This plane also served as the symmetrical axis of the drive system, running parallel to the vibrators' axis.

Measuring vibration amplitude values and their trajectories using piezoelectric accelerometers (in the future referred to as electronic measurements) during the startup of a vibrating machine can result in significant measurement errors. These errors may arise from the need for double integration of the measured acceleration (to obtain displacement values). To derive displacement values in the program, a regression integration method was applied, necessitating the imposition of a high-pass filter to eliminate the DC component. However, the filter value should be set as low as possible to ensure that significant displacement (amplitude) data during startup are not omitted. Therefore, to unambiguously indicate the actual amplitude values and their trajectories, it was decided to conduct measurements optically, allowing for the verification of the results obtained using electronic methods. The additional measurement method enables the validation of the accuracy of measurements made with piezoelectric accelerometers.

The optical measurement was made possible by using a device attached to the screen with the help of neodymium magnets. This device generates two light points precisely spaced 10.0 mm $\pm$ 0.1 apart from each other. Subsequently, by employing appropriate lighting conditions (darkened environment) and using the Panasonic DMC-FZ1000 camera (Panasonic: Osaka Prefecture, Japan) (mounted on a tripod), an appropriately long exposure time (exceeding the startup time) was set to capture the trajectory of the light points (screen) during startup. Figure 6 shows a view of such a device with a fixed distance between the light points (on the left side), while on the right side, an example of the captured image during startup is presented, with the dimensions indicated.

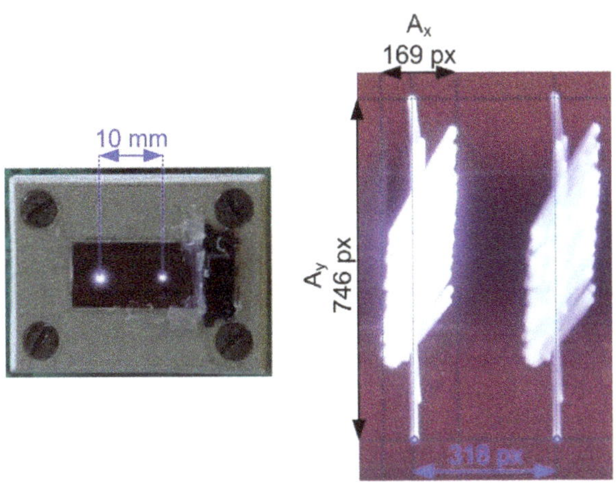

Figure 6. Measurement device with light points and a view of the captured image during measurement with overlaid measurements.

The figure's maximum range of horizontal vibrations is denoted as Ax, and the full range of vertical vibrations is marked as Ay. These range values, provided in pixels, were determined using GIMP software (version 2.10.30). The highlighted value in blue (318 px) represents the known distance between the light points on the measuring device (10 mm), which allowed for a straightforward conversion of measured vibration ranges in pixels into values in millimetres with an accuracy of 0.1 mm.

4. Study of Vibrational Motion Parameters during the Startup of a Screen

The analysis of the vibrating screen during its startup focused on mechanical analysis (analysis of vibration trajectories) and electrical analysis (determination of power consumption during startup). These experiments were conducted for predetermined startup times, which were 0 s, 0.1 s, 0.2 s, 0.5 s, 1 s, 2 s, 3 s, 4 s, and 8 s.

4.1. Mechanical Analysis

The research aimed to answer the question of what high-pass filter values should be set to obtain the real values of maximum displacements (amplitudes) during startup and vibration trajectories. For this purpose, the motion parameters of the vibrating screen were studied using piezoelectric accelerometers, and then during the analysis of the results of these studies, a high-pass filter with the following values was applied: 1.0; 1.5; 2.0; 2.5; 3.0; 3.5; 4.0; 4.5; 5.0; and 5.5 Hz. For each filter variant and for each of the specified startup times, a trajectory of vibrations in the displacement coordinate system in the x and y directions was generated. The value of the applied filter significantly impacts the resulting displacement value and the obtained vibration trajectory.

Applying the optical measurement method made generating an actual image of the vibration trajectory possible, allowing for a comparison with the trajectories obtained by the electronic method. The optical measurements' results, analysis (located in the top left corner), and plots of calculated trajectories for the applied high-pass filter values are presented in Figures 7–15.

The results of the vibrating screen's motion parameters for different startup time settings were analysed. The first stage of the analysis involved determining the maximum amplitude values in the x and y directions obtained by the electronic measurement method. These amplitude values represent the respective components of the generated vibration trajectory—the projection onto the coordinate system's horizontal (Ax) and vertical (Ay) axes. Subsequently, an analysis was conducted to determine the high-pass filter value that provided the best representation (both qualitatively and quantitatively) of the obtained vibration trajectory to the actual values (i.e., results obtained by the optical method).

The results of this analysis are presented in the form of a table (Table 1). This table includes the following information: the values of Ax and Ay measured optically; the range of amplitude values variation in the x and y directions (for the full range of analysed high-pass filter values); and the value of the high-pass filter for the best representation of the vibration trajectory, as well as the obtained amplitude values when using this filter. The accuracy of the recorded amplitude values for both measurement methods is ±0.1 mm.

Table 1. Compilation of obtained research results.

Startup Time	Real Value Optical Measurement		Electronic Measurement: Range of Obtained Values in the Full Measurement Range		Filter Value for the Best Representation	Amplitude Values Obtained for the Chosen Filter		Comparison of Methods Absolute Error	
	Ax	Ay	Ax	Ay		Ax	Ay	Ax	Ay
	[mm]	[mm]	[mm]	[mm]	[Hz]	[mm]	[mm]	[%]	[%]
0.0 [s]—Figure 7	5.3	23.3	4.0 ÷ 12.7	12.8 ÷ 29.1	2.0	5.0	23.5	6	1
0.1 [s]—Figure 8	5.4	23.4	3.8 ÷ 9.1	12.4 ÷ 24.8	2.0	4.7	23.9	13	2
0.2 [s]—Figure 9	5.9	22.9	4.0 ÷ 9.9	11.9 ÷ 28.7	2.0	4.7	22.3	20	3
0.5 [s]—Figure 10	6.2	22.8	4.4 ÷ 12.6	11.3 ÷ 30.3	2.0	5.6	22.4	10	2
1.0 [s]—Figure 11	8.1	21.9	6.3 ÷ 14.0	11.7 ÷ 26.0	2.0	7.5	21.6	7	1
2.0 [s]—Figure 12	7.6	20.5	7.0 ÷ 11.3	7.2 ÷ 22.9	1.5	7.2	20.5	5	0
3.0 [s]—Figure 13	6.6	12.0	6.1 ÷ 14.3	6.5 ÷ 22.1	1.5	7.2	12.0	9	0
4.0 [s]—Figure 14	7.6	18.7	6.5 ÷ 26.1	6.7 ÷ 41.1	2.0	7.4	16.7	3	11
8.0 [s]—Figure 15	9.7	13.2	5.4 ÷ 24.0	4.8 ÷ 38.5	2.0	7.9	12.7	19	4

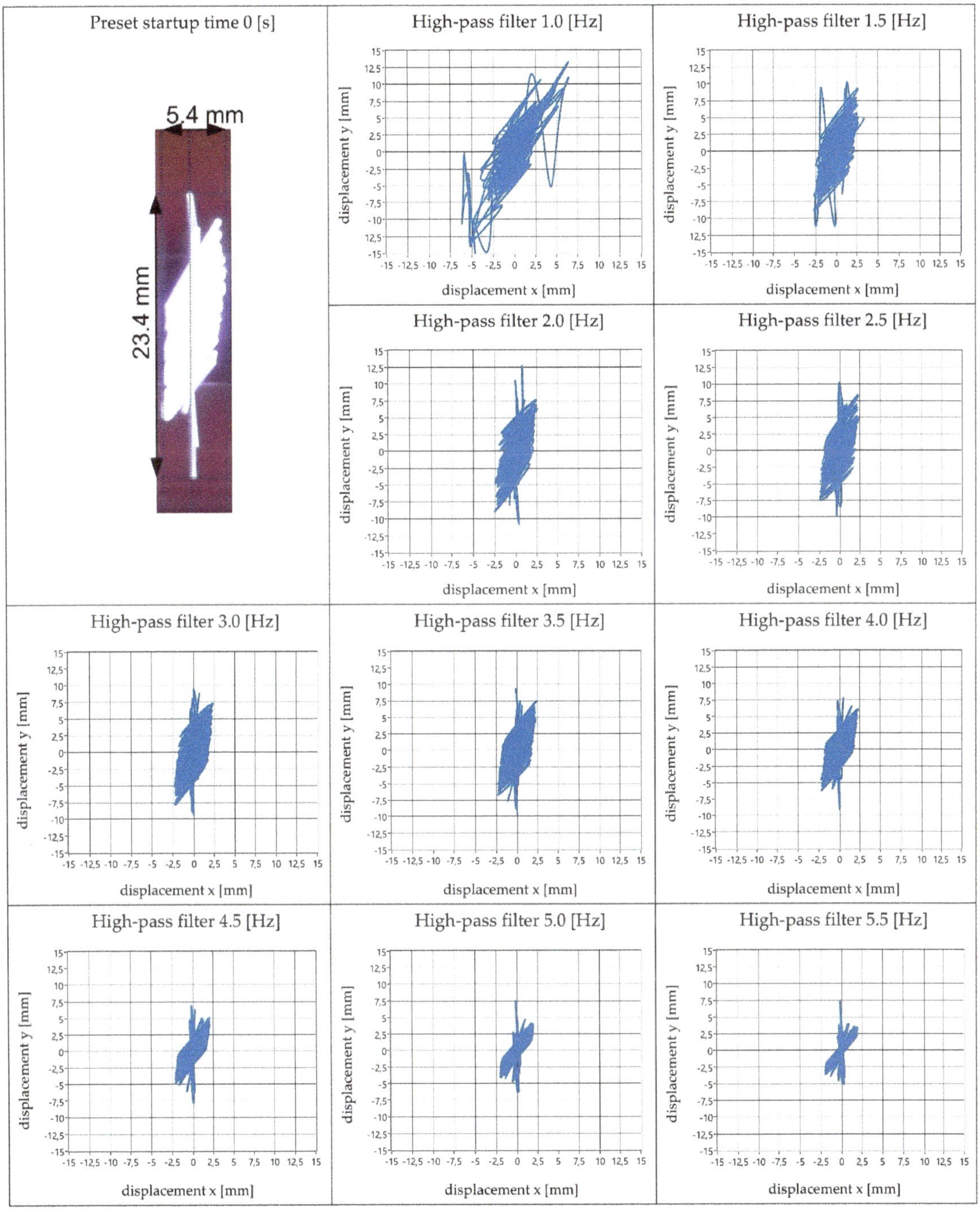

Figure 7. View of obtained amplitude values and vibration trajectory shapes—startup time 0.0 s.

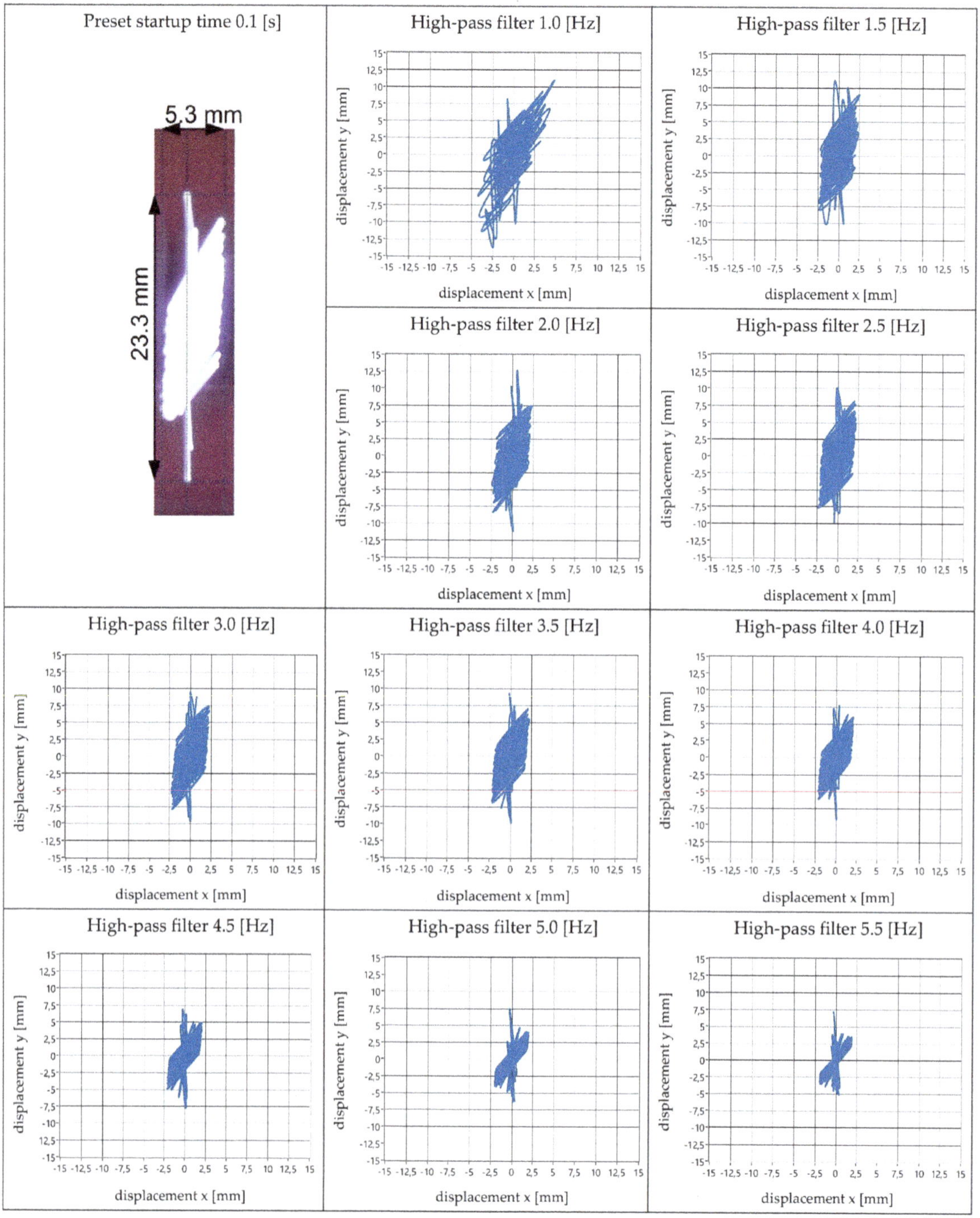

Figure 8. View of obtained amplitude values and vibration trajectory shapes—startup time 0.1 s.

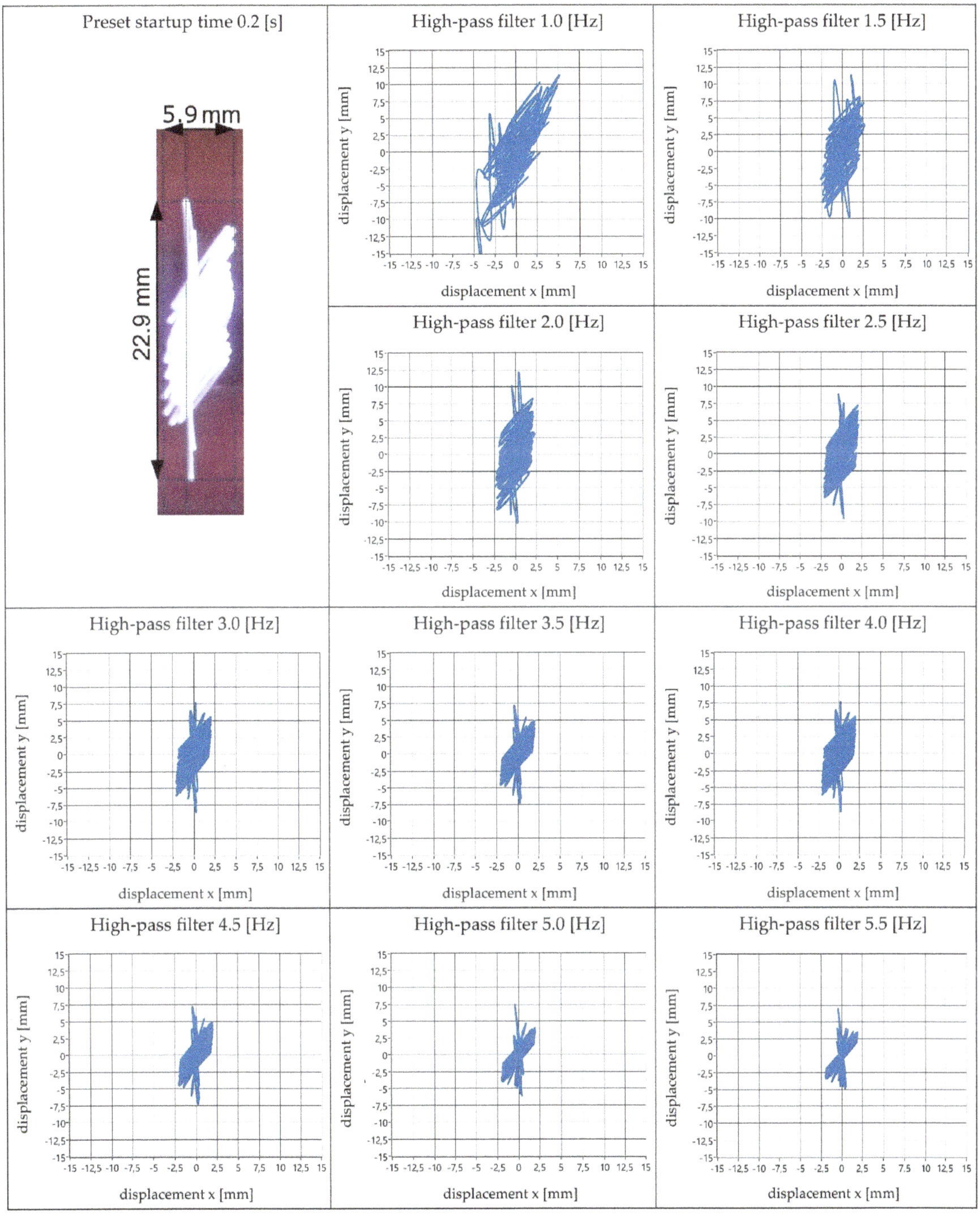

Figure 9. View of obtained amplitude values and vibration trajectory shapes—startup time 0.2 s.

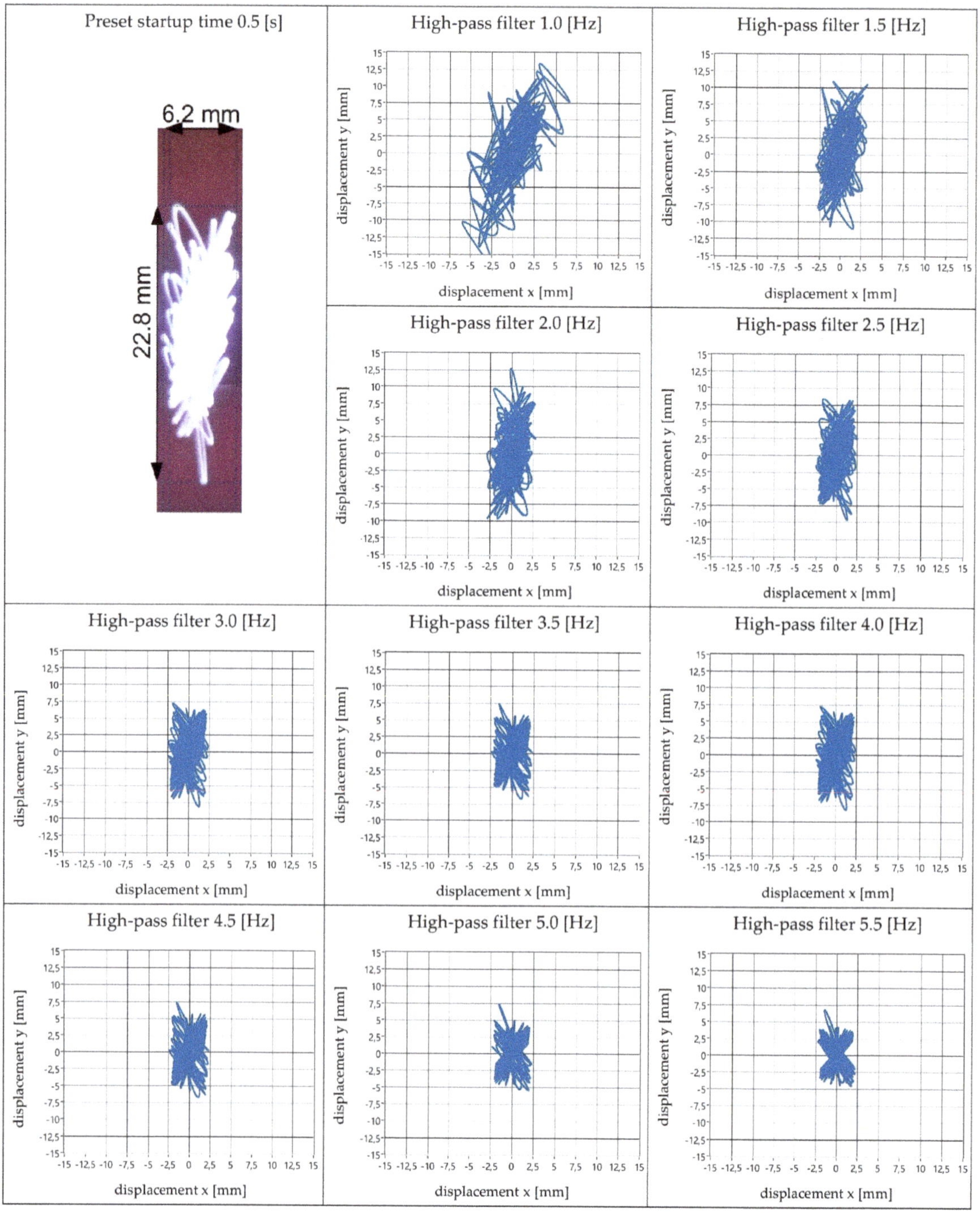

Figure 10. View of obtained amplitude values and vibration trajectory shapes—startup time 0.5 s.

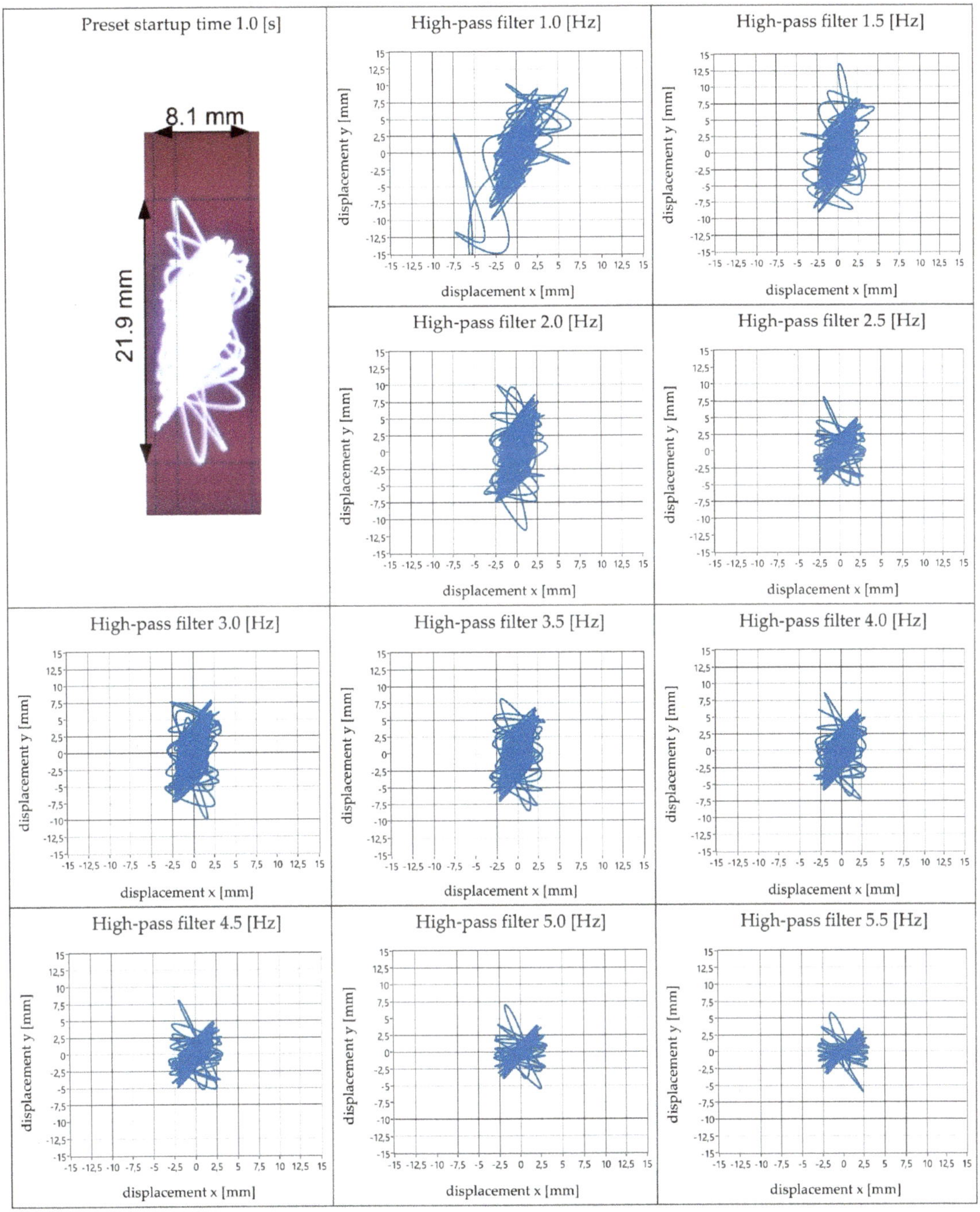

Figure 11. View of obtained amplitude values and vibration trajectory shapes—startup time 1.0 s.

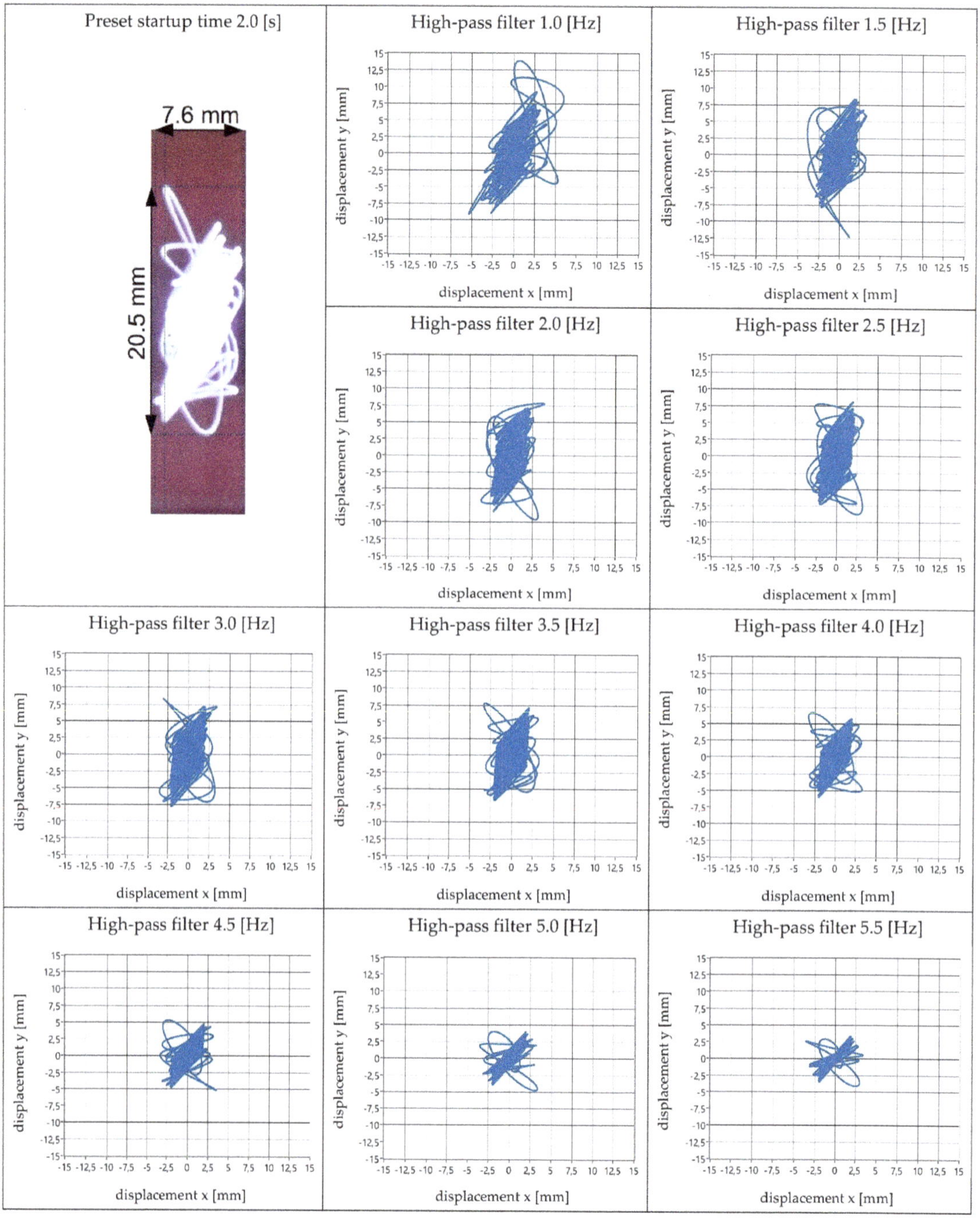

Figure 12. View of obtained amplitude values and vibration trajectory shapes—startup time 2.0 s.

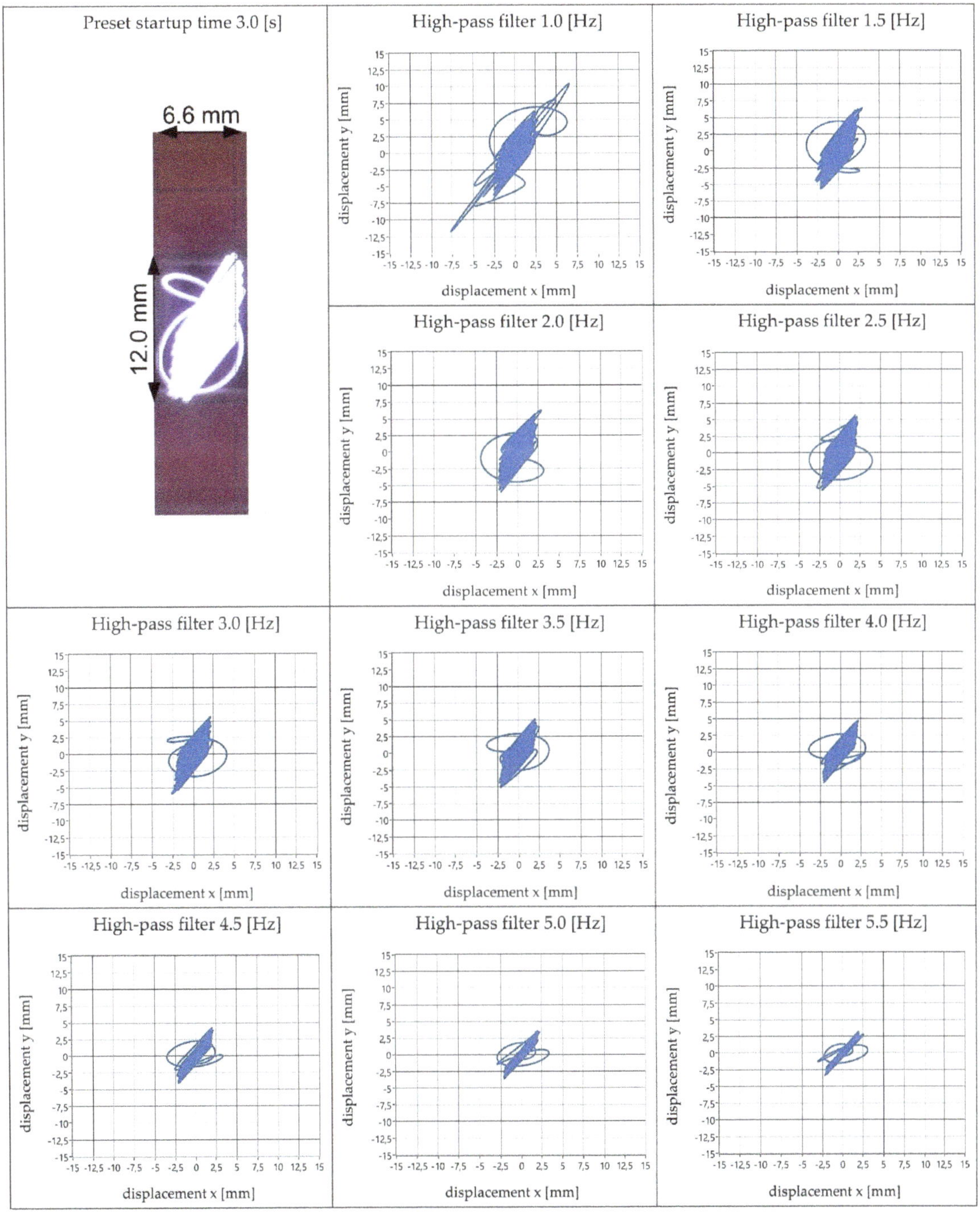

Figure 13. View of obtained amplitude values and vibration trajectory shapes—startup time 3.0 s.

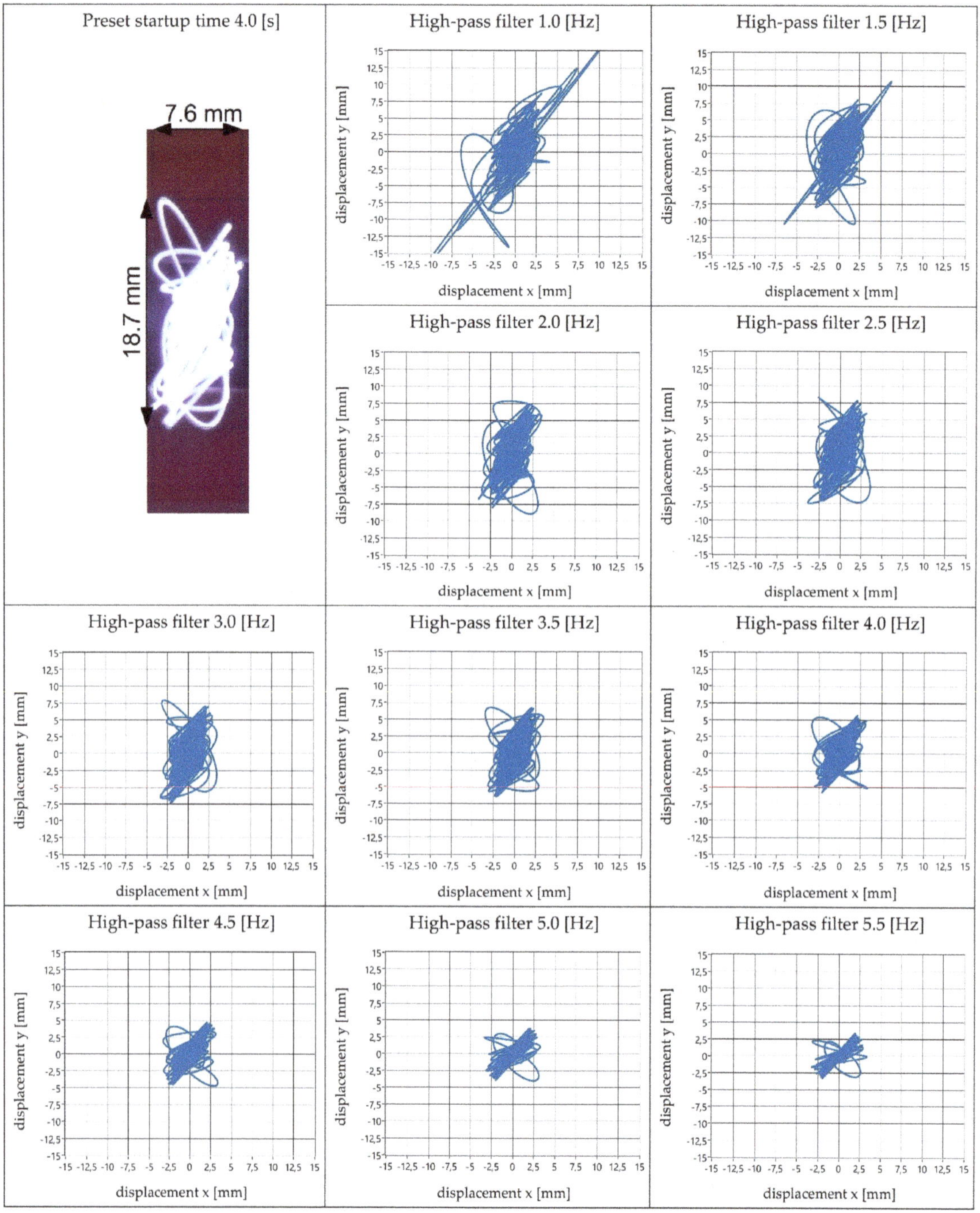

Figure 14. View of obtained amplitude values and vibration trajectory shapes—startup time 4.0 s.

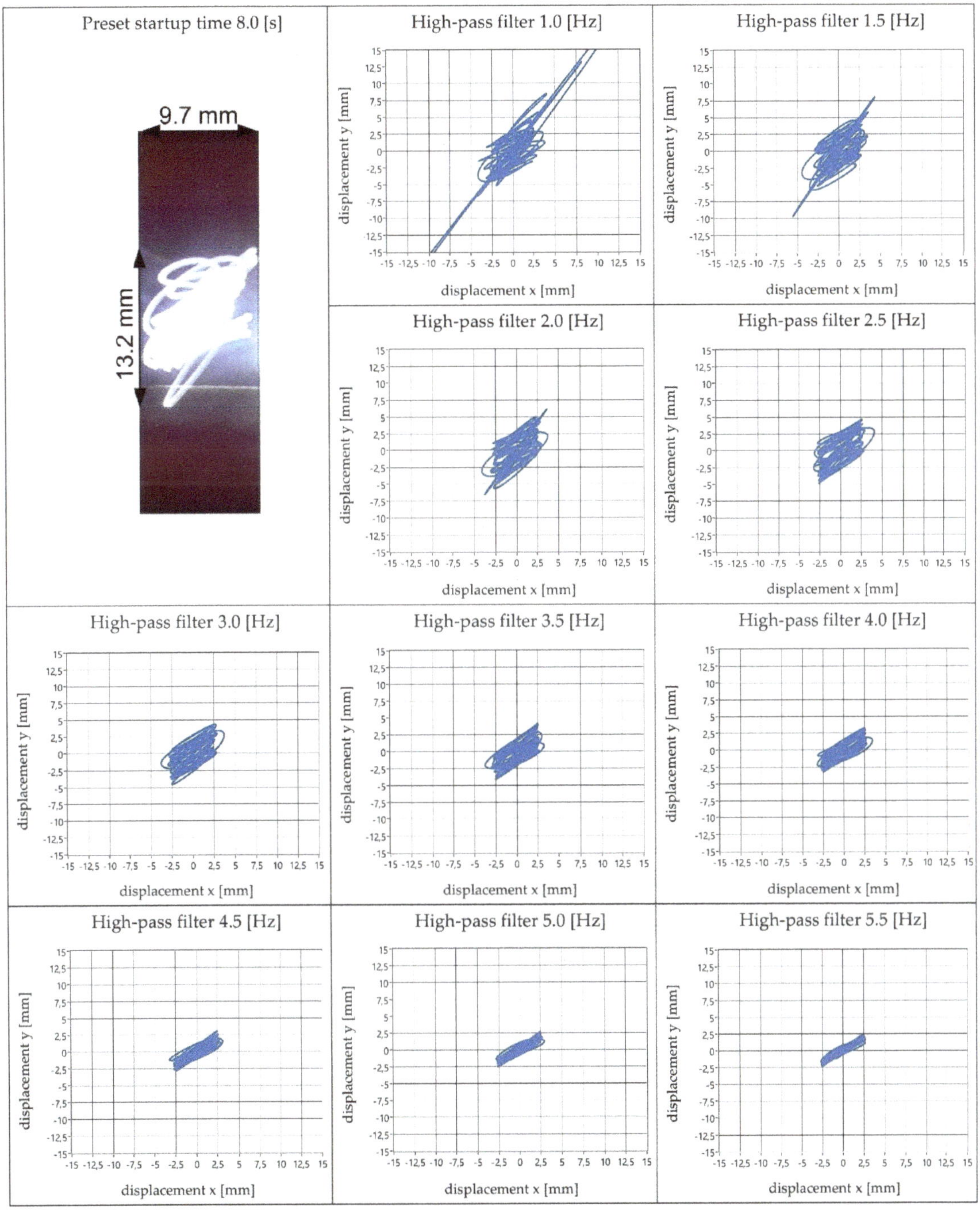

Figure 15. View of obtained amplitude values and vibration trajectory shapes—startup time 8.0 s.

The analysis of the obtained research results, which is presented in Table 2, showed that the best representation of the real values of vibration amplitudes of the screen during its startup by using the accelerometer measurement method can be achieved by applying a high-pass filter with values ranging from 1.5 to 2.0 Hz. The highest relative errors for the selected filter values that provided the best representation of real results were related to the amplitude in the x direction (ranging from 3 to 20%). For the y direction, perfect agreement of results was achieved, with errors mostly ranging from 1 to 4%, except for one measurement (for a startup time of 4.0 s), where the error was 11%.

Table 2. Maximum current and power values during start-up.

Time [s]	Maximum Current [A]	Maximum Power [kW]
0.0	4.55	5.11
0.1	4.32	4.88
0.2	4.63	5.19
0.5	3.55	4.09
1.0	2.42	2.93
2.0	1.23	1.73
3.0	1.22	1.71
4.0	1.16	1.65
8.0	0.96	1.45

4.2. Power and Current Values during Start-Up

The second aspect of the conducted research involved measuring the current and power consumption of the screen drive during startup. Power was continuously measured and recorded, while the maximum current value was determined for current measurement. An example of the power consumption variation during the screen drive startup is shown in Figure 16.

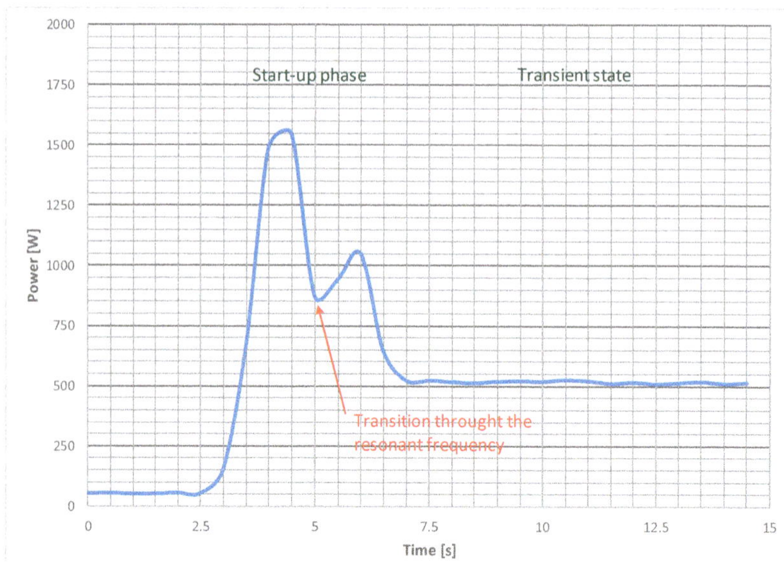

Figure 16. Example chart of instantaneous power variation for a 4-s startup.

During the startup process, we can observe a significant initial increase in instantaneous power to its maximum value, resulting from the necessity of obtaining the required starting torque. Subsequently, as the system approaches its resonant frequency, there is a decrease in power consumption (a local minimum of power). To surpass this resonant frequency and transition to steady-state rotation, it is necessary to increase the torque once again, increasing power consumption. A noticeable reduction in instantaneous power is observed for longer startup times (above 2 s), while for shorter startup times, passing through the resonant frequency results in a change in the rate of power decrease. Table 2 presents the maximum current values recorded during startup and provides the maximum instantaneous power value.

The current and power remain constant for the first three startup times of 0 s, 0.1 s, and 0.2 s, with values around 4.5 A and 5 kW, which represents approximately 4.5 times the rated power. For longer startup times, the measured values decrease and are only slightly higher than the rated values. In steady-state operation, the screen drive consumes approximately 520 W of energy, which is half of the rated power.

5. Conclusions

The results presented in the article regarding the startup time of the screen and its impact on vibration parameters have led to the following conclusions:

1. The measurements using three-axis acceleration sensors are subject to a small error only when the measurement method is calibrated correctly.
2. The measurements can be calibrated using an optical method, allowing for the determination of correct maximum amplitude values obtained by electronic measurement.
3. The initial calibration of the measurement system presented in the article (performed in the range from 1 to 5.5 Hz with a step of 0.5 Hz) allowed for obtaining measurement results with an accuracy ranging from 3 to 20% for the x-axis and up to 11% for the y-axis (up to 4% for eight out of nine analyzed startup times). The achieved measurement accuracy is very satisfactory, demonstrating the practical applicability of the proposed method in measuring vibrations of over-resonant vibrating screens. There is a possibility of improving the accuracy of the obtained results by conducting calibration tests with greater accuracy of the applied filter values (e.g., every 0.1 Hz).
4. The research has shown that the startup time significantly impacts the maximum amplitude value, which can be up to seven times greater than in steady-state operation.
5. Setting the screen startup time below 0.2 s does not affect the change in the screen's vibration trajectory during startup, which is related to the insufficient startup torque of the drive.
6. The most favourable startup profile for the screen was recorded for a startup time of 3 s, during which the lowest vibration amplitudes were obtained. This demonstrates that achieving the nominal rotational speed of vibrators as quickly as possible is not always the most advantageous approach in over-resonant vibrating machines.

The second part of the research focused on the impact of the set startup time on the energy parameters of the screen. The results of these studies indicate that the startup time significantly affects the maximum instantaneous power consumed by the screen. The following conclusions can be drawn from these studies:

1. For startup times below 0.2 s, the maximum instantaneous power remains constant at approximately 5 kW (phase current of 4.5 A). This represents 4.5 times the rated power and falls within the permissible motor overload range. The lack of differences in the power values is most likely due to the insufficient startup torque of the motor, preventing startup in less than 0.2 s.
2. Longer startup times significantly reduce the maximum instantaneous power consumption. A significant decrease was observed for startup times ranging from 0.5 to 2.0 s. Above this time range, the power consumption stabilizes at a level between 1.5 kW and 1.7 kW.
3. The analysis of power consumption changes during startup (Figure 16) also identifies two stages of the screen startup: startup from 0 to the resonant frequency and from the resonant frequency to steady-state operation. When rotating at the resonant frequency, local minima of power consumption by the motors can be observed. This minimum is

less pronounced for shorter startup times and is characterized by changes in the slope of the power curve (the rate of decrease).

4. Regarding energy efficiency, the most favourable startup times are above 2.0 s. Using such startup times would allow for a threefold reduction in power consumption by the drive. Consequently, it would be possible to select motors with much lower power ratings, as low as 0.173 kW, thanks to the allowable fivefold overload of the drive (two engines) during startup (1.73 kW/5/2 = 0.173 kW). This value is lower than the steady-state power consumption of 0.52 kW (approximately 0.26 kW per motor). For the tested screen, with a 2.0 s startup, the drive size could be minimized to 0.3 kW motors (a 45% reduction in rated power), allowing for safe startup and operation at a power level close to the rated power.

The practical application of the results presented in this article contributes to the development of new guidelines for selecting drives for over-resonant vibrating screens. Such a well-matched drive will reduce environmental impact by reducing reactive power consumption, limiting the value of the resonance amplitude, as well as reducing operational and purchasing costs for the screen.

Author Contributions: Conceptualization, J.F. and P.T.; methodology, J.F. and P.T.; software, P.T.; validation, formal analysis, investigation, J.F. and P.T.; resources, J.F. and P.T.; data curation, J.F. and P.T.; writing—original draft preparation, J.F. and P.T.; writing—review and editing, P.T.; visualization, P.T. and J.F.; supervision, J.F.; project administration, J.F. and P.T.; funding acquisition, J.F. and P.T. All authors have read and agreed to the published version of the manuscript.

Funding: Works financed by the AGH University of Krakow.

Data Availability Statement: The data presented in this study are available on request from the corresponding author.

Acknowledgments: Thanks for technical support to Paweł Pawlik.

Conflicts of Interest: The authors declare no conflict of interest.

References

1. Modrzewski, R. *Przesiewacze Sitowe Materiałów Uziarnionych*; Monografie Politechniki Łódzkiej, Nr 2339: Łódź, Poland, 2020. (In Polish)
2. Feliks, J.; Filipowicz, A. Przesiewacze w wytwórniach mas mineralno-bitumicznych. *Kopaliny* **2008**, *1*, 73–75. (In Polish)
3. Gawenda, T.; Stempkowska, A.; Saramak, D.; Foszcz, D.; Krawczykowska, A.; Surowiak, A. Assessment of Operational Effectiveness of Innovative Circuit for Production of Crushed Regular Aggregates in Particle Size Fraction 8–16 mm. *Minerals* **2022**, *12*, 634. [CrossRef]
4. Sidor, J.; Feliks, J.; Mazur, M.; Tomach, P. Technika wibracyjna w przeróbce surowców mineralnych. In *Mechanizacja, Automatyzacja i Robotyzacja w Górnictwie*; Krauze, K., Ed.; AGH w Krakowie: Kraków, Poland, 2015; pp. 101–111. (In Polish)
5. Makinde, O.A.; Ramatsetse, B.I.; Mpofu, K. Review of vibrating screen development trends: Linking the past and the future in mining machinery industries. *Int. J. Miner. Process.* **2015**, *145*, 17–22. [CrossRef]
6. Trumic, M.; Magdalinovic, N. New model of screening kinetics. *Miner. Eng.* **2011**, *24*, 42–49. [CrossRef]
7. Modrzewski, R.; Wodziński, P. Analysis of screening process of crushed basalt performed by a double-frequency screen. *Physicochem. Probl. Miner. Process.* **2013**, *49*, 81–89.
8. Linhares, T.B.; Vimieiro, C.B.S. Analysis of the dynamic forces acting on a vibrating screen and its support structure using a scale model. *Measurement* **2021**, *176*, 109179. [CrossRef]
9. Banaszewski, T. *Przesiewacze*; Wydawnictwo "Śląsk": Katowice, Poland, 1990. (In Polish)
10. Gursky, V.; Krot, P.; Korendiy, V.; Zimroz, R. Dynamic Analysis of an Enhanced Multi-Frequency Inertial Exciter for Industrial Vibrating Machines. *Machines* **2022**, *10*, 130. [CrossRef]
11. Modrzewski, R.; Obraniak, A.; Rylski, A.; Siczek, K. A Study on the Dynamic Behavior of a Sieve in an Industrial Sifter. *Appl. Sci.* **2022**, *12*, 8590. [CrossRef]
12. Yu, C.; Geng, R.; Wang, X. A Numerical Study of Separation Performance of Vibrating Flip-Flow Screens for Cohesive Particles. *Minerals* **2021**, *11*, 631. [CrossRef]
13. Zhao, G.; Wang, X.; Lin, D.; Xu, N.; Yu, C.; Geng, R. Study of Double-Deck Vibrating Flip-Flow Screen Based on Dynamic Stiffness Characteristics of Shear Springs. *Minerals* **2021**, *11*, 928. [CrossRef]
14. Yu, J.; Li, Z.; Zhang, Z.; Zhao, W.; Niu, Z.; Cheng, J. Vibration Test and Control of Factory a Building under Excitation of Multiple Vibrating Screens. *Buildings* **2022**, *12*, 607. [CrossRef]

15. Tomach, P.; Feliks, J. *Influence of the Type of Drive of the Laboratory Screen with Rectilinear Vibrations on the Parameters of Its Work in the Steady State*; Nauka-Technologia-Technika Seria; Wydawnicza AGH: Kraków, Poland, 2021; Volume 2, pp. 161–176, (In Polish). [CrossRef]
16. Feliks, J.; Tomach, P.; Foszcz, D.; Gawenda, T.; Olejnik, T. Research on the New Drive of a Laboratory Screen with Rectilinear Vibrations in Transient States. *Energies* **2021**, *14*, 8444. [CrossRef]
17. Nuraliyev, A.; Ibadullayev, M. Research and development of resonant electromagnetic vibration screen for intensive vibration technologies. *E3S Web Conf.* **2020**, *216*, 01114. [CrossRef]
18. Krot, P.; Shiri, H.; Dąbek, P.; Zimroz, R. Diagnostics of Bolted Joints in Vibrating Screens Based on a Multi-Body Dynamical Model. *Materials* **2023**, *16*, 5794. [CrossRef] [PubMed]
19. Kryszak, D.; Bartoszewicz, A.; Szufa, S.; Piersa, P.; Obraniak, A.; Olejnik, T.P. Modeling of Transport of Loose Products with the Use of the Non-Grid Method of Discrete Elements (DEM). *Processes* **2020**, *8*, 1489. [CrossRef]
20. Cheng, X.; Yang, H.; Yuan, L.; Lu, Y.; Cao, C.; Wu, G. Fault Feature Enhanced Extraction and Fault Diagnosis Method of Vibrating Screen Bearings. *Machines* **2022**, *10*, 1007. [CrossRef]
21. Delaney, G.W.; Cleary, P.W.; Hilden, M.; Morrison, R.D. Testing the validity of the spherical DEM model in simulating real granular screening processes. *Chem. Eng. Sci.* **2012**, *68*, 215–226. [CrossRef]
22. Chen, Z.; Tong, X.; Li, Z. Numerical Investigation on the Sieving Performance of Elliptical Vibrating Screen. *Processes* **2020**, *8*, 1151. [CrossRef]
23. Lyu, Z.; Zhou, S. Research on spatial Lissajous trajectory vibrating screen. *Adv. Mech. Eng.* **2019**, *11*, 1687814019872851. [CrossRef]
24. Peng, L.P.; Jiang, H.S.; Chen, X.H.; Liu, D.Y.; Feng, H.H.; Zhang, L.; Zhao, Y.M.; Liu, C.S. A review on the advanced design techniques and methods of vibrating screen for coal preparation. *Powder Technol.* **2019**, *347*, 136–147. [CrossRef]
25. Moncada, M.; Rodriguez, C.G. Dynamic Modeling of a Vibrating Screen Considering the Ore Inertia and Force of the Ore over the Screen Calculated with Discrete Element Method. *Shock. Vib.* **2018**, *2018*, 1714738. [CrossRef]
26. Gong, S.; Oberst, S.; Wang, X. An experimentally validated rubber shear spring model for vibrating flip-flow screens. *Mech. Syst. Signal Process.* **2020**, *139*, 106619. [CrossRef]
27. Duda-Mróz, N.; Anufriiev, S.; Stefaniak, P. Application of Wavelet Filtering to Vibrational Signals from the Mining Screen for Spring Condition Monitoring. *Minerals* **2021**, *11*, 1076. [CrossRef]
28. Ogonowski, S.; Krauze, P. Trajectory Control for Vibrating Screen with Magnetorheological Dampers. *Sensors* **2022**, *22*, 4225. [CrossRef]
29. Rossi, A.; Bocchetta, G.; Botta, F.; Scorza, A. Accuracy Characterization of a MEMS Accelerometer for Vibration Monitoring in a Rotating Framework. *Appl. Sci.* **2023**, *13*, 5070. [CrossRef]
30. Wu, T.; You, D.; Gao, H.; Lian, P.; Ma, W.; Zhou, X.; Wang, C.; Luo, J.; Zhang, H.; Tan, H. Research Status and Development Trend of Piezoelectric Accelerometer. *Crystals* **2023**, *13*, 1363. [CrossRef]

energies

MDPI

Article

Application of an Analytical Model of a Belt Feeder for Assessing the Load and Stability of Its Structure

Krzysztof Krauze [1,*], Tomasz Wydro [1], Ryszard Klempka [2] and Kamil Mucha [1]

[1] Faculty of Mechanical Engineering and Robotics, AGH University of Kraków, al. A. Mickiewicza 30, 30-059 Kraków, Poland; wydro@agh.edu.pl (T.W.); kmucha@agh.edu.pl (K.M.)
[2] Faculty of Electrical Engineering, Automatics, Computer Science and Biomedical Engineering, AGH University of Kraków, al. A. Mickiewicza 30, 30-059 Kraków, Poland; klempka@agh.edu.pl
* Correspondence: krauze@agh.edu.pl

Abstract: Belt conveyors, owing to their simple construction, high reliability and relatively low energy consumption, are the basic means of transporting loose and granular materials. Currently, thanks to continuous development, belt conveyors can reach a length of up to several kilometres, and their belt width can be more than two meters. Such possibilities are achieved thanks to increasingly better belts and drives. However, the most common are short belt conveyors with a length of up to 40 m and belt widths of up to 1 m, frequently referred to as belt feeders. Apart from the mining industry, they are widely used in power engineering, metallurgy and other industries (chemical plants, trans-shipment ports, storage yards, etc.). The design of machines, including belt feeders, is based on calculations. Modern design in technology is based on advanced computational methods and the possibilities of computer technology. Multi-variant simulation calculations are necessary, especially in the case of belt feeders, where none of the devices—despite the use of typical elements and subassemblies—are a repeatable solution. Only this procedure guarantees the selection of rational solutions already at the early stages of design. Therefore, in this article, an analytical model of a typical belt feeder was developed and its stability and forces in the supports were determined. This allowed the development of an application for testing the stability of the belt feeder at the design stage or when introducing structural changes.

Keywords: transport of loose materials; belt feeder; analytical model; stability; computer application

Citation: Krauze, K.; Wydro, T.; Klempka, R.; Mucha, K. Application of an Analytical Model of a Belt Feeder for Assessing the Load and Stability of Its Structure. *Energies* **2023**, *16*, 8111. https://doi.org/10.3390/en16248111

Academic Editors: Dameng Liu and Fernando Rubiera González

Received: 23 September 2023
Revised: 21 November 2023
Accepted: 14 December 2023
Published: 17 December 2023

1. Introduction

Belt conveyors are devices transporting various types of materials with granulation adapted to width b_t. The belt is an element that moves the material at a certain speed [1,2]. The efficiency of the conveyor results from the active cross-sectional area of the material on the belt F_p (trough) and its speed v_t. The characteristic parameters of the belt conveyor are therefore its length b_t and speed v_t [3–5]. These parameters, together with the type of transported material and the inclination of the conveyor α_p, determine the size of installed power N_p [6,7]. A typical belt conveyor with an upper conveyor belt consists of a return pulley, a supporting structure, usually in the form of a grid with idlers, and a drive with an electric motor or, less frequently, a hydraulic motor and a mechanical transmission. The movement of the belt, powered by the drive, at a certain speed v_t on idlers built on supports (trestles) from the return pulley to the drive, and causing the material to be transferred with the required efficiency, is a working movement. In contrast, the movement of the belt from the drive to the return pulley is a return (idle) movement; in this case, the belt also moves on idlers, but only single ones. As mentioned before, belt conveyors are devices that transport various types of materials in a continuous way. Therefore, they are mostly applied wherever constant feeding of material with a certain efficiency is required [3,4].

The above information applies to all belt conveyors with lengths ranging from several metres to even several kilometres, and belt widths from several centimetres to more than two meters [8,9]. This group of conveyors includes short belt conveyors with a length of up to 40 m and a belt width reaching up to 1 m, often referred to as belt feeders [10,11].

Due to their broad scope of work, belt feeders are eagerly chosen by investors in various industries. They are commonly used in reloading and transporting works, warehouses, landfills, construction sites, plants processing and exploiting materials and mineral resources (Figure 1), and for transporting excavated material and overlayers in the mining industry. These devices guarantee the stable and safe transport of materials. The entire process is convenient and quick, which brings significant benefits (saving time and money). Users are sure that the investment in high-class feeders will pay off quickly, and the profits will be felt for years. The use of belt feeders means easy assembly, disassembly, and transport because the structures are modular—assembled from ready-made elements. It allows the modernisation of production or distribution and adapting it to current standards. A large selection of parts and accessories allows users to modify the structure as needed. Additionally, the operation of belt feeders is quite quiet.

Figure 1. Belt feeder in the technological line of the processing plant of an aggregate mine.

For the mineral resources industry, belt feeders are, among others, used in opencast rock mining for transporting the crushed mineral. They are usually included in the process line, moving the mineral from one loading point (feeder, tank, screen, or crusher) to another discharge point (other means of transport, warehouse, or heap). Typically, the abovementioned feeders consist of a drive (Figure 2), a return pulley (Figure 3) and a route (Figure 4) with idlers and a belt. Such a feeder moves the material lying on the belt at angle α_{tp}, usually upwards. The maximum value of this angle results from the frictional coupling between the transported material and the belt μ_{tu} ($\alpha_{tp} + \Delta\alpha_{tp} \leq \mathrm{arctg}\mu_{tu}$). It can work horizontally ($\Delta\alpha_{tp} = 0°$) and on a downward negative slope ($\Delta\alpha_{tp} < 0°$) or on an upward positive slope ($\Delta\alpha_{tp} > 0°$) [3]. Additional equipment utilised with such feeders are a charging hopper (Figure 5) and a discharge hopper, external and internal scrapers, platforms (Figure 6), as well as various types of protection related mainly to health and safety regulations and control (Figure 7) [12,13].

Figure 2. Belt feeder's drive (gear motor).

Figure 3. Belt feeder's return pulley with a belt tensioning system.

Figure 4. Belt feeder's route with idlers, trestles, and a safety line.

Figure 5. Belt feeder's return pulley with a charging hopper and a safety net.

Figure 6. Belt feeder's route with a platform for the crew.

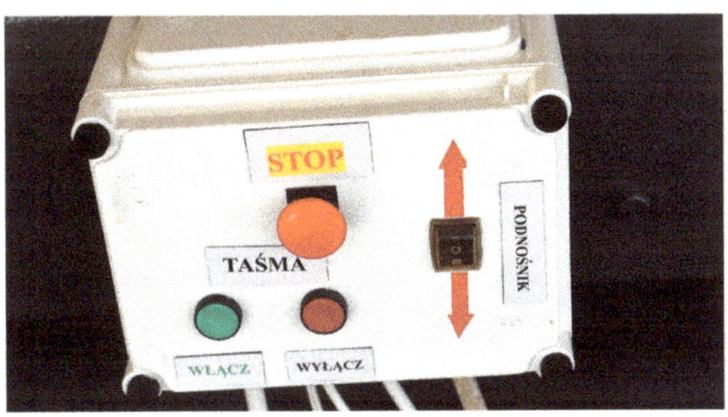

Figure 7. Local control of the belt feeder.

A characteristic feature of these feeders is their installation in the workplace, on stationary or mobile supports. Straight stationary supports (SSSs) are built on previously made foundations (Figure 8), and their number is dependent on the length of the feeder. The V-type stationary supports (VSSs), equipped with slides (skids), hold the feeder structure at the required angle α_{tp}, resting on the ground (Figure 9). If wheels are installed in place of skids, V-type mobile supports (VMS) are used (Figure 10). Belt feeders with V-type stationary supports are the most common [1–3], which poses a big challenge due to the stability of the structure, both during operation and at standstill. This issue requires the testing of the feeder's stability [14] at the design stage and subsequent operational tests.

Figure 8. Belt feeder with SSSs (straight stationary supports).

Figure 9. Belt feeder with VSSs (V-type stationary supports).

The available literature is rich in issues related to various aspects of belt conveyors. There are a few articles related to belt feeders; among others, Maton [15] and Bates [16] tested and calculated the load on such feeders. However, as yet, there is no description of the problem related to the stability of the belt feeder structure and its load. Therefore, an analytical model of a typical belt feeder with V-type supports was developed, and its stability and the forces in the supports were determined. It is worth emphasising that all the parameter values adopted for the calculations were obtained from belt feeder manufacturers and users. These are the actual values with which these feeders work. As a result, an application was developed using computer technology to test the feeder's stability at the design stage or while making changes.

Figure 10. Belt feeder with VMSs (V-type mobile supports).

2. Analytical Model of the Belt Feeder

The starting point for developing an analytical model of the VSS and VMS belt feeder in order to assess its stability is the reduction of masses [17]. This allows the determination of the coordinates of the centre of mass of the feeder's supporting structure (Figure 11). The continuous linear load $q_{jpi,}$, different for the three parts of the belt feeder (1), was assumed as rectangles with known coordinates of the centre of mass. The first part of the feeder's supporting structure is the section stretching from the hopper (return pulley) to the first support, the second part is the section between the supports, and the third one stretches from the second support to the discharge (drive). This model does not take into account the weights of the feeding hopper G_z (drum, belt tension, hopper, and covers) or the discharge G_w (drum, counterweight drive, and platform). These will be taken into account when determining the forces and moments for the entire feeder.

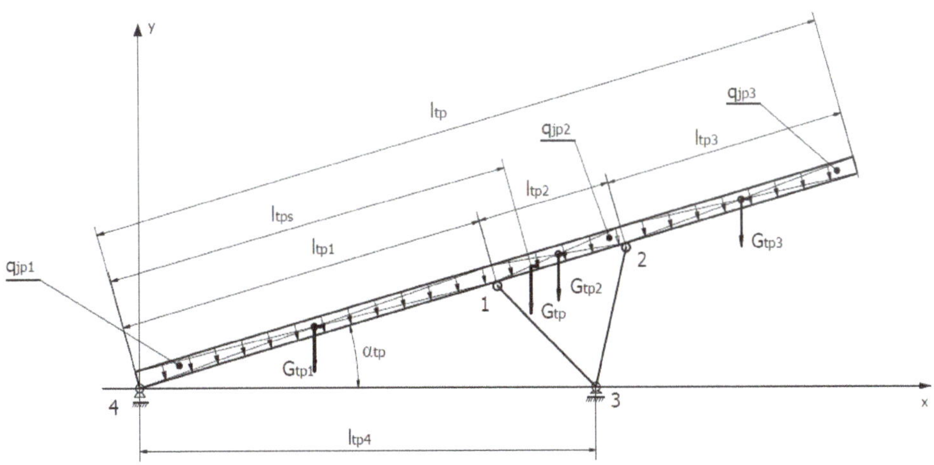

Figure 11. Diagram of the route (load-bearing structure) of the belt feeder for determining the centre of mass l_{tps}.

$$G_{tp} = \sum_{i=1}^{n} q_{tpi} \cdot l_{tpi} \quad (1)$$

where:

q_{tpi}—linear continuous load for the i-th section of the feeder;
l_{tpi}—length of the i-th section of the feeder;
n—number of feeder sections under consideration.

Concentrated weights for the three sections of the belt feeder can be determined from dependencies (2), (3), and (4). The total weight of the supporting structure of the belt feeder G_{tp} is the sum of three individual weights G_{tp1}, G_{tp2}, and G_{tp3} (5). Similarly, the total length of the belt feeder l_{tp} is the sum of its three sections l_{tp1}, l_{tp2}, and l_{tp3} (6). For the conditions of equilibrium relative to the y-axis, it is possible to determine the resultant moment M_{Gtp} as well as single moments for the three masses M_{Gtp1}, M_{Gtp2}, and M_{Gtp3}, which balance each other (7). The structural dimensions of the feeder's route and the locations of the concentrated weights G_{tp1}, G_{tp2}, and G_{tp3} (8), (9), (10), and the resultant mass (11) allow the determination of the coordinate of the centre of mass l_{tps} of the resultant mass M_{Gtp} for different (13) or identical continuous linear loads q_{tpi} (14).

$$G_{tp1} = q_{tp1} \cdot l_{tp1} \quad (2)$$

$$G_{tp2} = q_{tp2} \cdot l_{tp2} \quad (3)$$

$$G_{tp3} = q_{tp3} \cdot l_{tp3} \quad (4)$$

$$G_{tp} = G_{tp1} + G_{tp2} + G_{tp3} \quad (5)$$

$$l_{tp} = l_{tp1} + l_{tp2} + l_{tp3} \quad (6)$$

$$M_{Gtp} = M_{Gtp1} + M_{Gtp2} + M_{Gtp3} \quad (7)$$

$$M_{Gtp1} = G_{tp1} \cdot 0.5 \cdot l_{tp1} \cdot \cos(\alpha_{tp}) \quad (8)$$

$$M_{Gtp2} = G_{tp2} \cdot (l_{tp1} + 0.5 \cdot l_{tp2}) \cdot \cos(\alpha_{tp}) \quad (9)$$

$$M_{Gtp3} = G_{tp3} \cdot (l_{tp1} + l_{tp2} + 0.5 \cdot l_{tp3}) \cdot \cos(\alpha_{tp}) \quad (10)$$

$$M_{Gtp} = G_{tp} \cdot l_{tps} \cdot \cos(\alpha_{tp}) \quad (11)$$

On the basis of Formula (7) as well as (8), (9), and (10), the M_{Gtp} dependence can be expressed (12) as follows:

$$M_{Gtp} = [G_{tp1} \cdot 0.5 \cdot l_{tp1} + G_{tp2} \cdot (l_{tp1} + 0.5 \cdot l_{tp2}) + G_{tp3} \cdot (l_{tp1} + l_{tp2} + 0.5 \cdot l_{tp3})] \cdot \cos(\alpha_{tp}) \quad (12)$$

Based on Formula (11) and Formulas (2)–(5), the coordinate of the centre of mass l_{ps} describes dependence (13) or (14) if (6) is taken into consideration.

$$l_{tps} = \frac{M_{Gtp}}{G_{tp} \cdot \cos(\alpha_{tp})} == \frac{0.5 \cdot G_{tp1} \cdot l_{tp1} + G_{tp2} \cdot (l_{tp1} + 0.5 \cdot l_{tp2}) + G_{tp3} \cdot (l_{tp1} + l_{tp2} + 0.5 \cdot l_{tp3})}{G_{tp}} \quad (13)$$

$$l_{tps} = \frac{0.5 \cdot l_{tp1}^2 + l_{tp2} \cdot (l_{tp1} + 0.5 \cdot l_{tp2}) + l_{tp3} \cdot (l_{tp1} + l_{tp2} + 0.5 \cdot l_{tp3})}{l_{tp1} + l_{tp2} + l_{tp3}} \quad (14)$$

The above dependencies allowed the development of an analytical model of the VSS or VMS belt feeder (Figure 12), located in a horizontal excavation ($\Delta\alpha_{tp} = 0°$), inclined upwards ($\Delta\alpha_{tp} > 0°$) and inclined downwards ($\Delta\alpha_{tp} < 0°$). The belt feeder, with the load-bearing structure weight G_{tp}, the weight of the discharge (drive) G_w, and the feeding hopper (return pulley) G_z, is supported at points 3 and 4. The feeder's return pulley (point 4) exerts pressure on the ground with the force of reaction S_4, whereas reaction S_3 at point 3 comes from two forces S_1 and S_2 in the supports having lengths l_{st1} and l_{st2}. At point 3, friction force T_{s4} will occur. Attaching the supports in points 1 and 2 should ensure the stability of the feeder (permanent equilibrium); i.e., in this case, reaction S_4 must always be positive. This is possible when the moment about pole (point) 3 is zero or positive. By projecting the forces onto the x- and y-axes and determining the moments of these forces relative to point 3, dependencies describing reaction forces S_1, S_2, S_3, and S_4 can be obtained. However, prior to this, it is necessary to establish the relationships between the construction parameters of the feeder, especially with α_{tp}, $\Delta\alpha_{tp}$, and l_{tp}. Height H_{tp}, to which the feeder can transport the material, is described by dependence (15), whereas heights H_{tp1} and H_{tp2} can be determined from dependencies (16) and (17). The angles of inclination of support 1 α_{s1} having length l_{tp1} and of support 2 α_{s2} with length l_{tp2} are described by dependencies (18) and (19). The required support length 1 l_{ts1} results from the value of angle α_{s1} and height H_{tp1} (20). The same is true for support 2, where its length l_{tp2} also results from the value of angle α_{s1} and height H_{tp2} (21).

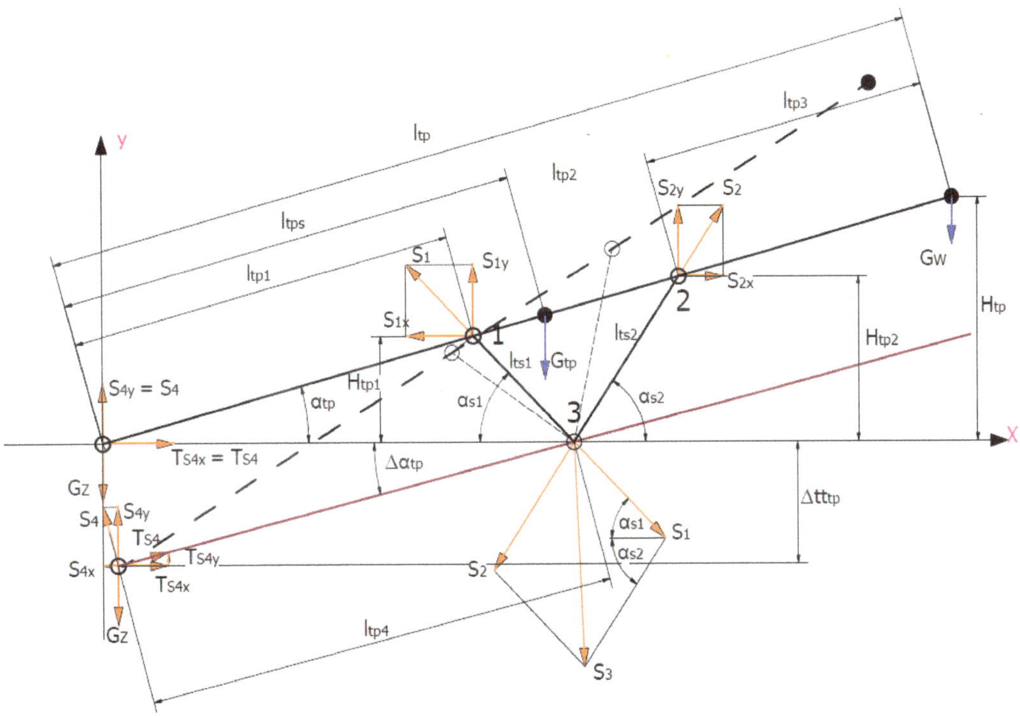

Figure 12. Diagram of the belt feeder load in a horizontal excavation ($\Delta\alpha_{tp} = 0°$)—continuous line; and in an excavation inclined upwards ($\Delta\alpha_{tp} > 0°$)—dotted line.

$$H_{tp} = l_{tp} \cdot \sin(\alpha_{tp} + \Delta\alpha_{tp}) - l_{tp4} \cdot \sin(\Delta\alpha_{tp}) \quad (15)$$

$$H_{tp1} = l_{tp1} \cdot \sin(\alpha_{tp} + \Delta\alpha_{tp}) - l_{tp4} \cdot \sin(\Delta\alpha_{tp}) \tag{16}$$

$$H_{tp2} = (l_{tp1} + l_{tp2}) \cdot \sin(\alpha_{tp} + \Delta\alpha_{tp}) - l_{tp4} \cdot \sin(\Delta\alpha_{tp}) \tag{17}$$

$$\alpha_{s1} = \arctan\left(\frac{H_{tp1}}{l_{tp4} \cdot \cos(\Delta\alpha_{tp}) - l_{tp1} \cdot \cos(\alpha_{tp} + \Delta\alpha_{tp})}\right) \tag{18}$$

$$\alpha_{s2} = \arctan\left(\frac{H_{tp2}}{(l_{tp1} + l_{tp2}) \cdot \cos(\alpha_{tp} + \Delta\alpha_{tp}) - l_{tp4} \cdot \cos(\Delta\alpha_{tp})}\right) \tag{19}$$

$$l_{st1} = \frac{H_{tp1}}{\sin(\alpha_{s1})} \tag{20}$$

$$l_{st2} = \frac{H_{tp2}}{\sin(\alpha_{s2})} \tag{21}$$

When the above analytical relationships between the design parameters of the belt feeder are known, it is possible, using the force balance equations (x- and y-axis) and moments (point 3), to obtain the dependencies describing the searched reaction forces S_1, S_2, S_3, and S_4 and to test the stability of the structure. The sum of the projections of forces on the x-axis should be zero (22).

$$\sum_i P_{ix} = T_{s4x} - S_{4x} - S_{1x} + S_{2x} = 0 \tag{22}$$

$$T_{s4x} = S_{4y} \cdot \mu_x = \mu_x \cdot S_4 \cdot \cos(\Delta\alpha_{tp}) \tag{23}$$

$$S_{4x} = S_4 \cdot \sin(\Delta\alpha_{tp}) \tag{24}$$

$$S_{1x} = S_1 \cdot \cos(\alpha_{s1}) \tag{25}$$

$$S_{2x} = S_2 \cdot \cos(\alpha_{s2}) \tag{26}$$

$$A_{s4x} \cdot S_4 - S_1 \cdot \cos(\alpha_{s1}) + S_2 \cdot \cos(\alpha_{s2}) = 0 \tag{27}$$

$$A_{s4x} = \mu_x \cdot \cos(\Delta\alpha_{tp}) - \sin(\Delta\alpha_{tp}) \tag{28}$$

Similarly, when these forces are projected onto the y-axis, their sum should also be zero (29).

$$\sum_i P_{iy} - G_z + S_{4y} + T_{s4y} + S_{1y} - G_{tp} + S_{2y} - G_w = 0 \tag{29}$$

$$S_{4y} = S_4 \cdot \cos(\Delta\alpha_{tp}) \tag{30}$$

$$S_{1y} = S_1 \cdot \sin(\alpha_{s1}) \tag{31}$$

$$S_{2y} = S_2 \cdot \sin(\alpha_{s2}) \tag{32}$$

$$T_{s4y} = S_{4x} \cdot \mu_y = \mu_y \cdot S_4 \cdot \sin(\Delta\alpha_{tp}) \tag{33}$$

$$G_z + G_{tp} + G_w = G_{zpw} \tag{34}$$

$$A_{s4y} \cdot S_4 + S_1 \cdot \sin(\alpha_{s1}) + S_2 \cdot \sin(\alpha_{s2}) - G_{zpw} = 0 \tag{35}$$

$$A_{s4y} = \mu_y \cdot \sin(\Delta\alpha_{tp}) - \cos(\Delta\alpha_{tp}) \tag{36}$$

For the structure to remain in equilibrium, the sum of the moments relative to point 3 should also be zero (37). As a result of transforming Equation (37) and introducing additional notations (38), (39), (40), (41), (42), and (43), the dependence describing force S_4 (44) was obtained.

$$\begin{aligned}\sum_i M_{i03} &= G_z \cdot l_{tp4} \cdot \cos(\Delta\alpha_{tp}) + T_{s4x} \cdot l_{tp4} \cdot \sin(\Delta\alpha_{tp}) - S_{4x} \cdot l_{tp4} \cdot \sin(\Delta\alpha_{tp}) - T_{s4y} \cdot l_{tp4} \cdot \cos(\Delta\alpha_{tp}) - S_{4y} \\ &\cdot l_{tp4} \cdot \cos(\Delta\alpha_{tp}) + S_{1x} \cdot H_{tp1} - S_{1y}\left[l_{tp4} \cdot \cos(\Delta\alpha_{tp}) - l_{tp1} \cdot \cos(\alpha_{tp} + \Delta\alpha_{tp})\right] \\ &+ G_{tp}\left[l_{tp4} \cdot \cos(\Delta\alpha_{tp}) - l_{tps} \cdot \cos(\alpha_{tp} + \Delta\alpha_{tp})\right] \\ &+ S_{2y}\left[(l_{tp1} + l_{tp2}) \cdot \cos(\alpha_{tp} + \Delta\alpha_{tp}) - l_{tp4} \cdot \cos(\Delta\alpha_{tp})\right] - S_{2x} \cdot H_{tp2} \\ &- G_w\left[l_{tp} \cdot \cos(\alpha_{tp} + \Delta\alpha_{tp}) - l_{tp4} \cdot \cos(\Delta\alpha_{tp})\right] = 0\end{aligned} \tag{37}$$

$$A_{s1y} = l_{tp4} \cdot \cos(\Delta\alpha_{tp}) - l_{tp1} \cdot \cos(\alpha_{tp} + \Delta\alpha_{tp}) \tag{38}$$

$$A_{s2y} = (l_{tp1} + l_{tp2}) \cdot \cos(\alpha_{tp} + \Delta\alpha_{tp}) - l_{tp4} \cdot \cos(\Delta\alpha_{tp}) \tag{39}$$

$$\begin{aligned}A_{zpw} = G_z \cdot l_{tp4} \cdot \cos(\Delta\alpha_{tp}) + G_{tp}\left[l_{tp4} \cdot \cos(\Delta\alpha_{tp}) - l_{tps} \cdot \cos(\alpha_{tp} + \Delta\alpha_{tp})\right] \\ -G_w\left[l_{tp} \cdot \cos(\alpha_{tp} + \Delta\alpha_{tp}) - l_{tp4} \cdot \cos(\Delta\alpha_{tp})\right]\end{aligned} \tag{40}$$

$$A_{s4} = 0.5(\mu_x - \mu_y) l_{tp4} \cdot \sin(2\Delta\alpha_{tp}) - l_{tp4} \tag{41}$$

$$A_{s1} = H_{tp1} \cdot \cos(\alpha_{s1}) - A_{s1y} \cdot \sin(\alpha_{s1}) \tag{42}$$

$$A_{s2} = A_{s2y} \cdot \sin(\alpha_{s2}) - H_{tp2} \cdot \cos(\alpha_{s2}) \tag{43}$$

$$S_4 = -\frac{A_{s1} \cdot S_1 + A_{s2} \cdot S_2 + A_{zpw}}{A_{s4}} \tag{44}$$

By transforming Equation (22) and introducing additional notations (45), (46), and (47), the force S_1 dependence (48) was obtained.

$$-\frac{A_{s4x}}{A_{s4}}(A_{s1} \cdot S_1 + A_{s2} \cdot S_2 + A_{zpw}) - S_1 \cdot \cos(\alpha_{s1}) + S_2 \cdot \cos(\alpha_{s2}) = 0 \tag{45}$$

$$B_{s2x} = \cos(\alpha_{s2}) - \frac{A_{s4x} \cdot A_{s2}}{A_{s4}} \tag{46}$$

$$B_{s1x} = \cos(\alpha_{s1}) + \frac{A_{s4x} \cdot A_{s1}}{A_{s4}} \tag{47}$$

$$S_1 = \frac{B_{s2x} \cdot S_2 \cdot A_{s4} - A_{s4x} \cdot A_{zpw}}{B_{s1x} \cdot A_{s4}} \tag{48}$$

By transforming Equation (29) and introducing additional notations (49), (50), and (51), the force S_2 (52) dependence (52) was obtained.

$$S_1\left(\sin(\alpha_{s1}) - \frac{A_{s4y}\cdot A_{s1}}{A_{s4}}\right) + S_2\left(\sin(\alpha_{s2}) - \frac{A_{s4y}\cdot A_{s2}}{A_{s4}}\right) - \frac{A_{s4y}\cdot A_{zpw}}{A_{s4}} - G_{zpw} = 0 \quad (49)$$

$$B_{s2y} = \sin(\alpha_{s2}) - \frac{A_{s4y}\cdot A_{s2}}{A_{s4}} \quad (50)$$

$$B_{s1y} = \sin(\alpha_{s1}) - \frac{A_{s4y}\cdot A_{s1}}{A_{s4}} \quad (51)$$

$$S_2 = \frac{\left(A_{s4y}\cdot A_{zpw} + G_{zpw}\cdot A_{s4}\right)B_{s1x} + A_{s4x}\cdot A_{zpw}\cdot B_{s1y}}{A_{s4}\left(B_{s2y}\cdot B_{s1x} + B_{s2x}\cdot B_{s1y}\right)} \quad (52)$$

Reaction S_3 at point 3 is the geometric sum of forces S_1 and S_2 (53).

$$S_3 = \sqrt{S_1^2 + S_2^2 - 2S_1\cdot S_2\cdot\cos(\alpha_{s1} + \alpha_{s2})} \quad (53)$$

The obtained analytical dependencies resulting from the adopted model of the belt feeder make it possible to determine the forces that are important for its structure load and stability. This is, of course, particularly important at the stage of creating the model, and designing and preparing documentation, but also when making changes to the feeder structure (modernising). As mentioned in the introduction, one should also remember the adhesion of the excavated material to the belt, expressed by the coefficient of friction μ_{tu}, which corresponds to friction angle ρ_{tw}. Then, the total angle of feeder inclination (α_{tp} + $\Delta\alpha_{tp}$) cannot exceed the value of angle ρ_{tw} (54). Of course, changing the angle of the feeder's inclination also requires checking its stability.

$$\alpha_{tp} + \Delta\alpha_{tp} \leq \arctan\mu_{tu} \quad (54)$$

The above-mentioned comments and requirements resulted in the development of an algorithm and a computer application that enable the load of the feeder structure and its stability to be calculated, without the need to use CAD programs.

3. Application for Designing and Analysing the Construction of a Belt Feeder

In order to support the design and analysis of the feeder load, an application in the Matlab environment has been developed. After compilation, it will be an independent application. Figure 13 shows the main application window.

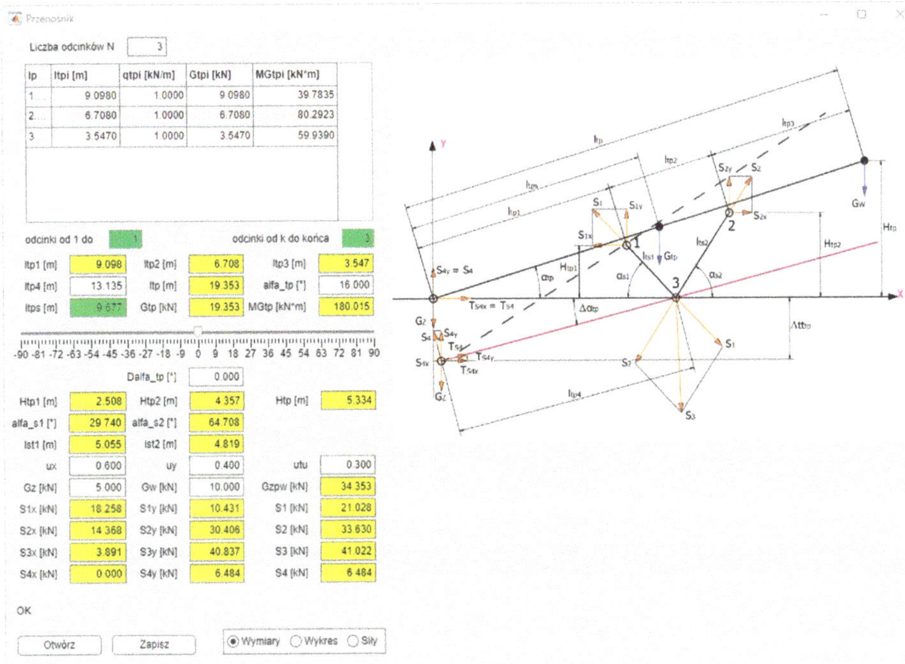

Figure 13. Main window of the application for designing and analysing the belt feeder.

In the left part of the application window, one can see the data entered by the user and the results of calculations based on the relationships presented in the previous section. The input data are the lengths and weights of the feeder elements (top right corner of the application), the distance l_{tp4}, the angle of the feeder inclination α_{tp}, the angle of the feeder's deviation from the base plane (lowering or elevating the starting point of the feeder), orthogonal components of the friction coefficients μ_x and μ_y, as well as the forces (of weights) G_z and G_w. All these quantities are entered in the white edit boxes. Yellow and green edit boxes are reserved for calculation results. Each change in the input quantity causes recalculation of all the dependencies. The input data entered in the feeder calculations can be saved in a text file with the "Save" button to enable quick loading of the data with the "Open" button without re-entering the data manually. An important element of the application is the ability to change the value of angle $\Delta\alpha_{tp}$, i.e., rotate the feeder around support point 3.

In the lower right part of the application, there is a three-position switch for selecting the drawings located on the right side of the application. Initially, a diagram of the feeder with its marked dimensions (parameters) and the forces acting on it is drawn. This considerably facilitates entering the input data and interpreting the obtained results, as they can be quickly located. The second option "Graph" is a visualisation of the feeder based on the entered and calculated data (Figure 14). In this option, the designed feeder is drawn in the neutral position ($\Delta\alpha_{tp} = 0°$) with a blue dashed line, and in the position taking into account the set angle $\Delta\alpha_{tp}$. In the case of the second option, the acting forces (weights) G_z, G_w, G_{zpw} (black), as well as forces S_1, S_2, S_3, and S_4 with their orthogonal components are drawn. A change in angle $\Delta\alpha_{tp}$ by means of a slider updates the calculations and the schematic drawing of the feeder. The third drawing option includes graphs of S_i forces with their orthogonal components as a function of the sum of angles α_{tp} and $\Delta\alpha_{tp}$ (Figure 15).

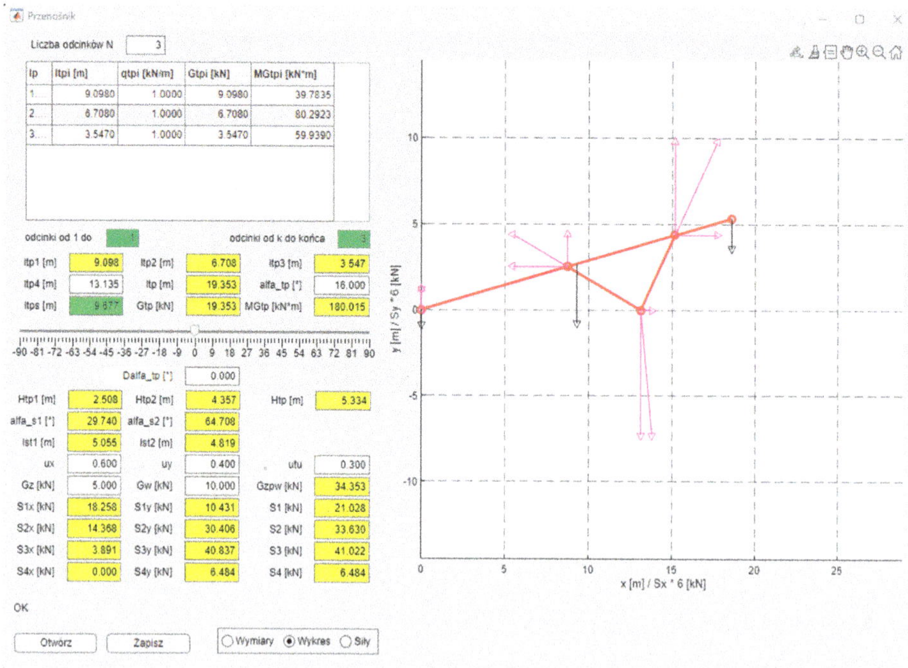

(a)

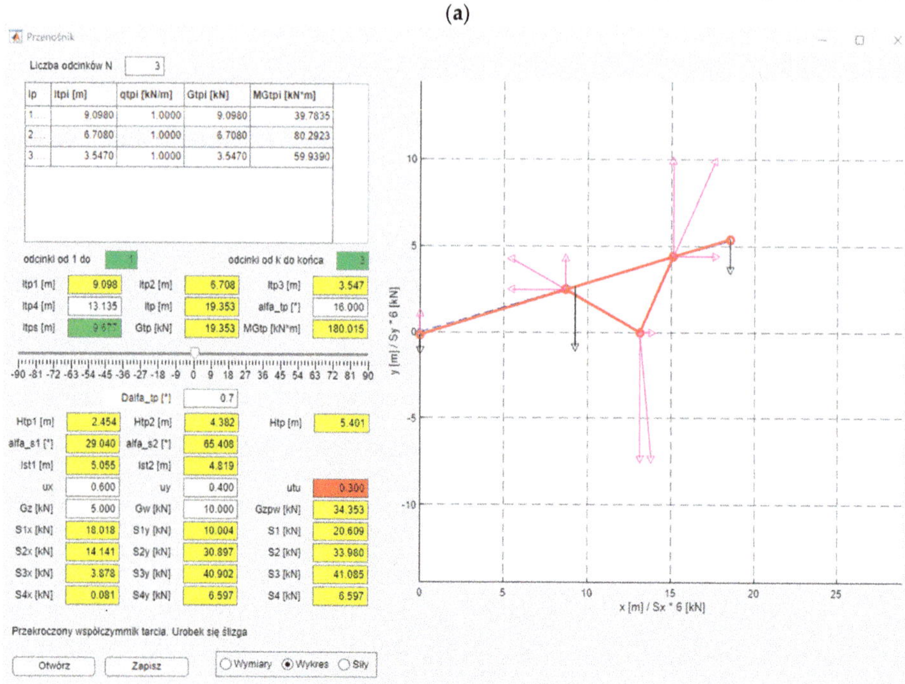

(b)

Figure 14. *Cont.*

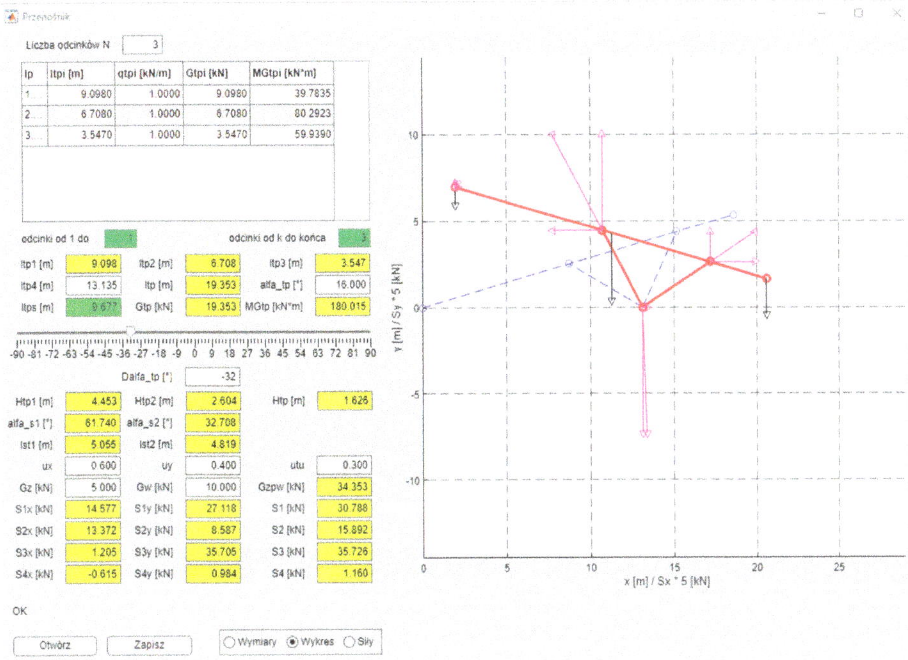

(**c**)

Figure 14. View of the application after schematic depiction of the designed feeder for different values of $\Delta\alpha_{tp}$: (**a**) $\Delta\alpha_{tp} = 0°$, (**b**) $\Delta\alpha_{tp} = 0.7°$, and (**c**) $\Delta\alpha_{tp} = -32°$.

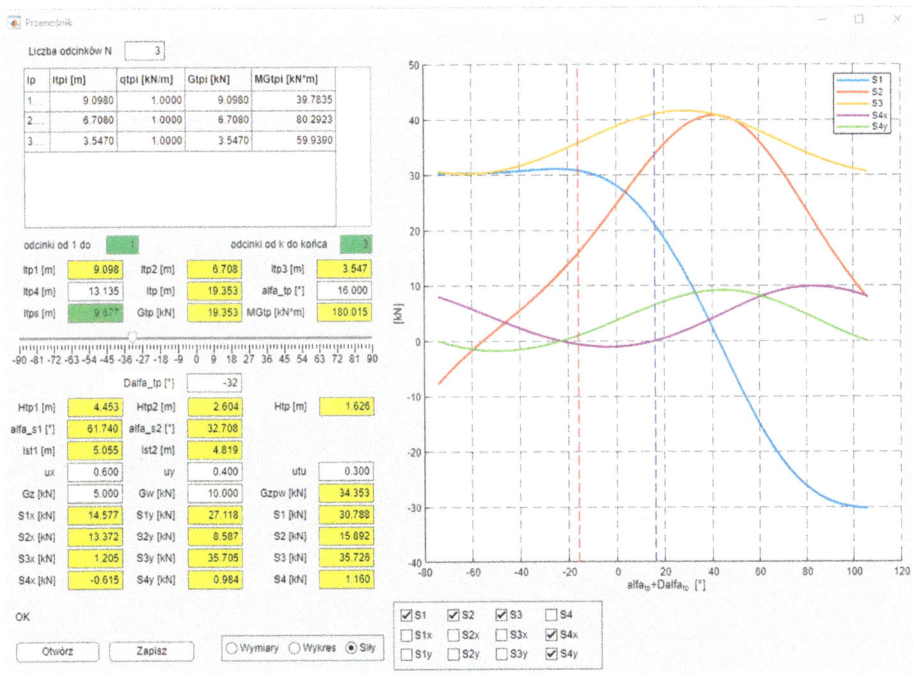

Figure 15. Graphs of forces as a function of the sum of angles α_{tp} and $\Delta\alpha_{tp}$.

The values of the forces acting on the feeder depend on its positioning. Graphs of these functions can be helpful in the process of designing the feeder. Twelve "checkboxes" have been placed under the graphs so that the designer can choose which forces they want to depict, also taking into account their orthogonal components. Two vertical dashed lines are also drawn on the graph. The blue line represents the feeder in the neutral position ($\Delta\alpha_{tp} = 0°$), whereas the red one defines the feeder's position after taking into account the rotation of the feeder relative to support point 3. Every change in the feeder parameters is updated in the calculations and in the diagrams. Figure 16 shows the same graphs as in the application (Figure 15) with an additional red vertical line marked for $\Delta\alpha_{tp} = 0.7°$.

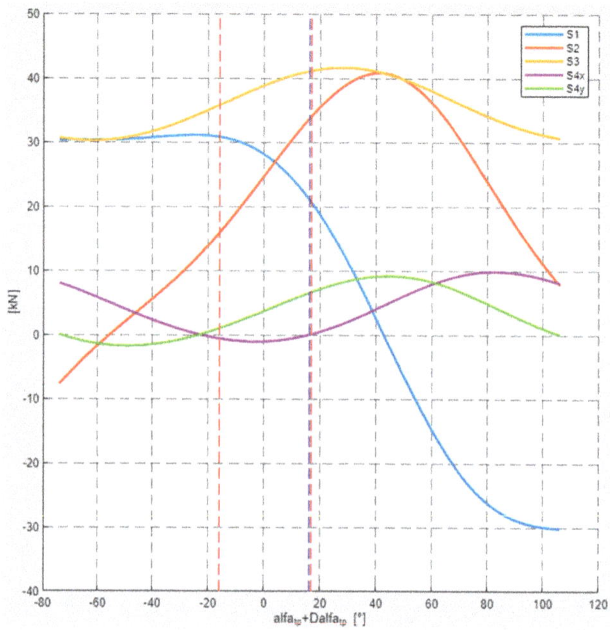

Figure 16. Graphs of forces as a function of the sum of angles α_{tp} and $\Delta\alpha_{tp}$ with marked dashed lines of positions $\Delta\alpha_{tp} = -32°$ (pink), $\Delta\alpha_{tp} = 0°$ (blue), and $\Delta\alpha_{tp} = 0.7°$ (red).

4. Usage of the Application

The results of the calculations (yellow and green edit boxes on Figure 13) together with Figures 14–16 were obtained with the use of the application in question and the data of a real feeder with a length of 20 m, inclination $\alpha_{tp} = 6°$, and a belt width of 1000, placed on V-type stationary supports. (Figure 10). This feeder is mainly designed to work on inclination $\alpha_{tp} = 16°$, in which case $\Delta\alpha_{tp} = 0°$. This is due to the coupling of the transported material with the belt, where angle $\rho_{tu} = 16.7°$ (54) for $\mu_{tu} = 0.3$, in which case $\Delta\alpha_{tp} = 0.7°$. Increasing the upward inclination for this type of feeder ($\Delta\alpha_{tp} > 0°$) is practically impossible in the case of a smooth belt. In the event that the feeder works downwards ($\Delta\alpha_{tp} < 0°$), its total inclination can reach $-32°$. The value and sense of force S_{4y} lead to the conclusion that the feeder's stability can also be maintained (Figure 16, green). It can be easily seen that the course of force S_{4y} is decreasing in the examined range. Conversely, force S_{4x} increases in value. For $\Delta\alpha_{tp} = 0°$, force S_{4x} has a zero value, which is consistent with reality. In the range of permissible belt inclination angles ($-32° < \alpha_{tp} < 0.7°$), force S_2 has a positive and increasing value. The opposite is true for force S_1, as its value is also positive but decreasing, whereas the resultant value of these forces, i.e., force S_3 for angle $\alpha_{tp} = 16°$, reaches its maximum. It should be noted that force S_{4y} also has a positive value in this range, which leads to the conclusion that the feeder has the required stability. It may change when the weight of the discharge G_w increases. In such a case, force S_{4y} has a negative value above weight G_w = 25.5768 kN, which indicates the loss of stability of the feeder when it is working on an upward slope ($\Delta\alpha_{tp} > 0°$), Figure 17a. However, in the case of a feeder working on a downward slope with weight G_w > 12.2258 kN, force S_{4y} also has a negative value ($\Delta\alpha_{tp} < 0°$), which causes loss of stability (Figure 17b). Of course, the application allows other cases related to the design parameters of the feeder to be considered.

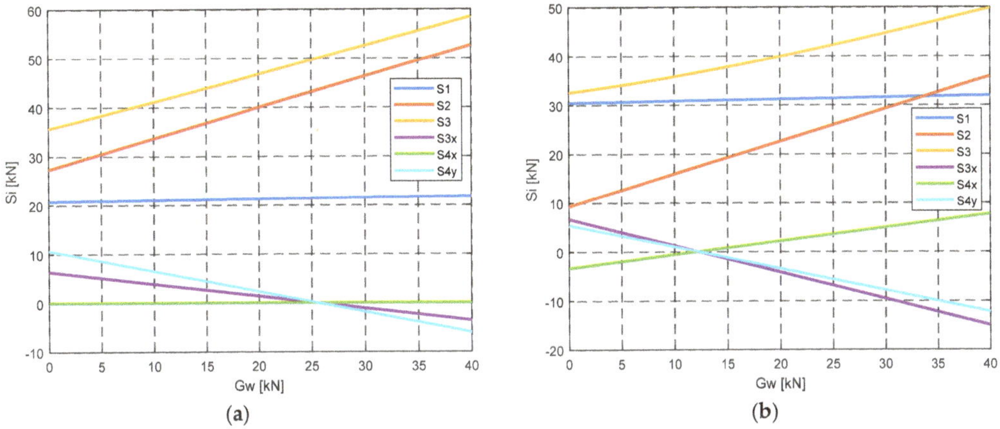

Figure 17. Graphs of forces as a function of discharge weight G_w, for α_{tp} = 16°: (**a**) $\Delta\alpha_{tp}$ = 0°; and (**b**) $\Delta\alpha_{tp}$ = −32°.

5. Conclusions

The previously described construction of belt feeders as well as their physical and mathematical models allowed the development of an algorithm and a computer application. The main goal of this project was to create a tool that would enable the structure of the designed and constructed feeder to be tested without the need for additional documentation. This is particularly useful at the stage of starting up a specific feeder in its place of work, where small changes in its construction and installation are often required. To sum up, it can be said that:

1. The model of a belt feeder, based on currently produced and used feeders of this type, enables analytical testing of their stability and structure load.
2. The developed algorithm and computer application enable and expedite computational processes with the possibility of their visualisation, especially when making structural and functional changes to existing objects.
3. The discussed application is used in practice by companies producing and installing this type of feeder. Users of these feeders also employ it, as it responds to their needs.

Author Contributions: Conceptualization, K.K.; methodology, K.K.; software, R.K.; validation, K.M., R.K. and T.W.; formal analysis, K.K. and R.K; resources, T.W.; data curation, R.K.; writing—original draft preparation, K.K., R.K. and K.M.; writing—review and editing, K.K. and K.M.; visualization, T.W. and R.K.; supervision, K.K.; project administration, K.M.; funding acquisition, K.K. All authors have read and agreed to the published version of the manuscript.

Funding: Works financed by the AGH University of Kraków.

Data Availability Statement: Data are contained within the article.

Conflicts of Interest: The authors declare no conflict of interest.

References

1. Yardley, E.D.; Stace, L.R. *Belt Conveying of Minerals*; Woodhead Publishing: Sawston, UK, 2008.
2. Subba Rao, D.V. *The Belt Conveyor: A Concise Basic Course*; CRC Press: Boca Raton, FL, USA, 2022.

3. Gładysiewicz, L. *Przenośniki Taśmowe Teoria i Obliczenia*; Wrocław University of Science and Technology Press: Wrocław, Poland, 2003. (In Polish)
4. Ananth, K.N.S.; Rakesh, V.; Visweswarao, P.K. Design and selecting the proper conveyor-belt. *Int. J. Adv. Eng. Technol.* **2013**, *4*, 43–49.
5. He, D.; Pang, Y.; Lodewijks, G. Speed control of belt conveyors during transient operation. *Powder Technol.* **2016**, *301*, 622–631. [CrossRef]
6. Kawalec, W.; Suchorab, N.; Konieczna-Fuławka, M.; Król, R. Specific Energy Consumption of a Belt Conveyor System in a Continuous Surface Mine. *Energies* **2020**, *13*, 5214. [CrossRef]
7. Kulinowski, P.; Kasza, P.; Zarzycki, J. Influence of Design Parameters of Idler Bearing Units on the Energy Consumption of a Belt Conveyor. *Sustainability* **2021**, *13*, 437. [CrossRef]
8. Cao, Y.; Yu, Q.; Yang, T.; Zhu, W.; Le, Z. Numerical Study on the Fracturing Mechanism of the Belt Conveyor Roadway in Dagushan Open-Pit Mine and Control Measures Evaluation. *Rock Mech. Rock Eng.* **2022**, *55*, 6663–6682. [CrossRef]
9. Mathaba, T.; Xia, X. A Parametric Energy Model for Energy Management of Long Belt Conveyors. *Energies* **2015**, *8*, 13590–13608. [CrossRef]
10. Strydom, E. The challenges and advances in belt feeder and hopper design. *Bulk Solids Handle* **2006**, *26*, 106–115.
11. Živanić, D.; Ilanković, N.; Zuber, N.; Đokić, R.; Zdravković, N.; Zelić, A. The analysis of influential parameters on calibration and feeding accuracy of belt feeders. *Eksploat. I Niezawodn.–Maint. Reliab.* **2021**, *23*, 413–421. [CrossRef]
12. Bałaga, D.; Kalita, M.; Siegmund, M.; Nieśpiałowski, K.; Bartoszek, S.; Bortnowski, P.; Ozdoba, M.; Walentek, A.; Gajdzik, B. Determining and Verifying the Operating Parameters of Suppression Nozzles for Belt Conveyor Drives. *Energies* **2023**, *16*, 6077. [CrossRef]
13. Bortnowski, P.; Kawalec, W.; Król, R.; Ozdoba, M. Types and causes of damage to the conveyor belt–Review, classification and mutual relations. *Eng. Fail. Anal.* **2022**, *140*, 106520. [CrossRef]
14. Santos, L.S.; Macêdo, E.N.; Ribeiro Filho, P.R.C.F.; Cunha, A.P.A.; Cheung, N. Belt Rotation in Pipe Conveyors: Failure Mode Analysis and Overlap Stability Assessment. *Sustainability* **2023**, *15*, 11312. [CrossRef]
15. Maton, A.E. Belt feeder design: Starting load calculations. *Bulk Solids Handle* **2009**, *29*, 454–457.
16. Bates, L. A Fundamental Approach to Belt Feeder Loads: How to assess loads on Feeders, (practically). *Bulk Solids Handle* **2015**, *35*, 14–18.
17. Fruet, G.; Miguel, L.F.F. Proposal of a methodology for mass optimization of realistic steel structural systems composed of columns and galleries for support of solid bulk conveyors. *Structures* **2023**, *53*, 833–847. [CrossRef]

energies

MDPI

Article

Development of Longwall Shearers' Haulage Systems as an Alternative to the Eicotrack System Used Nowadays

Krzysztof Kotwica [1,*], Grzegorz Stopka [1], Andrzej N. Wieczorek [2], Marek Kalita [3], Dominik Bałaga [3] and Michał Siegmund [3]

[1] Faculty of Mechanical Engineering and Robotics, AGH University of Science and Technology, 30-059 Kraków, Poland
[2] Department of Mining Mechanization and Robotisation, Faculty of Mining, Safety Engineering and Industrial Automation, Silesian University of Technology, 44-100 Gliwice, Poland
[3] Division of Machines and Equipment, KOMAG Institute of Mining Technology, 44-101 Gliwice, Poland
* Correspondence: kotwica@agh.edu.pl; Tel.: +48-60-746-7068

Abstract: Longwall shearers' haulage system of the Eicotrack type, used most often nowadays, is presented in this article. Its disadvantages, causing problems with a correct operation of mechanized longwall shearer systems, are discussed. The concept of the innovative Flextrack, which should reduce the occurrence of the disadvantages mentioned above, is described. A course and research results, connected with rig tests of the Flextrack haulage system functionality, are presented. Measurement results of wear and stresses, obtained for the Eicotrack and Flextrack haulage systems, are compared and presented. Based on an analysis of the obtained results, a modified version of the Komtrack haulage system was suggested and manufactured. This haulage system was tested in the field conditions, similar to underground conditions in a mine coal longwall. The obtained results confirmed its full functionality and lack of problems experienced in the case of the Eicotrack and Flextrack systems. At present, the Komtrack system is tested in a coal longwall in Piast coal mine, where comparative tests with the Eicotrack system are conducted. The big part of the results presented in this article was developed as part of the research project KOMTRACK jointly implemented by KOMAG Institute of Mining Technology, AGH University of Science and Technology, Łukasiewicz Research Network—Cracow Institute of Technology, Specodlew Innovative Foundry Company and Polish Mining Group Inc., co-financed by the European Regional Development Fund (Contract No. POIR.04.01.04-00-0068/17).

Keywords: longwall shearer haulage system; toothed segment; toothed wheel; stress; flexible collaboration; wear

Citation: Kotwica, K.; Stopka, G.; Wieczorek, A.N.; Kalita, M.; Bałaga, D.; Siegmund, M. Development of Longwall Shearers' Haulage Systems as an Alternative to the Eicotrack System Used Nowadays. *Energies* **2023**, *16*, 1402. https://doi.org/10.3390/en16031402

Academic Editor: Krzysztof Skrzypkowski

Received: 20 December 2022
Revised: 12 January 2023
Accepted: 26 January 2023
Published: 31 January 2023

1. Introduction

One of basic systems for a mechanical exploitation of hard coal in underground mining includes shearer longwall systems. Their application for mining hard coal started in mid-XX century. The first sharers were relocated along the longwall with use of pulling rope or chain haulage systems. They generated a very big hazard to the workers employed in the longwall, mainly caused by a possibility of strand break and so-called whipping, i.e., vibrations of the strand having a big amplitude. In the fifties of the XX century, a search for chainless haulage systems of shearers eliminating the abovementioned hazards [1–5] was started. The first trials were realized in semi-mechanized longwall faces in Germany and in Hungary, but the biggest development of this type of system started 10 years later in Great Britain [6]. The first chainless solutions such as Rackatrack, Peratrack or Dynatrack, after positive tests in British and French mines, became a thing of the past without their broader implementation in the abovementioned longwall systems [7–9].

In the seventies, the tests of chainless solutions of haulage systems with a drive wheel (gear wheel or pin) and different toothed bars were finished with a bigger success.

They were described in the literature [10–12], starting with the British Rollrack with the drive wheel and the toothed bar through similar solutions in Europe and China, with the vertical or horizontal rack in the form of toothed bar such as Powertrack, Eicotrack and Poltrack. Such a haulage system enables a relocation of a shearer along a coal longwall and a realization of its basic functions, i.e., cutting and loading coal on an armoured face conveyor.

Coal shearers, operated nowadays, are mainly equipped with chainless haulage systems, among which the Eicotrack system, together with its following versions, is most popular—e.g., Megatrack, Gigatrack. The Eicotrack system is composed of stiff toothed bars (racks) fixed to the conveyor line pans [1,10]. The teeth of the shearer drive wheel, situated vertically, gear with the rock pins positioned horizontally, forcing a change of the machine location.

A construction and operational principle of the Eicotrack haulage system (Figure 1) is broadly described in other publications [1–3,10]. Due to a stiff construction of toothed bars, on nonrectilinear sections of the armored face conveyor, a location change of the shearer drive wheel in relation to the axis of racks occurs. This leads to a local change of pitch between extreme pins of racks adjacent to each other and a change of distance between pins of rack segments and the axis of rotation of the drive wheel. Such a collaboration of the frictional pair leads to so called edging of teeth, due to which the values of permissible contact stresses between the collaborating surfaces are exceeded. Such a situation, in turn, causes an untimely wear of drive wheel surfaces (mainly the teeth of these wheels) and pins installed in racks [13–17]. In Figure 2, a view of wear of gear wheels, used in the Eicotrack haulage system and of one of the extreme rack pins, is shown as an example.

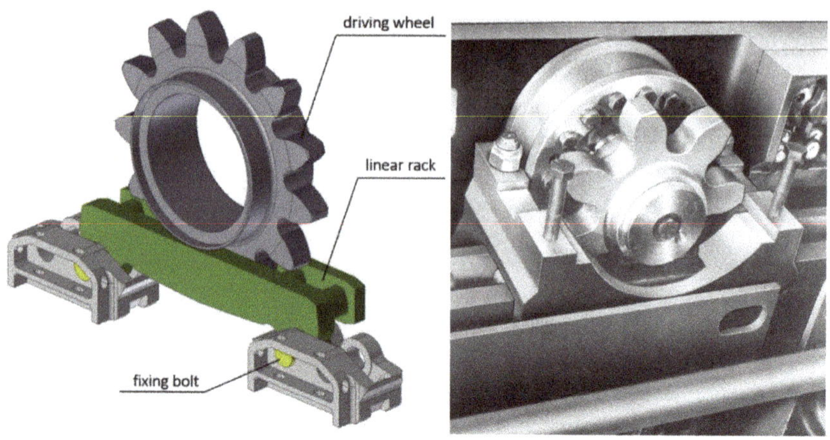

Figure 1. Model and view of the Eicotrack haulage system solution [18].

Figure 2. View of wear of gear wheels used in the Eicotrack haulage system of one of the extreme rack pins, (marked with red arrows).

2. Flextrack Haulage System—Results of Stand Tests, a Comparison with the Eicotrack System

A solution, eliminating above mentioned undesired results of collaboration of the frictional pair: a gear wheel-rack is an innovative haulage system of a longwall shearer with flexible toothed segments developed at the KOMAG Institute of Mining Technology within the framework of the FLEXTRACK project. The toothed segments are installed in a way that enables their mutual relocation in the guides. Replacing the Eicotrack rack, they contribute to a more advantageous collaboration of the drive wheel of the shearer haulage system and toothed segments, in particular in the situation of the conveyor line pan contraflexure both in the vertical plane, as well as in the horizontal one [19]. A model of the Flextrack haulage system, showing a layout of toothed elements in the guide, as well as a comparison of the Flextrack and Eicotrack haulage system elements, collaborating with the track wheel are shown in Figure 3.

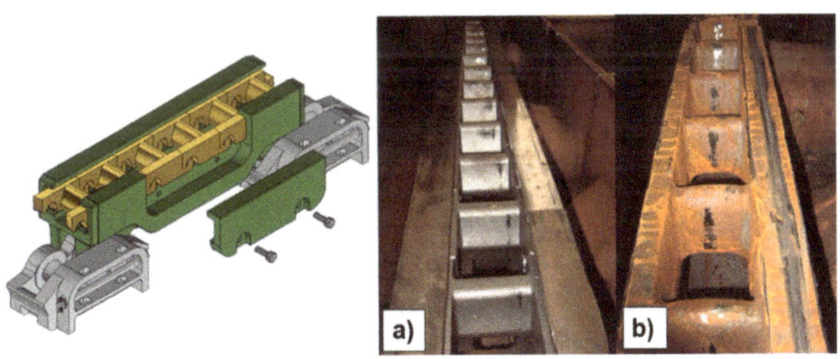

Figure 3. Model of the Flextrack haulage system with a layout of toothed elements in the guide and a comparison of elements collaborating with the track wheel: (a) Flextrack haulage system, (b) Eicotrack haulage system.

A special test rig, enabling the carrying out of comparative tests of wear of components of individual haulage systems, stresses generated in these components during a pass of the track wheel, resistances to haulage and vibrations, was elaborated and manufactured. This rig, shown in Figure 4, enabled about 15-metre of pass along the armored face conveyor line pan at its rectilinear position or contraflexure in the vertical and horizontal planes. A set of two pairs of rams enabled the generation of load resistances simulating cutting during the shearer passes.

Figure 4. View of the special test rig for conducting comparative tests of the Flextrack and Eicotrack haulage systems.

We planned to perform tests in the same operational conditions in the case of both systems; however, in the case of the Eicotrack haulage system, very big values of haulage load resistances, generated by this system, did not render it possible to conduct tests with a comparable value of the force simulating load resistances to cutting generated by rams. It was about 1/3 smaller. For the measurements the transducers of vibrations of 4507 B 005 type produced by Bruel and Kjaer, installed above the shearer haulage system (above each track wheel), were used. Average and maximum values of vibrations acceleration, significantly bigger (of about 40–50%) in the case of the Flextrack system where obtained, respectively, Eicotrack a_{med} = 2.5–4 m/s^2, a_{max} = 45–78 m/s^2, Flextrack a_{med} = 4–6.5 m/s^2, a_{max} = 75–105 m/s^2.

Moreover, different conditions of conducted tests had an impact on the measurement results of track wheels wear. The geometry of the wheels under testing, of new manufacture for each of the tested haulage systems was measured with use of the 3D scanner, HDI Advance, LMI Technology. A digitalization of the objects was carried out with use of the structural light method. A comparison of measurements, taken before and after the tests, enabled to determine quantitative and qualitative wear of track wheels for both systems. The obtained result of qualitative wear of wheels is shown in Figure 5. It is from 30 to 45% bigger in the case of the Flextrack system, however the form of wear is very uniform, and it is caused mainly due to abrasion of the wheels flanks. In the case of the track wheels, used in the Eicotrack system, an irregular wear, in a form of chippings on the edges of teeth, can be seen.

Measurements of stresses in the selected components of the Flextrack and Eicotrack systems were realized with use of an indirect method through measurements of deformations in these components using light pipe sensors with a Bragg net. These sensors, as measurement transducers of small dimensions lacking in sensitivity to the external electromagnetic field, enable the carrying out of noninvasive measurements in many branches of industry. Due to their sensitivity to temperature and stress, Bragg nets are often used as measurement sensors of these physical quantities. Most often, the phenomenon of relocating the wavelength of the Bragg resonance due to an impact of temperature of stress is used. Only the knowledge of the Young's modulus value of the material, from which the components under measurement are made, is required for an assessment of stresses. The applied measurement method is described in detail in the publications [20–23]. The light pipe sensors with Bragg net SC-01 and the optical interrogator FBG Scan 800 were used for measurements.

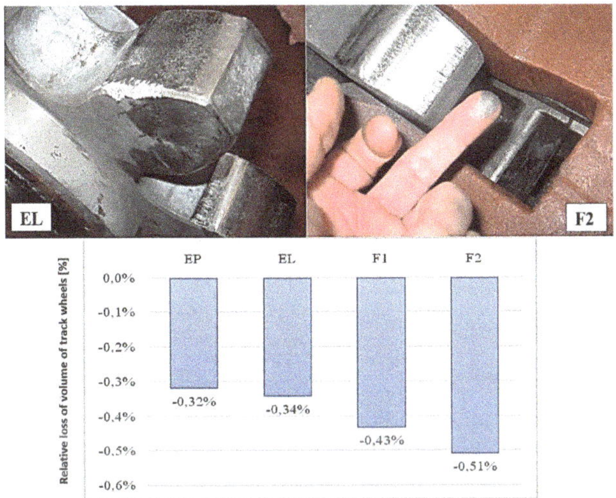

Figure 5. View of track wheel wear and graph of the relative volume loss of individual track wheels: EP and EL—right and left track wheel in the Eicotrack system, F1 and F2—right and left track wheel in the Flextrack system.

The measurements were taken for three elements: Flextrack—type 1, made of cast steel I; Flextrack—type 2, made of cast steel II and Eicotrack—the rack of the Eicotrack system. On the selected previously prepared components light pipe sensors were installed by their flooding. The sensors were placed in a vertical hole of 6 mm dimeter drilled in the selected components (Figure 6).

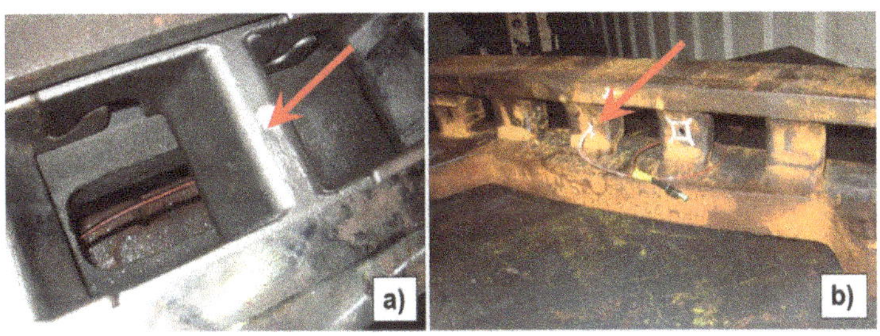

Figure 6. View of one of the toothed bar segments of the Flextrack system (**a**) and of the rack of the Eicotrack system (**b**), with a glued-in optoelectronic sensor with Bragg net for measurements of strains, (marked with red arrows).

The measurements were taken during a pass along a rectilinear conveyor section and along a deflected section both in the vertical plane and in the horizontal plane. A view of exemplary recorded courses for the toothed segments of Flextrack haulage system of type 1 and type 2 are shown in Figure 7. A deformation is caused in the first phase due to a compression of these components—a negative strain—but in the second phase when the wheel performs the proper work resulting from a transmission of the shearer haulage force to the toothed bar or the rack, it causes bending of the tooth or of the rack, i.e., a tension of the sensor (positive strain).

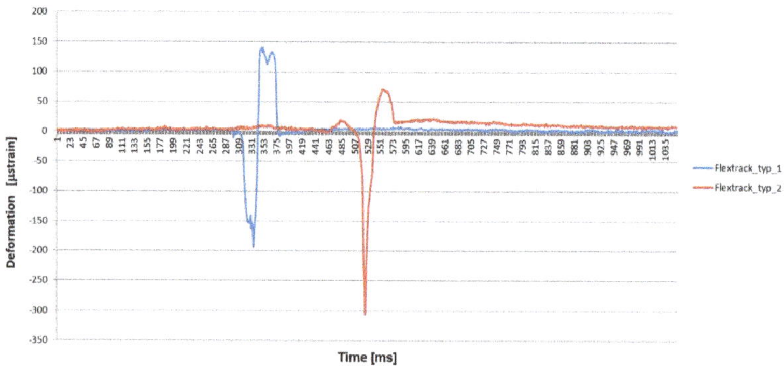

Figure 7. A course of the deformation value of the toothed segments of the Flextrack haulage system with the conveyor route curved in the horizontal plane.

Comparing the obtained results, it can be stated that a contraflexure of the conveyor line pan has an impact on increasing the values of strains and stresses, even several times greater, in particular in the case of the route bent in the vertical plane. The results obtained for the segments of the toothed bar—Flextrack type 1, made of cast steel I—were most advantageous. Not much bigger values, or in some cases comparable ones, were obtained in the case of the Eicotrack rack, whereas in the case of the toothed bar segments—Flextrack type 2, made of cast steel II—the obtained results are much higher in comparison with the toothed bar—Flextrack type 1—in relation to the configuration of the conveyor route, from 250% to even 550% (the route bent in the horizontal plane). It shows a big impact of the kind of material, from which the toothed bars were made, on the values of strains and stresses. The scope of measurement values of stresses is presented in Table 1.

Table 1. Scope of measured stress values in Flextrack and Eicotrack haulage system elements.

	Compressive Stresses (MPa)	Tensile Stresses (MPa)
Toothed segment—Flextrack—type 1	−2.2 to −7.2	4.3 to 7.2
Toothed segment—Flextrack—type 2	−6.5 to −55.6	6.7 to 29.6
Eicotrack rack	− 4.2 to −9.5	3.8 to 18.2

Analyzing all the obtained results of tests it is, however, possible to state that the innovative Flextrack system, subject to tests, is a more advantageous solution in comparison with the Eicotrack system used at present, despite the fact that an expected decrease in the track wheels' wear was not achieved in the new system. This might be caused by the fact that initial analyses concentrated on solving a problem of reducing contact stresses during a collaboration of the wheel with the toothed bar segment. To achieve this objective, the outline of teeth in the toothed bar was changed for a concave one. An analysis, enabling the determination of an impact of the second factor such as a slip connected with a collaboration of the pair: wheel–toothed bar, should be carried out again. A resignation from the meshing model of an involute outline could cause an increase of the slip share in the contact of the wheel tooth with the tooth of the toothed bar, which resulted in increased wear [24–26]. Comparative tests of noise measurements of the systems, subject to tests, confirm obtaining an effect of reducing contact stresses. However, the wear measurements indicate that a share of friction in the contact of the collaborating pair is significant, and the meshing requires an optimization. In the case of gears, it is difficult, and in practical life impossible, to optimize a collaboration of elements if one of them is subject to a modification. Thus, a conclusion was drawn that the following step should include an analysis and synthesis of the mesh pair based on the involute outline of the wheel tooth and an adaptation of the toothed bar to it, which in this case should have straight teeth or an involute outline.

Such tests and computer simulations were conducted, and their results are described in the publications [27,28].

3. Solution of the Komtrack Haulage System and Field Tests

The results of tests of the Flextrack system, conducted on the test rig, as well as the results of an impact analysis and the tooth outline of track wheel and of the flank of the toothed segment on a correctness of collaboration, were used in the design process of an innovative Komtrack haulage system, mainly of the toothed segment intended for an application in highly productive longwall systems of big power [28,29]. The toothed segment was designed with a bigger pitch of 151 mm and an innovative outline of the tooth flank. It reduced a number of segments installed in the guides to five. The 3D model of the guide and toothed segments, installed on the conveyor line pan, the 3D model of a new version of the Komtrack toothed segment and the model of collaboration of the shearer drive wheel with toothed segments are shown in Figure 8.

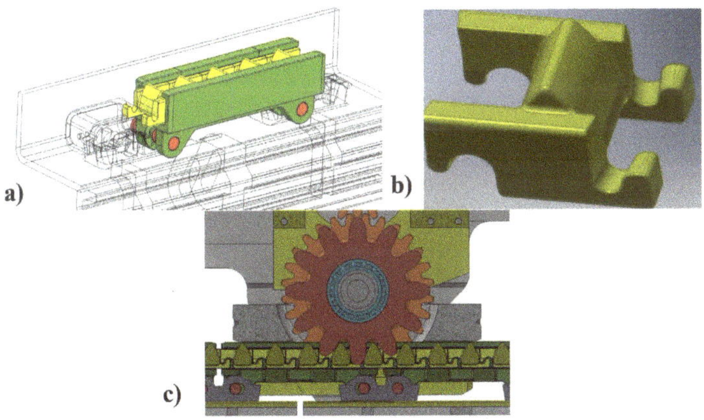

Figure 8. Komtrack haulage system: (**a**) 3D model of the guide and toothed segments installed on the conveyor line pan, (**b**) 3D model of a new version of the Komtrack toothed segments, (**c**) model of collaboration between the shearer drive wheel with toothed segments.

The developed Komtrack haulage system was subject to functional tests on the rig situated at the stacking yard of the Piast–Ziemowit, Mine Piast. Their objective was a verification of functionality of newly developed longwall shearer haulage system and of the wear degree of haulage system components, including toothed segments, on the field test rig simulating real conditions of loading the track wheel and of the toothed route. The prepared test rig of the length of about 50 m, with the installed Komtrack system, enabled a realization of the shearer passes under loading along a rectilinear route and deflected in the horizontal and vertical planes. A model of this rig is presented in Figure 9.

An armored face conveyor of the nominal length of line pans 1500 mm, in the number of 33 units, was mounted in the hardened base. A powered roof support unit was connected to every second line pan of the armored face conveyor to stabilize the conveyor position and enable its advance in a mechanized way, simulating contraflexures of the route in the horizontal plane.

The test rig enabled an assembly of the Komtrack haulage system guides in the existing seats on each line pan of the armored face conveyor. Toothed segments, blocked on both ends of the route with controlled hydraulic blocking, were placed inside the guides along the whole route length. The task of the blocking devices was an elimination of clearances occurring between toothed segments and a transmission of longitudinal loads generated by the longwall shearer moving along the armored face conveyor route.

Figure 9. Three-dimensional model and a view of the field test rig for conducting functional tests of the Komtrack haulage system: 1—coal shearer, 2—armored face conveyor, 3—Komtrack haulage system, 4—tensioning system of toothed segments, 5—hydraulic winch, 6—braking rope, 7—powered roof supports.

The KSW-1140E longwall shearer (made by FAMUR S.A.) with haulage units equipped with track wheels, collaborating with the toothed segments of the Komtrack haulage units and shoes adapted to a collaboration with the guides, were installed on the rig.

The KHT-6 hydraulic transport winch was used for loading the longwall shearer (simulation of load resistances resulting from cutting). The braking force was transmitted to the shearer frame with a steel rope. Due to an application of a driving pulley (pulley block), it was possible to obtain twofold higher values of forces loading the shearer in the result of braking the winch. A view of the rig, ready for conducting functional tests of the Komtrack haulage system, is shown in Figure 9.

4. Functional Verification of the Komtrack System

The elaborated test rig enabled to conduct functional tests of the Komtrack system on three trough alternatives of the armored face conveyer, simulating underground conditions, occurring during cutting the coal mass (Figure 10):

- "P" straight route reflecting a rectilinear route of the armored face conveyor;
- "S" route reflecting the route deflected in the horizontal plane by 800 mm simulating the conveyor advance;
- "G" route of the hill type reflecting the route deflected in the vertical plane by about 300 mm, which enables to obtain an angular contraflexure of 5°.

Figure 10. Three alternatives of situating the trough of the armored face conveyor simulating the conditions of underground operation: P—straight route of the conveyor, S—the route of the conveyor deflected in the horizontal plane by 800 mm, G—the route of the conveyor deflected in the vertical plane by about 300 mm.

The tests of the newly developed Komtrack haulage system were divided into three equal time periods, in which the tests were carried out on the straight route "P", on the route deflected in the horizontal plane "S" and on the route deflected in the vertical plane "G". On the route of type "P", 196 passes were made; on the route of type "S"—167 passes and on the route of type "G" altogether 172 passes were realized. In total, the Komtrack haulage system was subject to the loads resulting from the total number of longwall shearer passes reaching 535 passes. One pass consisted of covering the length of the conveyor one way and back, highlighting the fact than one way the shearer was slowed down with use of the winch and on the way back it returned at the neutral gear. Assuming the web depth of one cut on the level of 0.8 m, it was calculated that in the simulated conditions of the field test rig the shearer mined out the longwall of the panel length of nearly 430 m in the case of unidirectional cutting and twice the length in the case if bidirectional cutting.

During the conducted field tests, the following aspects were measured or subject to verification:

- Assembly technology of the haulage system;
- Collaboration correctness of the frictional pair: toothed track wheel—toothed segment;
- Strength of toothed segments;
- Operational correctness of tensioning system of toothed bars;
- Wear degree of individual components of the haulage system.

The realized work showed that the assembly and disassembly technology and an operation of the Komtrack system are simple by intuition, and they only require a short training of the miners' team in the scope of the system tensioning. A construction of guides of the haulage system enables their installation in the places, where the Eicotrack system racks have been installed up till now. An installation of toothed segments inside the guides results in getting a route flexible in all the planes, self-adjusting to a current position of the shearer track wheels. Conducted observations showed a correct character of operation of the frictional pair whose strength is not lower than in the case of the Flextrack system of comparable dimensions.

5. Verification of Wear Degree of Toothed Segments

The conducted functional tests enabled to determine the wear degree of the Komtrack toothed segments. The most visible form of the toothed segment wear was grinding of the tooth tip (Figure 11a), which occurred on the teeth of exceeded nominal height in the situation when a segment of the toothed bar was in the top zone of the guide. In the case of maintaining nominal overall dimensions, this phenomenon did not occur (Figure 11b).

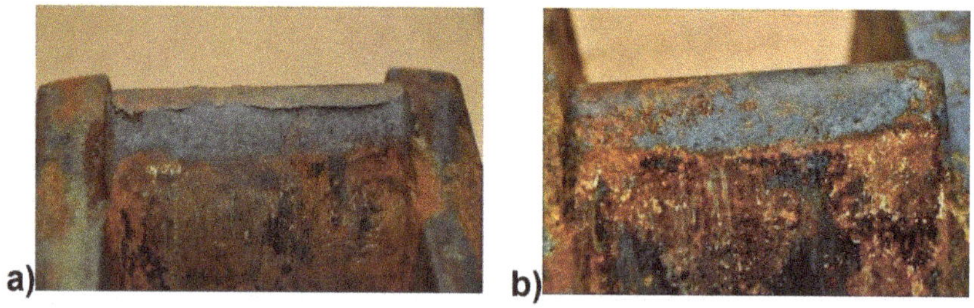

Figure 11. View of the top surface tooth of the Komtrack toothed segment: (**a**) damaged, (**b**) without any signs of damage.

The following form the toothed segment wear included indentations on the side walls of the keeps of these segments (Figure 12), caused by the pressure of the track wheel tooth in the situation of the toothed bar torsion in the horizontal plane. It was a proof of flexibility occurring in the haulage system, enabling a relocation of the toothed segments in relation to each other (tension). In extreme cases the above mentioned damages led to breaking off of keeping parts or resistance surfaces of toothed segments.

Figure 12. Damages of keeps of the Komtrack toothed segment, (marked with red arrows).

The most essential aspect, from the point of view of functioning of newly developed haulage system, was the wear degree of the tooth side surface of the toothed segment collaborating with the track wheel flank. Detailed digital data, indicating the wear degree of toothed segments, are presented in the further part of the article dedicated to a qualitative assessment of the toothed segments wear, based on the 3D scanning results. In the result of a visual inspection, it can be stated that the wear of the tooth flank is dependent on a position of the segments in the conveyor route. The toothed segment flank wear is distributed uniformly on its whole width (Figure 13a) in rectilinear conveyor section. In turn, in the places where the rectilinearity disturbances of the conveyor occur, the maximum flank wear moves towards the side wall of the toothed segment closer to the side wall (Figure 13b). Spherical load bearing surfaces of toothed segments, responsible for flexibility of the haulage system and maintaining its pitch, were also subject to a visual inspection. No forms of wear were identified on the above mentioned surfaces.

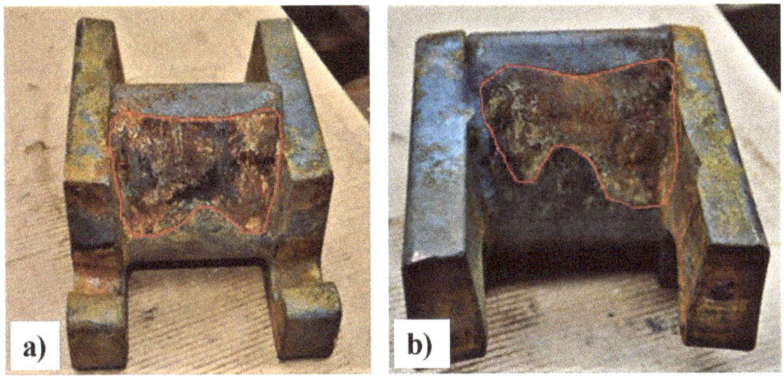

Figure 13. View of the wear of the tooth surface of the toothed segment, (marked with red frames). installed in the section: (**a**) rectilinear, (**b**) with disturbances of rectilinearity.

To assess the wear degree, the selected segments, before starting functional tests, were marked in a fast way and subjected to a process of weighing with accuracy of 0.5 g. Knowing the weight of individual specimen before starting the tests and after their ending, the wear degrees of individual segments were determined. They are presented in a form of weights differences and percentage wear. The results are presented in Table 2. In Figure 14, a graph of values measured before and after the tests of weights of individual toothed segments is shown.

Table 2. Measurement results of weights of individual toothed segments specimens.

Specimen	Weight before Test (g)	Weight after Test (g)	Difference of Weights (g)	Percentage Wear of Individual Toothed Segments (%)
K1	9578.0	9553	28.0	−0.26
K2	9441.5	9413	31.5	−0.30
K3	9394.0	9372	25.0	−0.23
K4	9608.0	9585	26.0	−0.24
K5	9358.0	9331	30.0	−0.29
K6	9685.5	9647.5	41.0	−0.39
K7	9589.5	9562.5	30.0	−0.28
K8	9422.5	9398.5	27.0	−0.25
Average	9509.63	9482.81	29.8125	−0.28

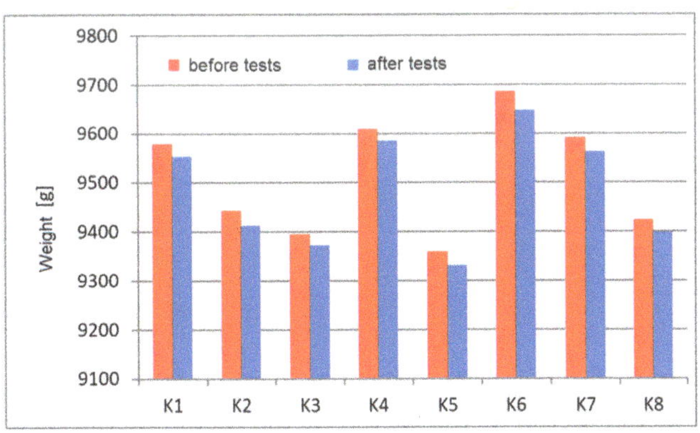

Figure 14. Weight value of individual specimens of toothed segments before and after functional tests.

An assessment of qualitative and quantitative wear degrees of the toothed segments was carried out with use of 3D scanning technology. Scanning was performed with use of the 3 DEINSCAN PRO 2 manual scanner made by the SHINING Company. Based on the 3D scans, comparative tests of the toothed segments surfaces, before and after functional tests, were realized. The measurements were taken with use of GOM Inspect 2019 software. Exemplary measurement results, in a form of the surface change deviation for selected segments, are shown in Figure 15. These segments are marked as:

- K1—toothed segment installed on the rectilinear route section before the contraflexure zone of the armored face conveyor;
- K3—toothed segment installed on the primary section of the conveyor contraflexure;
- K8—toothed segment installed on the rectilinear route section after the contraflexure zone of the armored face conveyor.

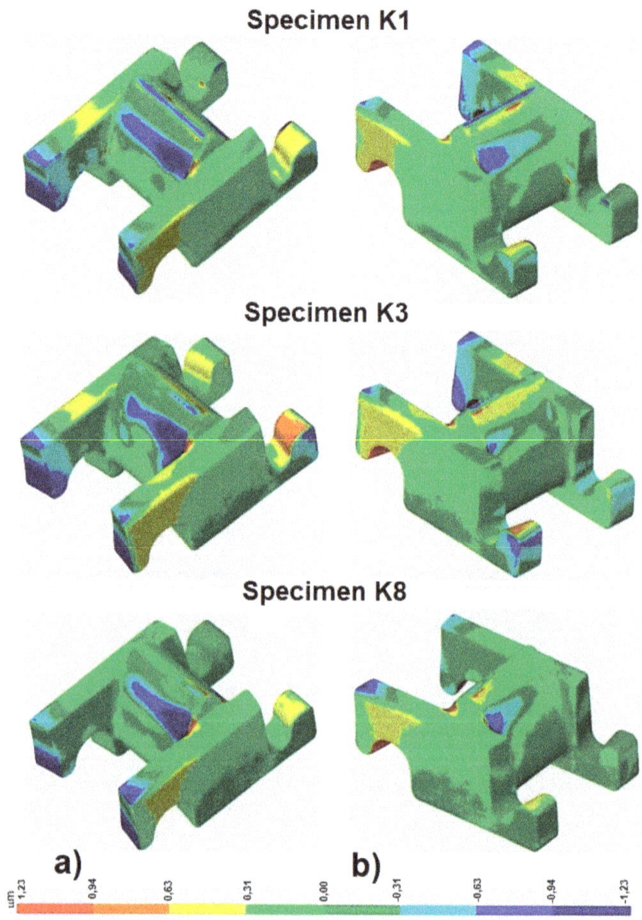

Figure 15. Measurements results of changes deviation of surfaces for selected segments, a comparison of scans of toothed segments models made before tests and after tests: (**a**) at loading caused by the shearer braking force, (**b**) at loading caused only by the shear pass.

The average deviation of surface changes before and after the tests for sample K1 was 0.1717 mm. The maximum upper deviation of the above mentioned specimen after tests, related to the scan made before tests, was 1.14 mm and for the lower deviation this value was −1.16 mm. Both flanks of the teeth collaborated asymmetrically with the track wheel teeth (i.e., bigger wear from the side of the coal mass). The loss is also visible in the tooth top part. Local plasticization of the material and relocations of compressed ends of catching keeps were identified as well.

For the K3 specimen, an average deviation of surface changes before and after tests was 0.2032 mm. Maximum upper deviation was 1.24 mm and the lower one −1.29 mm.

In turn, the average deviation of surface changes before and after tests for the K8 specimen was 0.1273 mm. Maximum upper deviation was 0.99 mm and the lower one −1.02 mm. Similarly, as for the K1 and K3 specimens, the increased wear on the tooth side collaboration during the shearer pass under load was stated.

In the case of all the specimens, maximum wear was identified on the tooth side corresponding to the shearer pass under load. On average, the deviation value between the reference model (before functional tests) and the models obtained in the result of 3D scanning of toothed segments after tests was 0.17 mm. Average values of deviations of model surfaces of all the toothed segments under analysis are shown in Figure 16.

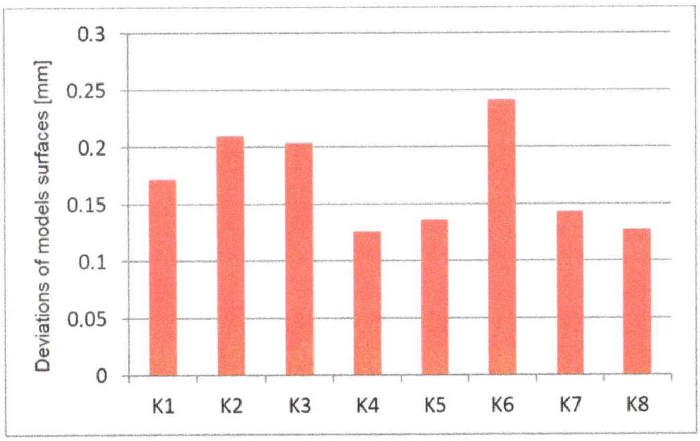

Figure 16. Value of the surface average deviation of the scanned model after tests in relation to the model before tests.

Analyzing the wear values of individual toothed segments, subject to functional tests in relation to their position on the conveyor route, it was spotted that the biggest wear occurred in the segments located in the zone of the armored face conveyor contraflexure. Based on the elaborated maps of deviations of surfaces of individual toothed segments, the following relationships were noticed:

- Visible increased flank wear of the toothed segment from the side of loading the shearer;
- Toothed segments characterized by positive tolerances of manufacture have damages in the tooth top part;
- Bigger wear of a part tooth flank of the tooted segment was stated from the side of the coal mass.

6. Verification of the State of Stresses in Toothed Segments

Within the framework of tests, also unique tests, oriented to a determination of the state of stress of the toothed segments due to a contact with the track wheel, were conducted. These tests were carried out with use of electric resistance strain gauges. A dozen or so special toothed segments, on which two, four and six strain gauges (Figure 17) were glued, were prepared. The strain gauges were glued in the pockets, also including the contact area of the toothed segment flank with the track wheel flank enabling to avoid their damage during a collaboration with the guide and the track wheel.

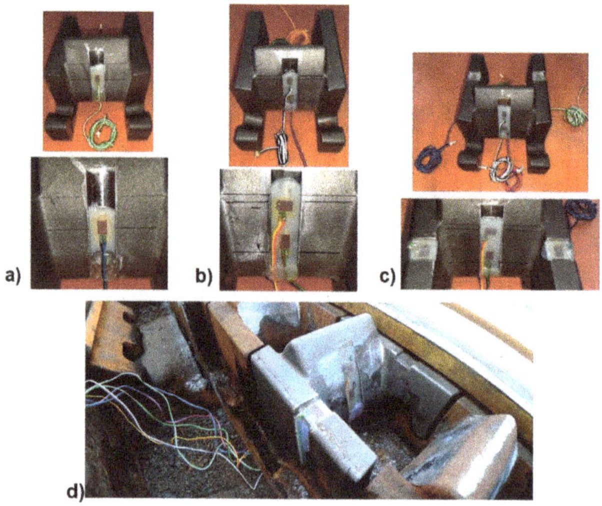

Figure 17. View of toothed segments prepared for tests with use of electric resistance strain gauges with glued strain gauges in the number: (**a**) two, (**b**) four, (**c**) six and (**d**) view of one segment with glued strain gauges installed in the guide.

After having installed toothed segments in the guides, the following longwall shearer passes were performed, recording current signals as reactions from the individual toothed segments. Collected data enabled to determine stresses in the place, where a strain gauge was glued in. In parallel with the strain gauge tests, the FEM numerical simulations, reflecting the conditions of the conducted strain gauge tests, were carried out. The exemplary result of conducting strain gauge tests and the results of the FEM simulations corresponding to them are shown in Figures 18 and 19.

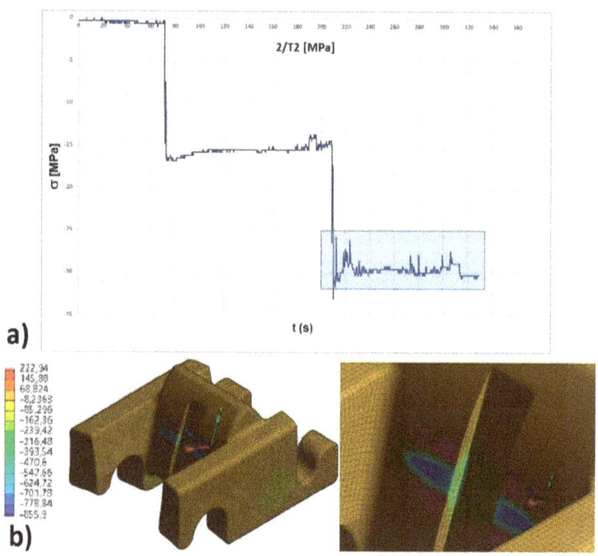

Figure 18. Values of main stresses in the contact place of the track wheel with the toothed segment with two glued strain gauges: (**a**) result of conducting strain gauge tests, (**b**) result of the FEM simulation.

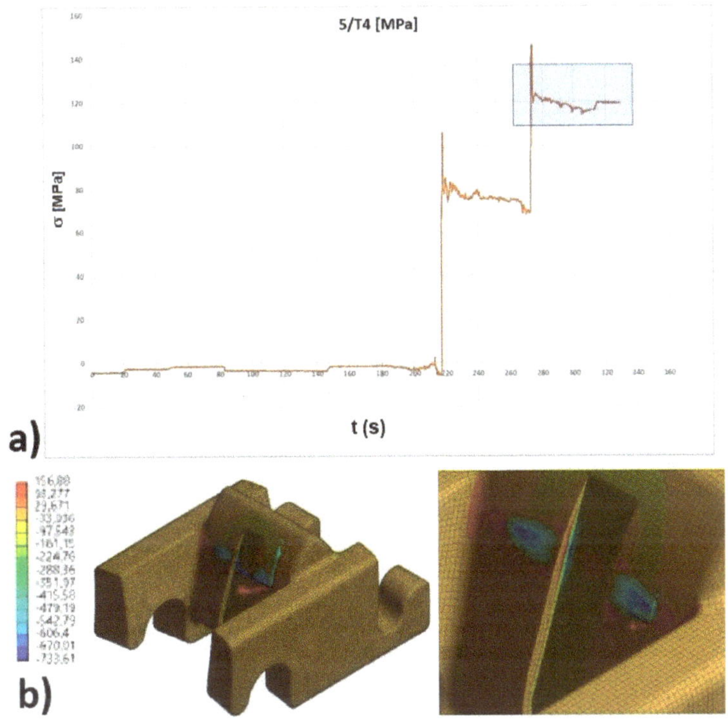

Figure 19. Values of main stresses in the contact place of the track wheel with the toothed segment with four glued strain gauges: (**a**) result of conducting strain gauge tests, (**b**) result of the FEM simulation.

In the case of the toothed segment with two glued strain gauges, maximum values of main stresses were in the range of 20–25 MPa, whereas the equivalent FEM simulations in this case showed the stresses on the level of 30 MPa. In turn, the values of stresses for the toothed segment with four glued strain gauges varied in the range of 115–120 MPa, whereas the values determined in the FEM simulations reached 130 MPa.

7. Modernization of Design Form of the Komtrack Toothed Segments

Based on the analysis of obtained results of the field tests, identified configurations of the toothed segments location inside the guide and their collaboration with the track wheels, some conclusions concerning a modification of the design form of the Komtrack toothed segment were formulated. The most essential ones are as follows:

1. Load-resistant and catching surfaces of the toothed segments, entering into a collision with the teeth of asymmetrically operating track wheel, are subject to damage (they break off), which in extreme cases can lead to a significant disturbance to the haulage system pitch.
2. Top surfaces of the tooth of the toothed segment, in relation to current position and tolerance of manufacture, entering into collision with a relocating shoe of the shearer, are subject to frictional wear; they introduce unwelcome resistances to motion and may cause a generation of disturbances in the haulage system pitch.
3. During slackening of the toothed route, the elements of keeps at the length of about ten meters enable pulling of the following toothed segments. A stroke of the cylinder of the tensioning system makes it impossible to eliminate backlash occurring among individual toothed segments along the whole route. To slacken the route completely, it is indispensable to move individual toothed segments manually in relation to the installed guides.

4. In the result of conducted tests and operational observations, as with regard to the Komtrack toothed segments, it was decided to modernize them through a removal of keeping parts (catches) due to the load-resistant surfaces gaining on the whole segment's height. Additionally, load-resistant surfaces, symmetrical in relation to the tooth axis, are subject to a collision with the track wheel teeth to a minimal extent. A reduction in the tooth height and an introduction of additional chamfering around the load-resistant surfaces decrease possibilities of abrasive wear of the toothed segment. The 3D model of the toothed segment, which was subject to field tests, and its modernized design are shown in Figure 20.

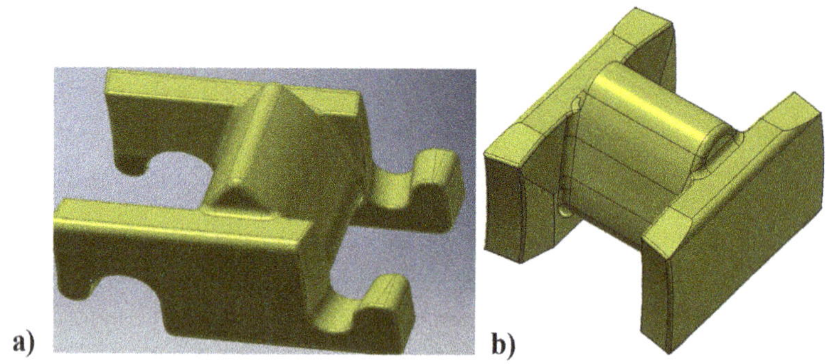

Figure 20. Three-dimensional model of the Komtrack toothed segment: (**a**) used in the field tests, (**b**) in the modernized version.

8. Summary

In the KOMTRACK project, realized by the authors, one of the stages included a verification of operational correctness of newly developed haulage system of highly productive longwall systems based on field tests. The tests were realized on the test rig, elaborated at the KOMAG Institute and constructed by the PGG S.A. Company. The test rig was located in the stacking yard of Piast–Ziemowit, Mine Piast. It enabled a simulation of operational conditions of a longwall system in underground conditions with regard to the equipment, loading and parameters characterizing the armored face conveyor with a chainless haulage system.

The results of tests and their analysis enabled, among others, answering the question of if a disadvantageous phenomenon of edging of track wheel tooth occurs in the newly developed haulage system. An analysis of films and photos, as well as of observations enables us to state that the developed Komtrack haulage system is characterized by flexibility resulting from design features of toothed segments. An analysis of the film material, in particular in the places, where disturbances of the route rectilinearity occur, indicated that a self-adjustment of collaborative surfaces of the frictional pair occurs in the result of a track wheel's single tooth acting on the tooth surface of the toothed segment. Due to this phenomenon, no durable, repeatable damage on the surface and on the edges of the track wheel occur.

Identified wear concerns both toothed segments, as well as the guide surfaces. The conducted analyses indicate that an average wear of toothed segments, in the results of tests, is smaller than differences in the tolerances of manufacture of new toothed segments. Thus, it can be concluded that the wear of individual toothed segments does not have any significant impact on functionality of the haulage system.

During a realization of tests, a few breakdowns were experienced, mainly regarding the shearer supply, but they did not have any negative impact on the scope and rate of conducting tests. Apart from some damage to the catching parts of toothed segments, the tests ran efficiently without any perturbations in the shearer's operation.

In the result of conducted functional tests, a modified design of toothed segments and guides of the Komtrack system were elaborated. These changes should increase, to a more significant extent, the operational reliability of the developed Komtrack system.

In the next stage of work in the KOMTRACK project, the newly developed haulage system was subjected to comparative tests with the Eicotrack system in in situ conditions in a longwall located in the Piast–Ziemowit mine. It was the last testing stage before preparing the haulage system for implementation on an industrial scale. The view of the shearer during the underground tests with the Komtrack haulage system is shown in Figure 21. The preliminary results of the tests showed that the shearer with the Komtrack system moves more smoothly along the conveyor route, with 2–3 times lower value of the vibration amplitude. In addition, thermal analyses have shown over 30% less heating of the cooperating system components.

Figure 21. View of the shearer during the underground tests with the Komtrack haulage system in Piast–Ziemowit coal mine.

On the other hand, in the case of wear of the cooperating components, very slight and even wear of the gang wheels and toothed segments was observed. In the case of the Eicotrack system, the wear of the cooperating components was higher and, especially in the case of rack pins, very uneven. The advantage of the Komtrack system is also lower by about 35% of volumetric unit consumption (cm^3/10,000 Mg of coal extracted from the longwall). A complete analysis of the results will be made after the end of the underground tests. Their completion is estimated in the second quarter of 2023. However, preliminary results confirm that the Komtrack shearer haulage system with modernized toothed segments can be successfully used in mechanized shearer longwall systems in coal mines. The obtained results of underground research will also be published and will be a continuation of and supplement to the article in question, in addition to the article that has already been published.

Author Contributions: Conceptualization, writing—original draft preparation, writing—review and editing, supervision, M.K., A.N.W. and K.K.; methodology, software, validation, formal analysis, visualization, K.K., G.S. and M.K.; investigation, data curation, resources, G.S., M.S. and D.B.; project administration and funding acquisition, M.K., A.N.W. and K.K. All authors have read and agreed to the published version of the manuscript.

Funding: This article was written based on the results of the tests carried out within the research project entitled New generation high-performance haulage system of the longwall system co-financed by the European Regional Development Fund (Contract no. POIR.04.01.04-00-0068/17).

Data Availability Statement: The data is not publicly available until the end of the project Contract no. POIR.04.01.04-00-0068/17.

Conflicts of Interest: The authors declare no conflict of interest.

References

1. Sikora, W.; Dolipski, M.; Osadnik, J.; Remiorz, E. Analiza geometryczna wybranych elementów układu ciągnienia 2BP. *Zesz. Nauk. Górnictwo/Politech. Śląska* **2005**, *269*, 543–551. (In Polish)
2. Korski, J. Capacity Losses Factors of Fully Mechanized Longwall Complexes. *Min. Mach.* **2020**, *3*, 12–21.
3. Fries, J. Rozwój bezcięgnowych systemów posuwu kombajnów ścianowych. *Uhli Rudy Geol. Pruzk.* **2003**, *10*, 22–24.
4. Gondek, H.; Marasova, D.; Fries, J. Możliwości zwiększenia trwałości zazębienia koła napędowego z zębatką mechanizmu bezcięgnowego posuwu kombajnu ścianowego. In Proceedings of the TEMAG 2000 "Trwałość elementów i węzłów konstrukcyjnych maszyn górniczych", VIII Konferencja Naukowo-Techniczna, Gliwice—Ustroń, Poland, 25–27 October 2000; pp. 73–80. (In Polish).
5. Matsui, K.; Shimada, H. Control of stability of retreat longwall gate road. *Min. Sci. Technol.* **1996**, *7*, 709–713.
6. Olson, J.J.; Tandanand, S. *Mechanized Longwall Mining: A Review Emphasizing Information Circular 8740: Foreign Technology;* United States Department of the Interior—Bureau of Mines: Washington, DC, USA, 1977.
7. Korski, J. Longwall shearer haulage systems—A historical review. Part 2—First cordless haulage systems solutions. *Min. Mach.* **2021**, *39*, 24–33.
8. Fennelly, S.D. *Chainless Haulage Systems for Power Loaders: Information Circular as of April 1978*; United States Department of Energy: Washington, DC, USA, 1978.
9. Braun, G. Chain Having Longitudal Stiffness Used with Hauling and Cutting Equipment. U.S. Patent 4,372,619, 8 February 1983.
10. Korski, J. Longwall shearer haulage systems—A historical review. Part 3—Chainless haulage systems with drive wheel and rack bar. *Min. Mach.* **2021**, *39*, 58–69.
11. Mackie, K.; Fimemme, T.E. Design & Development of Longwall Shearers for Overseas Application. *Adv. Min. Sci. Technol.* **1987**, *1*, 169–188.
12. Langefeld, O.; Paschedag, U. Longwall Mining—Development and Transfer. Mining Report—Glückauf. 1/2019, vol. 155. Available online: https://mining-report.de/wp-content/uploads/_pda/2019/02/MRG_1901_Longwall_Mining_Langefeld_Paschedag_190201.pdf (accessed on 17 October 2022).
13. Sikora, W.; Giza, T.; Mann, R. Wpływ przegięcia trasy przenośnika zgrzebłowego na współpracę koła napędowego z zębatką bezcięgnowego systemu posuwu typu BP. In Proceedings of the TUR 2003, III Międzynarodowa Konferencja Techniki Urabiania 2003, Kraków—Krynica, Poland, 23–26 September 2003; pp. 151–159. (In Polish)
14. Giza, T.; Mann, R. Wpływ zużycia zębów koła napędowego i zmiany podziałki zębatki na charakter zazębienia i prędkość posuwu kombajnu. *Zesz. Nauk. Politech. Śląskiej Górnictwo* **2005**, *269*, 407–416. (In Polish)
15. Piekło, J.; Maj, M.; Żuczek, R.; Pysz, S. Assessment of Durability of the Toothed Segment Based on FEM Analysis and Low Fatigue Cycle Test. *Arch. Foundry Eng.* **2016**, *16*, 113–118. [CrossRef]
16. Giza, T.; Sobota, P.; Mann, R. Wpływ zazębienia koła napędowego z zębatką sworzniową mechanizmu posuwu BP na parametry posuwu kombajnu ścianowego. In Proceedings of the Konferencja KOMTECH "Nowoczesne, Niezawodne i Bezpieczne Systemy Mechaniczne w Świetle Wymagań Unii Europejskiej", Szczyrk, Poland, 17–19 November 2003; pp. 189–197. (In Polish)
17. Twardoch, K.; Grzesica, P.; Nawrat, A. Analiza współpracy koła napędowego z zębatką o owalnym przekroju sworzni w mechanizmie posuwu EICOTRACK (2BP). In Proceedings of the TEMAG 2016, XXIV Międzynarodowa Konferencja Naukowo-Techniczna "Trwałość Elementów i Węzłów Konstrukcyjnych Maszyn Górniczych", Ustroń, Poland, 3–5 November 2016; pp. 291–304. (In Polish)
18. Kotwica, K.; Stopka, G.; Kalita, M.; Bałaga, D.; Siegmund, M. Impact of Geometry of Toothed Segments of the Innovative KOMTRACK Longwall Shearer Haulage System on Load and Slip during the Travel of a Track Wheel. *Energies* **2021**, *14*, 2720. [CrossRef]
19. Zachura, A.; Żuczek, R. *System Posuwu Flextrack Zwiększający Trwałość i Niezawodność Napędu Kombajnu Ścianowego. W: Innowacyjne Techniki i Technologie dla Górnictwa. Bezpieczeństwo—Efektywność—Niezawodność. KOMTECH 2014*; ITG KOMAG: Gliwice, Poland, 2014; pp. 151–161. (In Polish)
20. Bragg, W.H.; Bragg, W.L. The reflection of X-rays by crystals. *Proc. R. Soc. A* **1993**, *88*, 428–438.
21. Chehura, E.; James, S.W.; Tatam, R.P. Temperature and strain discrimination using a single tilted fibre Bragg grating. *Opt. Commun.* **2007**, *275*, 344–347. [CrossRef]
22. Hill, K.O.; Fujii, Y.; Johnson, D.C.; Kawasaki, B.S. Photosensitivity in optical fiber waveguides: Application to reflection fiber fabrication. *Appl. Phys.* **1978**, *32*, 647–649. [CrossRef]
23. Kisała, P. Metrological conditions of strain measurement optoelectronic method by the use of fibre Bragg gratings. *Metrol. Meas. Syst.* **2012**, *19*, 471–480. [CrossRef]
24. Si, L.; Wang, Z.B.; Tan, C.; Liu, X.H.; Xu, X.H. A feature extraction method for shearer cutting pattern recognition based on improved local mean decomposition and multi-scale fuzzy entropy. *Curr. Sci.* **2017**, *112*, 2243–2252. [CrossRef]
25. Linnik, Y.N.; Linnik, V.Y.; Zhabin, A.B.; Polyakov, A.V. Theoretical framework for the efficiency evaluation of coal mining machines. *Eurasian Min.* **2020**, *1*, 61–64. [CrossRef]
26. Morshedlou, A.; Dehghani, H.; Hoseinie, S.H. A data driven decision making approach for long-wall mining production enhancement. *Min. Sci.* **2019**, *26*, 7–20. [CrossRef]

27. Kotwica, K.; Stopka, G.; Gospodarczyk, P. Simulation tests of new solution of the longwall shearer haulage system. In Proceedings of the Scientific and Technical Conference on Innovative Mining Technologies (IMTech) 2019, Szczyrk, Poland, 14–16 October 2019. Part 2.
28. Kalita, M. Designing process of a toothed segment of the KOMTRACK haulage system. *New Trends Prod. Eng.* **2019**, *2*, 121–129. [CrossRef]
29. Kalita, M.; Mazurkiewicz, A.; Pieczora, E.; Tarkowski, A. *KOMTRACK—Nowej Generacji System Posuwu Wysokowydajnych Kompleksów Ścianowych—Wstępne Prace Projektowe Mechanizacja, Automatyzacja i Robotyzacja w Górnictwie*; Monografia; Wydawnictwa AGH: Kraków, Poland, 2020; pp. 23–31. (In Polish)

energies

MDPI

Article

Sorption–Dilatometric Properties of Coal from a High-Methane Mine in a CO_2 and CH_4 Atmosphere

Paweł Baran [1,*], Stanisław Kozioł [1], Katarzyna Czerw [1], Adam Smoliński [2] and Katarzyna Zarębska [1]

1 Faculty of Energy and Fuels, AGH University of Science and Technology, Al. Mickiewicza 30, 30-059 Cracow, Poland
2 Department of Energy Saving and Air Protection, Central Mining Institute, Pl. Gwarków 1, 40-166 Katowice, Poland
* Correspondence: baranp@agh.edu.pl; Tel.: +48-12-617-20-79

Abstract: Although highly developed countries are trying to diversify away from coal-based energy, many economies rely on this resource. Its consumption results in the production of carbon dioxide, which promotes global warming, necessitating its sequestration. This paper presents the sorption–dilatometric relationships of hard coal samples differing in vitrinite and inertinite content. The studies were carried out under isothermal conditions (298 K) at a free pressure drop complemented by measurements under non-isothermal conditions (298 K to 323 K). The tests were performed on an original apparatus, based on the operation of an Arduino microcontroller. For the natural porosity to be preserved and for a better representation of the behaviour of the coal–gas system, samples in the form of cuboidal blocks were used, making this apparatus unique worldwide. Based on the study, it appears that the difference in petrographic composition affects the behaviour of the coal structure, influencing differences in the sorption–dilatometric properties. In the case of the sample with higher vitrinite content, the amount of adsorbed gases is higher.

Keywords: coal; methane; carbon dioxide; sorption; swelling

Citation: Baran, P.; Kozioł, S.; Czerw, K.; Smoliński, A.; Zarębska, K. Sorption–Dilatometric Properties of Coal from a High-Methane Mine in a CO_2 and CH_4 Atmosphere. *Energies* **2023**, *16*, 1785. https://doi.org/10.3390/en16041785

Academic Editor: Krzysztof Skrzypkowski

Received: 28 December 2022
Revised: 4 February 2023
Accepted: 8 February 2023
Published: 10 February 2023

1. Introduction

Despite the transition away from fossil fuels, hard coal is still in the top three fuels in the world's energy industry [1,2]. With increasing urbanization and population, the demand for coal is still high. As a result, there is a continuous increase in anthropogenic carbon dioxide emission, which is one of the causes of global warming. Therefore, measures are being taken to reduce emissions of this gas, with particular emphasis on point source emitters [3]. In addition, 195 countries have accepted the so-called Paris Agreement. By doing so, they have committed to reducing greenhouse gas emissions, with a 2 °C reduction in temperature relative to the pre-industrial era as one of the main goals. In addition, more than 20 countries around the world have pledged to achieve a mission of net-zero emissions by investing in renewable energy and other technologies with reduced carbon dioxide emissions [4,5]. In terms of the method of carbon dioxide capture, post-combustion technology, pre-combustion technology, and oxygen combustion technology can be distinguished [6]. One method is geological sequestration [7]. Underground sites with storage potential include oil fields, gas fields, and subeconomic coal seams. In this case, 65 commercial CCS facilities are operating worldwide [8]. Among them, 26 are operating, 13 are in an advanced stage of development reaching front-end engineering design, 21 installations are at very early development stages, and 2 have suspended operations (mainly due to the economic downturn). In 2020, worldwide commercial facilities captured and stored ~40 Mt CO_2 per year.

Due to the global tendency to increase the diversification of energy sources, coalbed methane (CBM) has become an area of scientific interest. Its greatest advantages include its calorific value and its high quality as an energy carrier [9]. An additional advantage is the use of this gas as an input product in the production of hydrogen, which is often referred to as the fuel of the future. Apart from natural geological traps, which prevent methane from reaching the surface, methane is stored in coal beds inside the coal matrix due to the adsorption mechanism. To extract methane trapped in geological traps, the permeability of rock formations must be altered. The most common solution is the process of hydraulic fracturing. This process involves injecting a fracturing fluid, consisting of water, sand, and other chemical additives into a wellbore. As a result, natural fractures within the gas-bearing horizon are widened and new ones are created, allowing gas to escape through gaps in the bed [10]. To release the methane adsorbed inside the coal matrix, a series of physicochemical processes must be used. Hard coal, as a naturally porous material, has a system of micro- and macropores inside its structure, forming a biporous system [11]. Bituminous coal is composed of organic matter mixed with inorganic matter, and its elemental and petrographic composition may vary significantly. For this reason, understanding the process of methane adsorption on hard coal has a significant impact on subsequent attempts to extract methane gas from subeconomic hard coal seams. Some studies have suggested that the sorption capacity of coals is influenced by parameters such as the degree of coalification, the content of maceral moisture, rank and mineral content, and elemental composition [12–16]. When analysing the differences in methane and carbon dioxide adsorption on carbon of different origins, the most important parameter is the vitrinite reflectance (R_0). As R_0 increases, the sorption capacity towards CO_2 increases [17]. Jian et al. have shown in analyses of hard coal samples with high R_0 that changes in sorption capacity towards CO_2 are very small. Conversely, as the R_0 value decreases, changes in the sorption capacity of coal towards methane increase. Differences in the sorption capacity of hard coal towards CH_4 are also related to the maceral composition. Research has proven that coal with high content of vitrinite is characterised by higher sorption capacity than coal with higher content of inertinite [18,19]). The reason for such behaviour is a difference in the structure of particular macerals. Vitrinite shows a predominance of micropores, which therefore have the highest total volume, inertinite shows a predominance of mesopores, and liptinite is the least porous maceral [20]. Laxminarayana and co-workers [21] analysed coal samples from Australian mines, differing in the content of individual macerals. Vitrinite-rich coals were characterised by a fast methane desorption process. Coals with high inertinite content were characterised by slower sorption. Moreover, with increasing rank for inertinite-rich coals, the diffusion rate decreased. Similar observations occurred when vitrinite-rich coals were examined. In this case, the changes were not as pronounced. However, the trend was similar. When analysing the changes in the structure of hard coals accompanying increases in rank, one can observe a decrease in the number of meso- and macropores, which translates directly into a decrease in the rate of methane diffusion in the coal matrix [21–23].

By combining the process of geological sequestration of carbon dioxide with the process of extraction of methane adsorbed in the coal matrix, the ECBM process was developed [24]. It exploits the phenomenon of preferential sorption and pore occupation in the coal structure by carbon dioxide relative to methane, promoting the simultaneous desorption of methane to enable its extraction. Thus, this process is an alternative to traditional methane extraction and allows for the geological sequestration of large amounts of CO_2 in off-balance coal seams [25]. Laboratory studies conducted on cuboidal samples of hard coal show that for coals of similar rank, the sorption kinetics of methane and carbon dioxide have a similar relative rate, while the sorption capacity of CO_2 is twice as high as for methane [26]. One of the reasons for this behaviour may be the difference in molecule diameter. Carbon dioxide is characterised by a linear and centrosymmetric molecule with sp-hybridisation of central atom. Methane, on the other hand, is characterised by a tetrahedral structure with sp3 hybridisation, which makes it occupy more space than CO_2. Additionally, there is an observable difference between the diameters of the two molecules. Methane has a kinetic diameter of 380 pm, while carbon dioxide has a kinetic diameter of 330 pm. Therefore, some pores with small diameters inside the coal matrix can be preferentially occupied by carbon dioxide molecules. A second reason for the preferential sorption of CO_2 is the characteristic interaction of this gas with the coal matrix, which allows the gas to dissolve inside the matrix [27].

The process of sorption of carbon dioxide and methane on hard coal is linked with the process of swelling/shrinking of the coal matrix. This phenomenon can cause the coal to crack, changes the durability, and thus releases stored CO_2. Research carried out on samples in the form of solid, cuboid blocks [28] as well as cylindrical samples [29] has shown that as the number of adsorbed vapours and gases increases, the linear dimensions of the samples change for both methane and carbon dioxide. The magnitude of deformation, however, varies depending on the elemental and maceral composition of hard coal. Hard coal, characterised by a higher vitrinite content, undergoes greater deformation than samples with a higher inertinite content. The cited publications focused mainly on the study of the effect of gases such as methane or carbon dioxide on the behaviour of coal material, with an emphasis on the elemental composition of coal. It should be remembered, however, that the content of individual macerals may vary significantly despite a very similar elemental composition.

When analysing the physical basis of the sorption process, the influence of temperature must be borne in mind. The coal bed is exposed to certain temperature fluctuations, which translate into changes in the behaviour of the coal–gas system. For this reason, performing studies on the sorption of vapours and gases under isothermal conditions is insufficient, and it is necessary to carry out studies under changing temperature conditions. As the temperature increases, some of the vapours and gases may be desorbed [30], which may pose certain risks in the process of the geological storage of carbon dioxide. Some studies have shown that the opposite is true. As the temperature increases by 1 K, a 2% increase in the rate of sorption kinetics is observed (e.g., Wang et al. [31]). Due to the lack of data in the literature, this article discusses the sorption-induced swelling properties of hard coal in a methane and carbon dioxide atmosphere.

The vast majority of mines in the Upper Silesian Coal Basin are mined under conditions of high methane concentration [32]. Due to the fact that Polish hard coal is usually mined from a depth of 700–1000 m, the seams characterized by one of the highest methane-bearing capacities are mined. The coal used for the research described in this article comes from one of the most methane-affected mines [33]. It produces high-quality coking coal, a critical raw material, and the company that owns the mine is the largest producer in the European Union. Over the past 20 years, in Polish mines, there have been three significant disasters related to methane outbursts at the mine. In the first, during the tunnelling of a pipe lunette at the level of 1000 m, an additional methane and rock outburst occurred after setting off explosive charges. The concentration of methane was 84% at the face of the wall. Shortly after this disaster, another incident occurred. During the ventilation of one of the pit walls, the methane in the goaf ignited and then exploded. As a result, methane and burning gases ended up in the working space of the longwall [34]. The last catastrophe took place in April 2022. There was a series of three explosions that killed nine miners and rescuers, and seven people were trapped at a depth of 1000 m.

The studies were carried out on samples of similar elemental composition but different petrographic composition, where the main difference was the vitrinite-to-inertinite ratio. In the first part, the properties of hard coal were studied under isothermal conditions, with a free pressure drop caused by sorption processes. The second part of the experiment was carried out under non-isothermal conditions, where the temperature change was 25 °C.

2. Materials and Methods

Coal samples from the Upper Silesian Coal Basin were used. Both samples were obtained from the same coal mine. However, they came from two different seams that differed in depth. This is important because the potential CBM seams are not homogeneous and differ in their properties throughout their occurrence and depth. The elemental and petrographic composition are presented in Table 1. The ultimate and proximate analysis was performed in the Central Mining Institute in Katowice.

Table 1. Specification of the tested coal samples.

Sample	C^{daf}	S^{daf}	H^{daf}	N^{daf}	O^{daf}	W^{a}	A^{a}	V^{daf}	Vitrinite	Liptinite	Inertinite
	(%)	(%)	(%)	(%)	(%)	(%)	(%)	(%)	(%)	(%)	(%)
P1	84.24	0.39	4.58	1.52	4.58	1.75	3.01	27.12	73	7	20
P2	84.96	0.58	4.60	1.70	3.76	0.68	3.78	25.50	53	8	39

The elemental analysis was performed in the Central Mining Institute in Katowice. The moisture content was determined in accordance with the procedure set forth in the standard PN-80/G-04511, while ash content was established in accordance with PN-80/G-04512. The oxygen content was computed as the remainder of 100%, taking into account the moisture and ash content. W: moisture content, A: ash content, V: volatile matter content, C, H, N, O, S: content of element C, H, N, O, S, respectively, a: analytical basis, daf: dry–ash–free basis.

As shown in Table 1, the main difference between the two samples is the result of the petrographic analysis, specifically vitrinite and inertinite content. A strength of the conducted research is the use of solid hard coal samples in the shape of rectangular blocks. Studies in the literature on sorption phenomena of hard coal describe tests on samples in powder and grain form. The use of cuboidal coal blocks preserves the natural porosity and enables strain measurements to be performed. To show the hard coal samples studied under magnification, SEM images are shown in Figures 1 and 2. They show differences in grain size and external surface texture. The apparatus used in the study has been constructed for this experiment so that sorption and dilatometric changes can be measured simultaneously. It is possible to record changes in the linear dimensions of a sample in a given time interval. This type of measurement is based on changes in the resistance of strain gauges placed perpendicularly and parallel to the layering of the coal sample. In the first stage of the experiment, sorption capacity, together with linear deformations, was measured under isothermal conditions at a temperature of 298 K. In the second part of the experiment, a manostat was introduced into the system. The modification of the measuring system made it possible to determine the dilatometric changes of the sample under constant pressure. The temperature at which the measurement was carried out was in the range of 298–323 K.

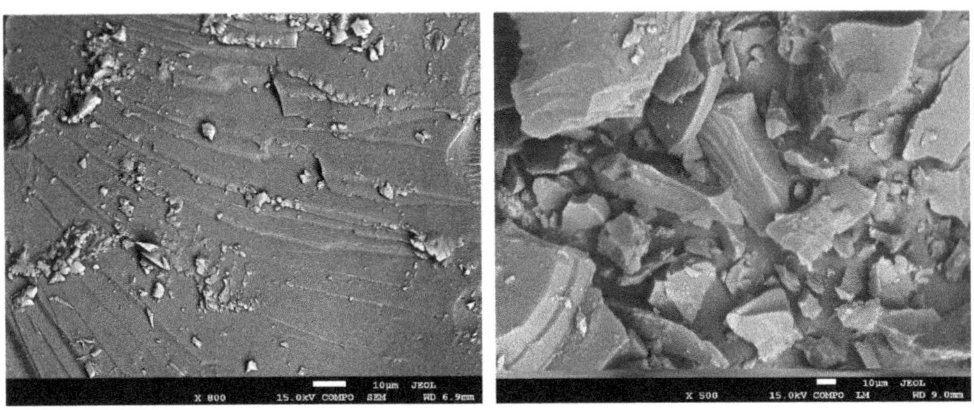

Figure 1. SEM images of the surface of sample P1.

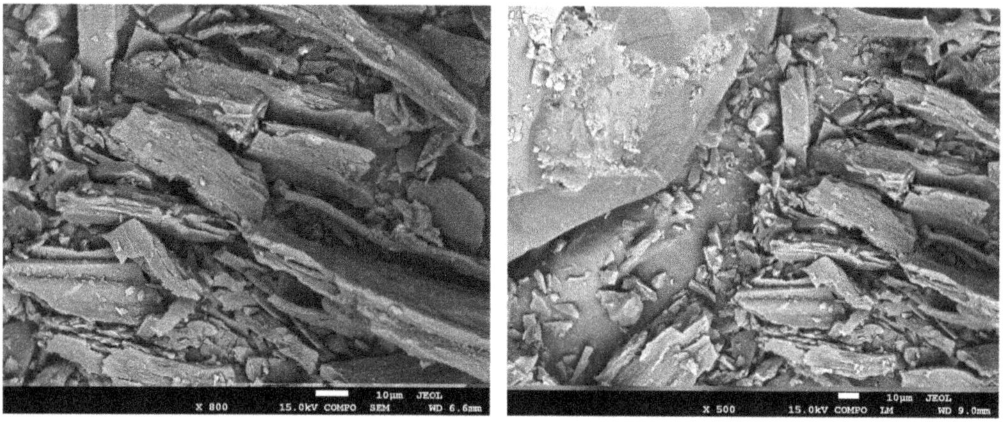

Figure 2. SEM images of the surface of sample P2.

3. Test Procedure

The measurements were made on a unique apparatus based on an Arduino microcontroller, which allows multiple parameters such as strain, pressure changes, and temperature measurements to be recorded simultaneously. It was built by the authors at the AGH University of Science and Technology in Kraków, Poland, and the proposed construction makes it adaptable to modifications and further improvements. In addition, a very wide library of software and the number of additional sensors allow the apparatus to be precisely adapted to the needs of the experiment performed. The implementation of the author's apparatus, combined with the hard coal samples subjected to analysis in the form of cuboidal solid blocks of coal, which is also not often found in the literature, makes the research performed unique. The wires coming out of the measuring ampoule were connected to a microcontroller. Tensometers were placed in the circuit based on the principle of a Wheatstone bridge. The experiment used resistance strain gauges with a resistance of $120 \pm 0.2\%\ \Omega$ and a strain sensitivity factor of k = 2.1. The tolerance of the k factor was 0.5%. From it, the signal is passed to a signal amplifier, based on the HX711 module, where the signal was converted to digital. Furthermore, the signal was sent to the Arduino UNO microcontroller. The results were recorded using the SD card reader module, and the current result is shown on the LCD1602 2×16 display with an I2C module. The pressure was measured using voltage S-20 transducers, which were operated in front of an additional Arduino UNO board with an LCD display. The schematic design of the device is shown in Figure 3, which includes the most important components of the device.

Cuboidal specimens were cut from the bulks of the bedrock. The linear dimensions of the samples were base 18×18 mm, walls 18×40 mm. The compensation sample had similar linear dimensions but was made of granite, a non-porous material. The weight of the samples used was approximately 20 g. Before the actual measurement, the samples were degassed with a vacuum pump. The vacuum value was 10^{-5} Pa, and the duration of degassing was 8 h. In addition, helium at a pressure of 10 kPa was passed through the sample to remove adsorbed vapours and gases inside the porous structure. The first point of the experiment was the introduction of the test gases into the ampoules with samples P1 and P2 and the compensation sample. In the second part of the experiment, the initial pressure was defined as the final pressure from the first part of the experiment. Next, the gas was introduced into the manostat and then into the measuring apparatus. Through the use of an electronic controller, the pressure in the apparatus was kept constant by gradually dosing the sorbate. The measurement temperature for the isothermal process was 298 K. Measurements under non-isothermal conditions, were conducted over the full temperature range of 298 K to 323 K.

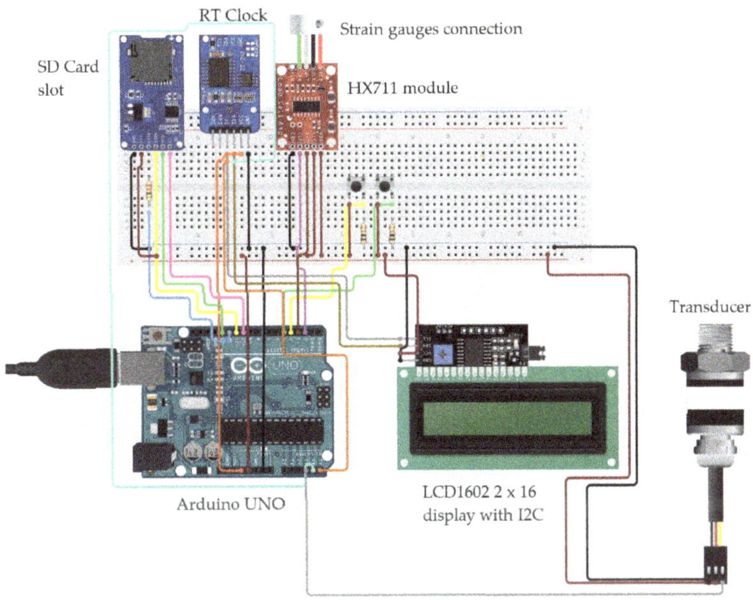

Figure 3. Schematic of the measurement system.

4. Results and Discussion

The sorption–dilatation results, for both samples, are presented in Figure 4a,b in the form of sorption kinetics. Both samples have a higher sorption capacity towards carbon dioxide than for methane. The final pressures for the individual samples are shown in Figure 4a, and are 6 bar for P1 under a CO_2 atmosphere, 8.4 bar for a P1 CH_4 atmosphere, 8.76 bar for a P2 CO_2 atmosphere, and 10.12 bar for a P2 CH_4 atmosphere. For both samples, there was an observable drop in pressure inside the measuring ampoule, which indicates the occurrence of sorption phenomena in both methane and carbon dioxide atmospheres. Due to the specific shape of the samples, i.e., cuboid blocks, the measurements are very time-consuming. The use of samples of this shape also causes the equilibrium state to be reached much later than for samples in the form of powders. Observing the literature data [26,35], it can be observed that it takes a considerable amount of time to reach equilibrium, so experiments in this form must be staggered. As shown in Figure 4a,b, in the initial phase of the experiment, the adsorption increases rapidly before the sorption equilibrium is reached. This is evidenced by the fact that after about 20 h of measurements, there is little change in sorption.

The sorption capacity of P1 towards carbon dioxide was about 0.93 mmol/g, while for methane, the adsorbed amount was close to 0.68 mmol/g. The diameter of the particles has a significant effect on the value of adsorption on hard coal. CO_2, having a smaller diameter, can better penetrate the small diameter pores present in the coal material (see Table 2). Another parameter that can influence the adsorption value is the critical temperature of the sorbate and thus the state in which the particles are found. For carbon dioxide, the critical temperature is 304 K, so in the first stage of the experiment, under isothermal conditions, it exists in the form of vapour [36]. For methane, the critical temperature is 191 K, so it always appears as a gas during the experiments. Due to the occurrence of this phenomenon at the investigated temperature of 298 K, it is possible that a part of the pores is inaccessible to methane but is preferentially occupied by carbon dioxide particles thanks to the lower energy required to expand the pores. A similar phenomenon is also observed for P2. The maximum sorption for carbon dioxide is about 0.5 mmol/g, and for methane it is about 0.33 mmol/g.

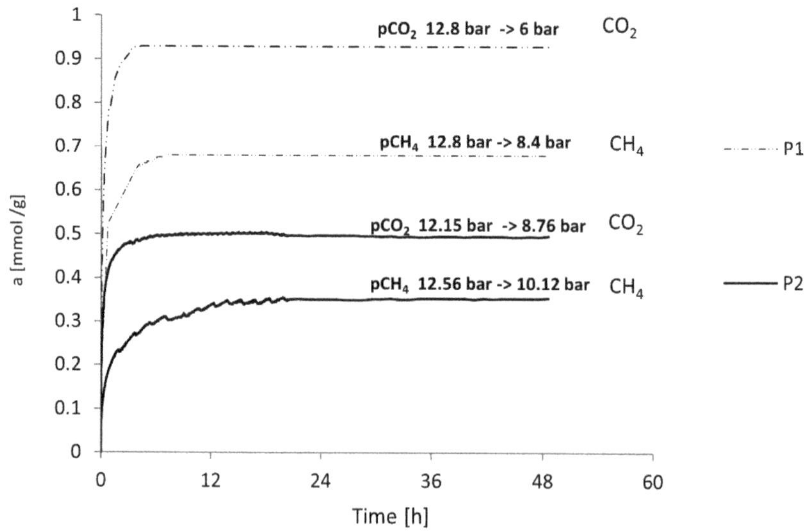

(a)

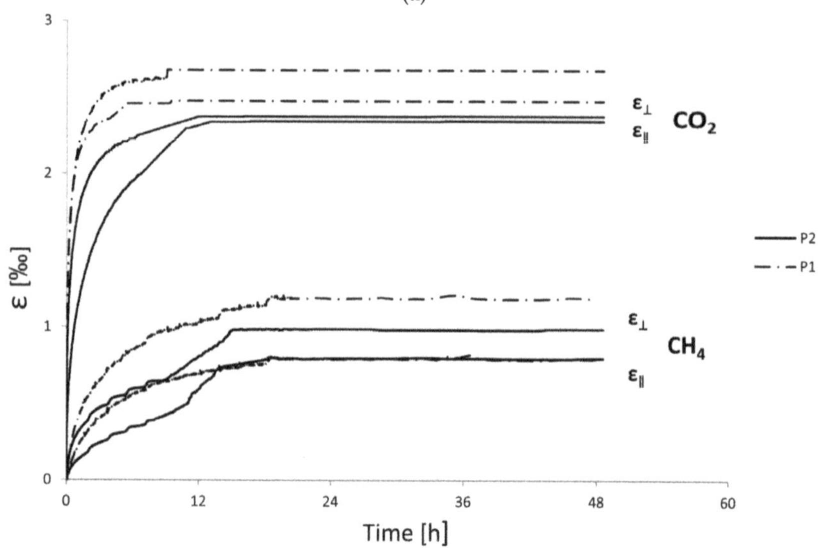

(b)

Figure 4. (**a**) Kinetics of CO_2 and CH_4 sorption. (**b**) Linear strains at 298 K for sample P1 and sample P2.

Table 2. Specification of the tested sorbates.

Sorbate	Melting Point (K)	Boiling Point (K)	Density at 1 atm, 273 K (kg/m^3)	Hybridization (-)	Kinetic Diameter (A)
CO_2	84.24	194.7 (sublimes)	1.977	sp	3.3
CH_4	90.694	111.6	0.717	sp^3	3.8

A course of the change in linear expansion is shown in Figure 4b. Based on other researchers' observations [26,37], significant differences can also be observed in the behaviour of the coal matrix in dioxide and methane atmospheres. In Figure 4b, there is a clear difference in the strain values in the perpendicular and parallel directions. For P1, the differences in strain are much more pronounced than for P2 for both carbon dioxide and methane atmospheres. To convert values from linear expansion to volumetric expansion, the following formula (1) was used:

$$\varepsilon_v = 2\varepsilon_{\parallel} + \varepsilon_{\perp} \quad (1)$$

ε_v—volumetric strain
$\varepsilon_{\parallel}$—deformations parallel to the layering
$\varepsilon_{\perp}$—deformations perpendicular to the layering

For both samples, carbon dioxide sorption involving volumetric expansion was observably higher. For P1, the maximum volumetric strain was nearly 7.5% for CO_2, and nearly 4% for methane. For P2, the volumetric strain for carbon dioxide was 5%, while the strain for methane was close to 2%. The relationship between the amount of sorbed CO_2 and CH_4 and the sorption-induced swelling is presented in Figure 4a,b.

The results presented in Figure 4b show a relatively fast expansion of the coal matrix in the presence of carbon dioxide, and after 12 h, no more changes were observed for both samples. For the CH_4–coal system, it took about 20 h to reach an equilibrium state.

The data shown in Figure 4a are similar to those available in the literature and similarly describe the behaviour of the coal matrix. Research conducted by scientists has shown that hard coals have a higher affinity for the sorption of carbon dioxide than methane [26,35,38]. These values, however, may vary depending on the petrographic composition. Vitrinite is a maceral characterised by higher porosity than inertinite, which means that it can sorb a higher amount of vapour and gases within its structure [39]. P1, which has a higher vitrinite content, generally has a higher sorption capacity than P2. This was observed in the whole range of pressures and temperatures investigated. The sorption–dilatation properties of coal are also affected by such parameters.

Due to the correlation between the coal-swelling process and the amount of sorbed vapours and gases, these parameters should not be considered separately. When analysing the sorption kinetics graphs for P1, it can be observed that a state close to the sorption equilibrium is reached after about 12 h of measurement and the sample reaches a saturated state with both methane and carbon dioxide. Compared with another study [26], this was relatively fast. In that case, it took as long as 50 h to reach equilibrium. In addition, note the relationship between the expansion of coal and its sorption capacity shown in Figure 4a,b. For P1, the course of change over the time range studied was nearly linear for carbon dioxide sorption and nonlinear for methane sorption. The linear course of these changes was also shown in another publication [40]. In the pressure range similar to the tests presented in this publication, the course was the same. However, differences started to appear at higher pressures. From the data obtained, it can be concluded that the expansion kinetics is slower than the sorption kinetics compared to the situation if the phenomena had occurred evenly.

By analysing the data in Figure 4b, it is possible to compare them with the data available in the literature [35]. The hard coal in these studies differed in elemental composition from the coal presented in this article. Therefore, it is not possible to compare these results directly with each other. Maximum deformations were reached in the first 24 h from the start of the measurements in both methane and carbon dioxide atmospheres. As in this paper, deformations induced by the influence of carbon dioxide were several times higher than deformations induced by the presence of methane. For P1, it is less than 2 times more, for P2 about 2.5×, more while in Majewska's study [35], it was 3× more. The difference between Ceglarska-Stefańska's study and the study presented in this article was the behaviour of the sample after the maximum expansion was reached [41]. In our case, the sample, after being saturated with the tested gases, did not change its shape until the end of the measurements. Similar behaviour was observed in another paper [28], where the petrographic composition was similar to that presented in this work. In Ceglarska-Stefańska's experiments, after reaching the maximum expansion, the sample slowly contracted. In our study, the relaxation phenomenon did not occur. Similar studies on sub-bituminous coals with vitrinite reflectance equal to 0.57% were carried out by Espinoza [29]. The main difference between the two experiments was the shape of the coal samples tested. In our case, they were cuboidal blocks, while Espinoza used cylindrical blocks of hard coal. In his case, higher expansion values were obtained for axial deformation, i.e., 0.006, while lower values were obtained for radial deformations, i.e., 0.004. These values are without a unit and are the result of the formulas proposed by the author.

An important parameter affecting sorption processes on hard coal is temperature, the effect of which on sorption–dilatometric properties is very evident under non-isothermal conditions. As is well known, the kinetic energy of sorbate molecules increases with increasing temperature. As the adsorption process is exothermic, an increase in temperature during the process can decrease the sorption capacity, which corresponds to the behaviour of P1 in a carbon dioxide atmosphere. As can be clearly seen in Figure 5a,b, P1 and P2 hardly increase their sorption capacity. The exception is P1, where a clear decrease in sorption capacity towards carbon dioxide is observed. A similar phenomenon can be found in the literature [26], in which researchers tested a wider range of temperatures and pressures, and both methane and carbon dioxide sorbed on the coal surface. These scientists identified the oxygen content as one of the reasons for this behaviour of the carbon material, which affects the number of active centres in the carbon matrix. An important parameter allowing discussion of the results is the diffusion coefficient, which can describe the decrease or increase in the sorption capacity in terms of temperature. With increasing temperature, the effective diffusion coefficient increases for methane [42,43]. Higher values of this parameter can be obtained for coals low in vitrinite, which may indicate differences in sorption capacity for P1 and P2 in the context of carbon dioxide. Figure 6a,b has been prepared based on the expansion kinetics and the sorption capacity of samples under non-isothermal conditions.

It was observed (Figure 7) that for the P1 sample and the carbon dioxide test, the sorption kinetics was slower than the expansion kinetics, because for a relatively small decrease in the sorption capacity, there was a clear change in the volumetric dimensions of the sample. In other cases, the change in sorption capacity due to a change in the volume dimensions of the sample was not so spectacular. As the temperature increases, samples treated with carbon dioxide undergo contraction. A completely opposite property was observable when the sample was tested in a methane atmosphere. Both P1 and P2 swelled with increasing temperature. Similar results were reported in another paper [44]. The purpose of this study was to investigate the behaviour of cuboidal coal samples under varying temperature conditions. In the case of dilatometric kinetics both for a vacuum and for helium, the changes in linear dimensions had exactly the same values, so when analysing the results of coal expansion, the thermal expansion parameter should be taken into account.

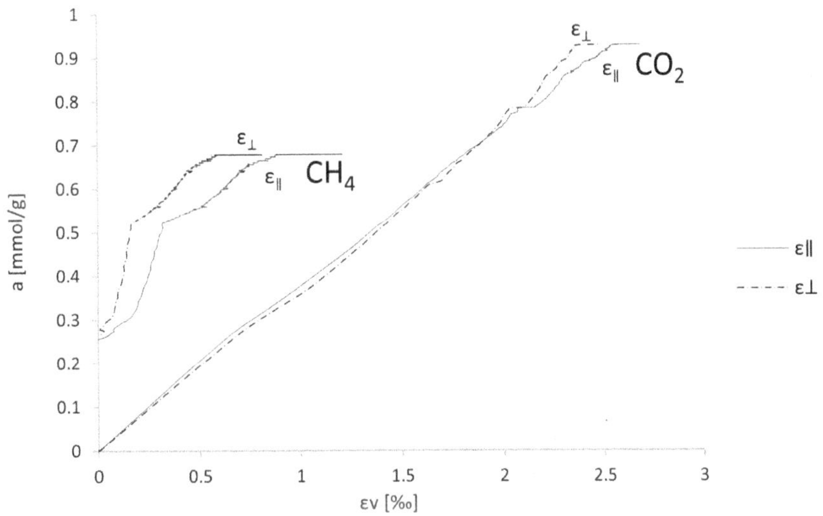

(a)

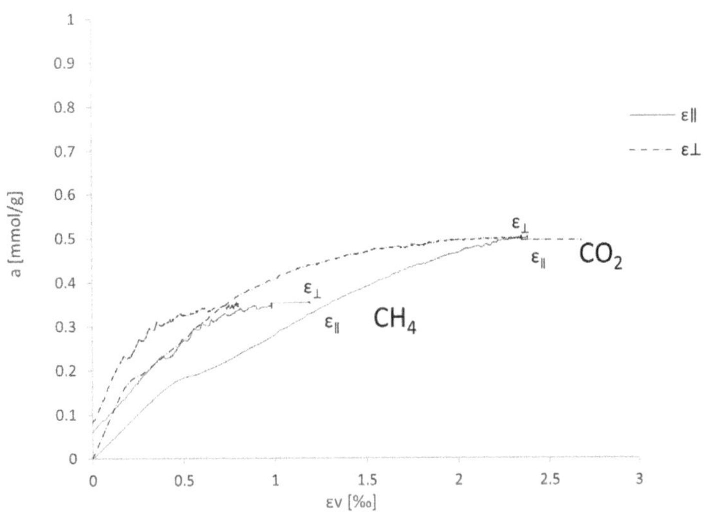

(b)

Figure 5. (**a**) Relationship between the amount of sorbed vapours and gases and the volumetric strain for sample P1. (**b**) Relationship between the amount of sorbed vapours and gases and the volumetric strain for sample P2.

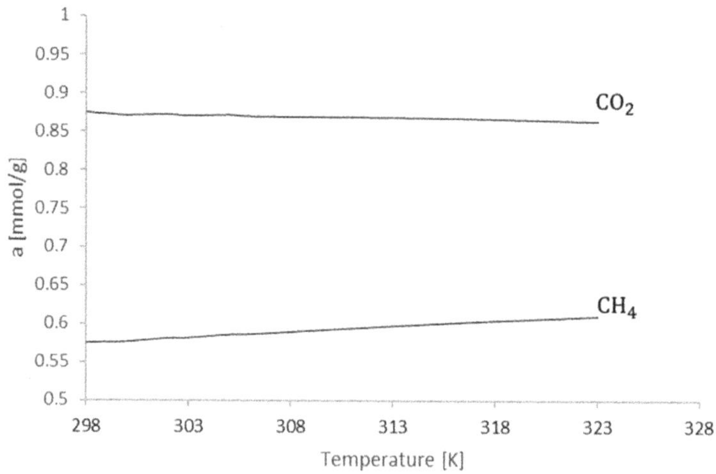

(a)

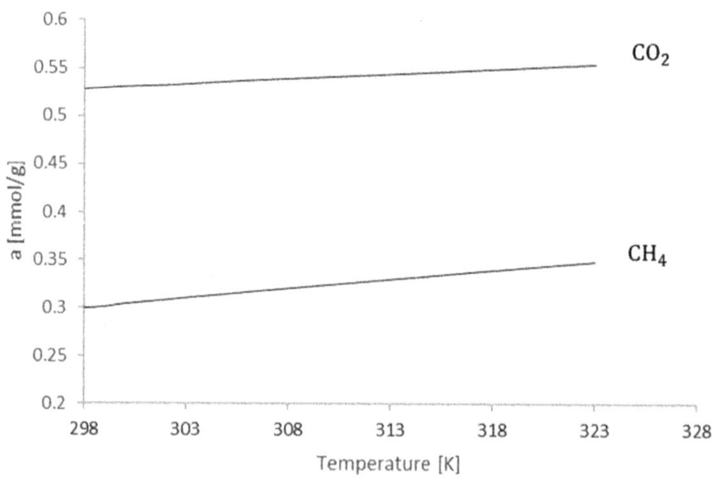

(b)

Figure 6. (**a**) Change in sorption capacity for CO_2 and CH_4 depending on the temperature for P1. (**b**) Change in sorption capacity for CO_2 and CH_4 depending on the temperature for P2.

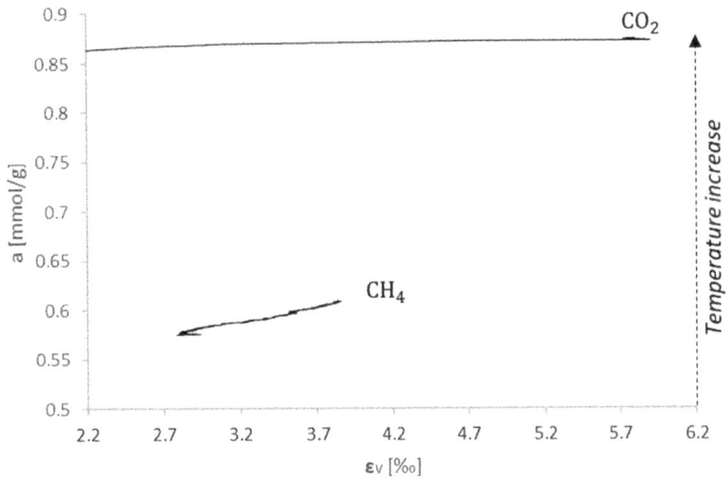

(**a**)

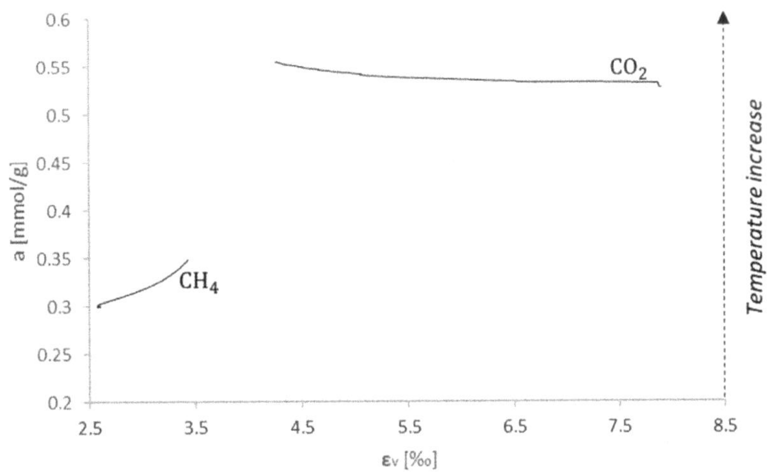

(**b**)

Figure 7. (**a**) Relationship between the amount of sorbed vapours and gases and the volumetric strain for sample P1 under non-isothermic conditions. (**b**) Relationship between the amount of sorbed vapours and gases and the volumetric strain for sample P2 under non-isothermic conditions.

5. Conclusions

- Change in the linear dimensions of the sample in the isothermal process due to carbon dioxide sorption is 2.5× greater than for methane sorption, with a noticeable difference between the two samples due to the different vitrinite content.
- The relationship between sorption capacity and volumetric expansion in the isothermal process for carbon dioxide for P1 is nearly linear, while for methane it is increasing. For P2, the graph has a curved shape. In the analysis of the later phase of the experiment, the rate of increase in volumetric strain is higher than the rate of increase in sorption capacity.

- Investigations carried out with increasing temperatures provided new data on the behaviour of the coal matrix. As the temperature increased, a slight increase in the sorption capacity of the sample was observed, with the sample shrinking in a carbon dioxide atmosphere and expanding in a methane atmosphere.
- The analysis of sorption capacity changes as a function of volume expansion under non-isothermal conditions provides relevant information about the carbon–gas system. As the temperature increases, no significant change in sorption capacity is observed, while a change in the volume dimensions of the sample is observed, indicating that the parameter related to the movement of vapours and gases within the carbon matrix has a greater influence on this process.

Author Contributions: Conceptualization, P.B.; methodology, P.B. and K.Z.; validation, P.B. and K.C.; formal analysis, P.B., S.K., K.C., A.S. and K.Z.; investigation, S.K.; resources, P.B. and K.Z.; data curation, S.K.; writing—original draft preparation, S.K. and K.C.; writing—review and editing, P.B. and K.Z.; visualization, S.K.; supervision, P.B.; project administration, K.Z.; funding acquisition, K.Z. All authors have read and agreed to the published version of the manuscript.

Funding: The authors gratefully acknowledge financial support from the "Initiative of Excellence—Research University—IDUB" program of AGH University of Science and Technology in Krakow, Poland.

Institutional Review Board Statement: Not applicable.

Informed Consent Statement: Not applicable.

Conflicts of Interest: The authors declare no conflict of interest.

References

1. International Energy Agency. *Coal 2019*; Market Report Series: Coal; OECD: Paris, France, 2019; ISBN 9789264391604.
2. Kowalska, N.; Brodawka, E.; Smoliński, A.; Zarębska, K. The European Education Initiative as a Mitigation Mechanism for Energy Transition. *Energies* **2022**, *15*, 6633. [CrossRef]
3. Kelemen, P.; Benson, S.M.; Pilorgé, H.; Psarras, P.; Wilcox, J. An Overview of the Status and Challenges of CO_2 Storage in Minerals and Geological Formations. *Front. Clim.* **2019**, *1*, 9. [CrossRef]
4. Hassanpouryouzband, A.; Yang, J.; Tohidi, B.; Chuvilin, E.; Istomin, V.; Bukhanov, B.; Cheremisin, A. CO_2 Capture by Injection of Flue Gas or CO_2–N_2 Mixtures into Hydrate Reservoirs: Dependence of CO_2 Capture Efficiency on Gas Hydrate Reservoir Conditions. *Environ. Sci. Technol.* **2018**, *52*, 4324–4330. [CrossRef] [PubMed]
5. Hassanpouryouzband, A.; Joonaki, E.; Edlmann, K.; Haszeldine, R.S. Offshore Geological Storage of Hydrogen: Is This Our Best Option to Achieve Net-Zero? *ACS Energy Lett.* **2021**, *6*, 2181–2186. [CrossRef]
6. Markewitz, P.; Kuckshinrichs, W.; Leitner, W.; Linssen, J.; Zapp, P.; Bongartz, R.; Schreiber, A.; Müller, T.E. Worldwide Innovations in the Development of Carbon Capture Technologies and the Utilization of CO_2. *Energy Environ. Sci.* **2012**, *5*, 7281–7305. [CrossRef]
7. Zhang, D.; Song, J. Mechanisms for Geological Carbon Sequestration. *Procedia IUTAM* **2014**, *10*, 319–327. [CrossRef]
8. Page, B.; Stern, N.; Hameister Oam, J. *Global Status of CCS 2020*; Global CCS Institute: Melbourne, Australia, 2020.
9. Karacan, C.Ö.; Ruiz, F.A.; Cotè, M.; Phipps, S. Coal Mine Methane: A Review of Capture and Utilization Practices with Benefits to Mining Safety and to Greenhouse Gas Reduction. *Int. J. Coal Geol.* **2011**, *86*, 121–156. [CrossRef]
10. Mangi, H.N.; Detian, Y.; Hameed, N.; Ashraf, U.; Rajper, R.H. Pore Structure Characteristics and Fractal Dimension Analysis of Low Rank Coal in the Lower Indus Basin, SE Pakistan. *J. Nat. Gas Sci. Eng.* **2020**, *77*, 103231. [CrossRef]
11. Moore, T.A. Coalbed Methane: A Review. *Int. J. Coal Geol.* **2012**, *101*, 36–81. [CrossRef]
12. Gürdal, G.; Yalçın, M.N. Gas Adsorption Capacity of Carboniferous Coals in the Zonguldak Basin (NW Turkey) and Its Controlling Factors. *Fuel* **2000**, *79*, 1913–1924. [CrossRef]
13. Corrente, N.J.; Zarębska, K.; Neimark, A. V Deformation of Nanoporous Materials in the Process of Binary Adsorption: Methane Displacement by Carbon Dioxide from Coal. *J. Phys. Chem. C* **2021**, *125*, 21310–21316. [CrossRef]
14. Guo, D.; Guo, X. The Influence Factors for Gas Adsorption with Different Ranks of Coals. *Adsorpt. Sci. Technol.* **2017**, *36*, 904–918. [CrossRef]
15. Zarebska, K.; Baran, P.; Cygankiewicz, J.; Dudzinska, A.; Zarçbska, K.; Baran, P.; Cygankiewicz, J.; Dudzińska, A. Sorption of Carbon Dioxide on Polish Coals in Low and Elevated Pressure. *Fresenius Environ. Bull.* **2012**, *21*, 4003–4008.
16. Macuda, J.; Baran, P.; Wagner, M. Evaluation of the Presence of Methane in Złoczew Lignite: Comparison with Other Lignite Deposits in Poland. *Nat. Resour. Res.* **2020**, *29*, 3841–3856. [CrossRef]
17. Jian, X.; Guan, P.; Zhang, W. Carbon Dioxide Sorption and Diffusion in Coals: Experimental Investigation and Modeling. *Sci. China Earth Sci.* **2012**, *55*, 633–643. [CrossRef]
18. Hildenbrand, A.; Krooss, B.M.; Busch, A.; Gaschnitz, R. Evolution of Methane Sorption Capacity of Coal Seams as a Function of Burial History—A Case Study from the Campine Basin, NE Belgium. *Int. J. Coal Geol.* **2006**, *66*, 179–203. [CrossRef]

19. Wojtacha-Rychter, K.; Howaniec, N.; Smoliński, A. Effect of Porous Structure of Coal on Propylene Adsorption from Gas Mixtures. *Sci. Rep.* **2020**, *10*, 11277. [CrossRef]
20. Clarkson, C.R.; Marc Bustin, R. Variation in Micropore Capacity and Size Distribution with Composition in Bituminous Coal of the Western Canadian Sedimentary Basin: Implications for Coalbed Methane Potential. *Fuel* **1996**, *75*, 1483–1498. [CrossRef]
21. Laxminarayana, C.; Crosdale, P.J. Role of Coal Type and Rank on Methane Sorption Characteristics of Bowen Basin, Australia Coals. *Int. J. Coal Geol.* **1999**, *40*, 309–325. [CrossRef]
22. Sadasivam, S.; Masum, S.; Chen, M.; Stańczyk, K.; Thomas, H. Kinetics of Gas Phase CO_2 Adsorption on Bituminous Coal from a Shallow Coal Seam. *Energy Fuels* **2022**, *36*, 8360–8370. [CrossRef]
23. Zhang, W.; Jiang, S.; Wang, K.; Wang, L.; Xu, Y.; Wu, Z.; Shao, H.; Wang, Y.; Miao, M. Thermogravimetric Dynamics and FTIR Analysis on Oxidation Properties of Low-Rank Coal at Low and Moderate Temperatures. *Int. J. Coal Prep. Util.* **2015**, *35*, 39–50. [CrossRef]
24. Talapatra, A. A Study on the Carbon Dioxide Injection into Coal Seam Aiming at Enhancing Coal Bed Methane (ECBM) Recovery. *J. Pet. Explor. Prod. Technol.* **2020**, *10*, 1965–1981. [CrossRef]
25. Talapatra, A.; Karim, M.M. The Influence of Moisture Content on Coal Deformation and Coal Permeability during Coalbed Methane (CBM) Production in Wet Reservoirs. *J. Pet. Explor. Prod. Technol.* **2020**, *10*, 1907–1920. [CrossRef]
26. Czerw, K.; Baran, P.; Zarębska, K. Application of the Stretched Exponential Equation to Sorption of Mine Gases and Sorption Induced Swelling of Bituminous Coal. *Int. J. Coal Geol.* **2017**, *173*, 76–83. [CrossRef]
27. Ranathunga, A.S.; Perera, M.S.A.; Ranjith, P.G.; Rathnaweera, T.D.; Zhang, X.G. Effect of Coal Rank on CO_2 Adsorption Induced Coal Matrix Swelling with Different CO_2 Properties and Reservoir Depths. *Energy Fuels* **2017**, *31*, 5297–5305. [CrossRef]
28. Czerw, K. Methane and Carbon Dioxide Sorption/Desorption on Bituminous Coal—Experiments on Cubicoid Sample Cut from the Primal Coal Lump. *Int. J. Coal Geol.* **2011**, *85*, 72–77. [CrossRef]
29. Espinoza, D.N.; Vandamme, M.; Pereira, J.M.; Dangla, P.; Vidal-Gilbert, S. Measurement and Modeling of Adsorptive-Poromechanical Properties of Bituminous Coal Cores Exposed to CO_2: Adsorption, Swelling Strains, Swelling Stresses and Impact on Fracture Permeability. *Int. J. Coal Geol.* **2014**, *134*, 80–95. [CrossRef]
30. Gaowei, Y.; Chunlin, Z.; Liupeng, H.; Xinjun, Z. Measurement and Modeling of Temperature Evolution during Methane Desorption in Coal. *Sci. Rep.* **2020**, *10*, 3146. [CrossRef]
31. Wang, Z.; Tang, X.; Yue, G.; Kang, B.; Xie, C.; Li, X. Physical Simulation of Temperature Influence on Methane Sorption and Kinetics in Coal: Benefits of Temperature under 273.15 K. *Fuel* **2015**, *158*, 207–216. [CrossRef]
32. Jakubów, A.; Tor, A.; Tobiczyk, S. Wyrzut Metanu i Skał w Drążonej Lunecie Rurowej Do Szybu II Na Poziomie 1000m w KRK Pniówek- Okoliczności, Przyczyny i Skutki. In Proceedings of the Szkoła Ekspoloatacji Podziemnej, Szczyrk, Poland, 7–21 February 2003.
33. *Ocena Stanu Bezpieczeństwa Pracy, Ratownictwa Górniczego Oraz Bezpieczeństwa Powszechnego w Związku z Działalnością Górniczo-Geologiczną w 2015r*; WUG: Katowice, Poland, 2016.
34. Wasilewski, S.; Jamróz, P. Wybrane Katastrofy i Wypadki w Górnictwie Polskim—Zebranie Danych. *Pr. Inst. Mech. Górotworu PAN* **2018**, *20*, 197–206.
35. Majewska, Z.; Ceglarska-Stefańska, G.; Majewski, S.; Ziętek, J. Binary Gas Sorption/Desorption Experiments on a Bituminous Coal: Simultaneous Measurements on Sorption Kinetics, Volumetric Strain and Acoustic Emission. *Int. J. Coal Geol.* **2009**, *77*, 90–102. [CrossRef]
36. Zarębska, K.; Ceglarska-Stefańska, G. The Change in Effective Stress Associated with Swelling during Carbon Dioxide Sequestration on Natural Gas Recovery. *Int. J. Coal Geol.* **2008**, *74*, 167–174. [CrossRef]
37. Li, J.; Li, B.; Cheng, Q.; Gao, Z. Characterization of Anisotropic Coal Permeability with the Effect of Sorption-Induced Deformation and Stress. *Fuel* **2022**, *309*, 122089. [CrossRef]
38. Nikoosokhan, S.; Vandamme, M.; Dangla, P. A Poromechanical Model for Coal Seams Saturated with Binary Mixtures of CH4 and CO_2. *J. Mech. Phys. Solids* **2014**, *71*, 97–111. [CrossRef]
39. Guo, P. A Theoretical Model for Coal Swelling Induced by Gas Adsorption in the Full Pressure Range. *Adsorpt. Sci. Technol.* **2020**, *38*, 94–112. [CrossRef]
40. Ceglarska-Stefanska, G.; Zarebska, K. Expansion and Contraction of Variable Rank Coals During the Exchange Sorption of CO_2 and CH_4. *Adsorpt. Sci. Technol.* **2002**, *20*, 49–62. [CrossRef]
41. Ceglarska-Stefańska, G.; Zarębska, K. The Competitive Sorption of CO_2 and CH4 with Regard to the Release of Methane from Coal. *Fuel Process. Technol.* **2002**, *77–78*, 423–429. [CrossRef]
42. Wierzbicki, M. Changes in the Sorption/Diffusion Kinetics of a Coal-Methane System Caused by Different Temperatures and Pressures. *Gospod. Surowcami Miner.—Miner. Resour. Manag.* **2013**, *29*, 155–168. [CrossRef]
43. Dutka, B.; Kudasik, M.; Pokryszka, Z.; Skoczylas, N.; Topolnicki, J.; Wierzbicki, M. Balance of CO_2/CH4 Exchange Sorption in a Coal Briquette. *Fuel Process. Technol.* **2013**, *106*, 95–101. [CrossRef]
44. Baran, P.; Czerw, K.; Samojeden, B.; Czuma, N.; Zarębska, K.; Baran, P.; Czerw, K.; Samojeden, B.; Czuma, N.; Zarębska, K. The Influence of Temperature on the Expansion of a Hard Coal-Gas System. *Energies* **2018**, *11*, 2735. [CrossRef]

energies

MDPI

Article

Determination of the Salt-Dust Emission and the Efficiency of the Dedusting Installation in the Wieliczka Salt Mine

Mariusz Kapusta * and Krzysztof Skrzypkowski *

Faculty of Civil Engineering and Resource Management, AGH University of Science and Technology, Mickiewicza 30 Av., 30-059 Kraków, Poland
* Correspondence: kapustam@agh.edu.pl (M.K.); skrzypko@agh.edu.pl (K.S.)

Abstract: This article presents measurements of the emission of salt dust discharged into the atmosphere in the Wieliczka Salt Mine. Industrial tests have been carried out cyclically since 2004. The research methodology included six measurements of the salt dust's mass increments on the filters, including three before the inlet to the desalination installation and three measurements at the emitter outlet. In order to limit errors resulting from the diversity of concentrations in the mixture of dosed brine, three series of measurements are carried out each time. The proposed research methodology uses the Aspirator Stationary (AS-50) gravimetric aspirator for measurements with a designed probe adjusted to the characteristics of the dedusting installation used. The conducted tests allowed calculation of the efficiency of the scrubber and confirm the usefulness of the measuring probe used. Moreover, long-term and cyclical measurements make it possible to observe the effectiveness of the installation and salt production using the wet system from brine obtained in the mine. The proposed method is an innovative solution for monitoring salt-dust emissions in mining applications.

Keywords: salt mine; salt-dust emission; air dust; gravimetric dust meter; efficiency of the dedusting installation

Citation: Kapusta, M.; Skrzypkowski, K. Determination of the Salt-Dust Emission and the Efficiency of the Dedusting Installation in the Wieliczka Salt Mine. *Energies* **2022**, *15*, 8122. https://doi.org/10.3390/en15218122

Academic Editors: Ben McLellan and Nikolaos Koukouzas

Received: 7 September 2022
Accepted: 28 October 2022
Published: 31 October 2022

1. Introduction

Direct measurements of salt-dust emissions to the atmosphere and their variability over time are rare but necessary to attempt to explain their environmental impact. Based on the long-term research conducted, the results can be used to develop models determining the variability of transport routes and the amount of the emissions from the dust source area based on environmental conditions [1]. Various properties, different from other dusts, make it difficult to classify the salt dust into the appropriate group of industrial dusts. The positive inhalation properties and medical effects of sodium chloride (NaCl) are well known and proven by providing tourists and patients with routes and chambers in underground salt mines [2]. In this type of research, the authors focus on the assessment of mass concentration and determination of the chemical composition of particulate matter (PM), mainly in terms of their possible relationship with the therapeutic effect of underground air in a mine [3]. The hygroscopic behavior of salt as a function of relative humidity (RH), taking into account the thermodynamic model, is of significance [4]. Among the various studies, attempts are made to find the correlation between the deposition rate of solid particles and the parameters of the aerosol [5–7]. This dependence can be used to parameterize mathematical models of air–dust mixture flow in mine ventilation networks [8]. The exploitation of rock-salt deposits is carried out by underground or borehole method from the surface. The underground method of extracting rock salt is carried out in a dry way by mechanical mining or with the use of explosives [9]. Such a system is known as a "dry" underground mining method. In the case of a lower NaCl content in the rock-salt deposit, underground mining can be carried out using the "wet" system by mining the salt solids with water and producing evaporated salt from the obtained brine [10]. This type of salt extraction system has been designed in the Wieliczka Salt Mine, which obtains brine solutions as a result of drainage of mining excavations. These activities are necessary to ensure the safe functioning of the mining plant employees as well as people traveling along the mine's tourist routes [11]. An analysis of the world literature about salt dust showed that the conducted research concerns mainly the influence of salt on the pollution of lakes, rivers, seas and soils [12–14]. The long-term effects of improper mining activities often caused degradation of the natural environment [15–17]. Only a few researchers try to determine the factors influencing soil erosion and its impact on dust formation around mining enterprises [18,19]. Currently, there is no mining in the Wieliczka Salt Mine, and its main activity is to provide tourists with tourist routes with numerous chambers. The work of miners in a mining plant is to control and maintain the mine's workings and to control water leaks, which in the case of salt pose a catastrophic threat to the existence of the mine. The water is constantly pumped out and discharged to the water tanks on the eighth level. Then, the brine is pumped to the surface where the brewing process takes place in the Saline Water Treatment Plant. As a result of this process, salt is produced in the amount of approx. 20 thousand tons annually, the production of which is beneficial for the whole of Poland, especially for the Krakow region. Considering the very limited amount of information on the emission of salt dust, the article presents the results of tests of the efficiency of the installation in the saltwork in the Wieliczka Salt Mine in the years 2004–2021. On the basis of industrial research, it was found that the proposed measurement method with the use of the AS-50 gravimetric dust meter as well as the selection of filters to work with the selector constitute an original research methodology adapted to salt dust. The main aim of the research was to determine the efficiency of the applied installation. On the other hand, the secondary goal was to design a measuring probe and carry out measurements for such a designed installation. The tested installation was specially designed individually and does not allow for reliable measurements using traditional methods and devices. Therefore, one of the goals presented was to design and manufacture a measuring probe.

2. Methodology of Air Dustiness Tests

Dust measurement has a large number of different types of dust meters, working on the basis of using different physical properties of solid particles. Classic air pollution analysis consists of three stages: collecting a pollutant sample at the test stand; quantitative analysis of the collected pollutant sample; converting analysis results into the required concentration units. Sampling methods with the use of gravimetric, conimetric or optical dust meters are most often used to determine the value of dust pollutants in mining. In Polish mines, gravimetric dust meters "Barbara-3a" (Central Mining Institute, Katowice, Poland) and Aspirator Stationary type 50 (AS-50) (Cooperative Two-Met, Zgierz, Poland) are commonly used, as well as individual PM10 (particulate matter meter for airborne particles with a diameter of less than 10 μm), P25 (particulate matter meter for airborne particles with a diameter of less than 25 μm) and PO-01 (personal dust meter series number 01), which operate on the basis of aspiration pollutant collection methods. Moreover, the AS-50 aspirator used in the research, due to the way it works, is classified as a direct operating gravimetric dust meter. The sampling principle is based on filtration using measuring filters. The methodology of measuring dust emissions in the saltworks of the "Wieliczka" salt mine was developed by AGH (University of Science and Technology) employees on the basis of applicable standards and measurement procedures [20–23]. Due to the need to obtain large flows of sucked air, an aspirator powered from a 230 V network, type AS-50, manufactured by TWO-MET in Zgierz, was used for measurements (Figure 1a). The technical and operational parameters of the AS-50 aspirator are presented in Table 1. A RADWAG XA 82/220/2X laboratory balance was used to measure the mass of salt dust retained on the measuring filter. To calculate the air-flow velocity in the scrubber installation (salt scrubber), a Pitot–Prandtl tube was used, and the dynamic pressure was read using a liquid U-tube manometer filled with distilled water (Figure 1b). Temperature measurements were carried out with the EMT-02 temperature meter with a TP01 sensor, manufactured by ELMECH, with a range from −100 to +500 °C.

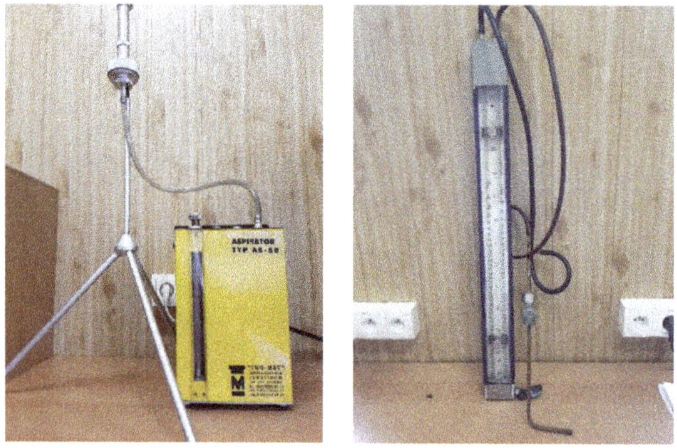

(**a**) (**b**)

Figure 1. Salt-dust measuring instruments: (**a**) AS-50 gravimetric dust meter; (**b**) U-tube liquid manometer with Pitot–Prandtl probe.

Table 1. AS-50 aspirator parameters.

Parameters	Value
Drive motor power	90 (W)
Nominal suction capacity	50 (L/min)
Nominal suction pressure	10 (kPa)
Measuring range of the rotameter	15–60 (L/min)
The range of the flow rate regulation	15–60 (L/min)
Flow tolerance for changes in pressure drop on the measuring filter 0.1–30.0 (kPa)	+/−5%
Overall dimensions (width, length, height)	155 × 200 × 330 (mm)
Weight	8.2 (kg)
Type of work	continuous

Dust emission to the atmosphere was determined on the basis of Formulas (1)–(8). The mass of the collected salt dust on the filters for individual measurement series is determined according to Equation (1):

$$\Delta m_i = m_{fzi} - m_{fci} \quad (1)$$

where:

Δm_i—weight gain on the filter (g).

m_{fci}—filter mass before measurement (g).

m_{fzi}—filter weight after measurement, (g).

The volume of gas taken by the gravimetric dust meter is calculated according to Equation (2):

$$V_i = q_s \cdot t_i \quad (2)$$

where:

V_i—gas volume taken (m^3).

q_s—flow rate of the gas recorded on the instrument (L/min).

t_i—sample measurement time (min).

The mass concentration of dust was determined according to Equation (3):

$$Z_i = \frac{\Delta m_i}{V_i} \quad (3)$$

The air-flow velocity was calculated according to Equation (4):

$$\omega_g = \sqrt{\frac{2P_d}{\rho_p}} \quad (4)$$

where:

ω_g—air-flow velocity (m/s).

P_d—dynamic pressure (Pa).

ρ_p—air density (kg/m^3).

In the tests, the total and static pressure were determined using a Pitot–Prandtl tube. Based on the readings of these pressures, the dynamic pressure was determined.

The air density is read from the tables on the basis of measurements of the temperature of the properties of moist air saturated in the duct [24]. The air volume flow (output) was calculated according to Equation (5):

$$Q_g = F_e \cdot \omega_g \quad (5)$$

where:

Q_g—gas volume flow (output) (m^3/h).

F_e—emitter cross-sectional area (m^2).

The total efficiency of the dedusting devices is determined from the mass flow rate of dust in the pipe upstream and downstream of the device, according to Equation (6):

$$\eta = \left(1 - \frac{Z_2}{Z_1}\right) \cdot 100 \quad (6)$$

where:

η—efficiency of the installation (%).

Z_1—dust concentration in front of the dust collector (g/m^3).

Z_2—dust concentration behind the dust collector (g/m^3).

The amount of salt-dust emissions to the atmosphere is calculated according to Equation (7):

$$E = Z_2 \cdot Q_g \quad (7)$$

where:

E—salt-dust emission (g/h).

Based on the statistical analysis, it was assumed that the distribution of dust in air with salt dust was log-normal. The cumulative distribution function of this distribution is determined by Equation (8):

$$F(Z) = \frac{-1}{\sigma_1 \sqrt{2\pi}} \cdot \int_{-\infty}^{\log Z} \exp \cdot \left[- \frac{(\log Z - \log Zn)^2}{2\sigma_1} \right] d \log Z \quad (8)$$

where:

logZ—logarithm of the dust value.

lgZn—arithmetic mean value from the logarithm of the dustiness value in a given population.

σ_1—standard deviation of the dust logarithm value.

The cumulative distribution function F(Z) determines the probability that the dustiness in the tested sample will be less than Z. The cumulative distribution value is also treated as the time interval in which the dustiness is lower than Z.

3. Industrial Tests

The salt-dust emission to the atmosphere was tested in the saltwork in the Wieliczka Salt Mine. The layout of the dust concentration measurement points is shown in Figure 2.

Measurements of the tested parameters were carried out in two places. The measuring point P1 is located upstream of the scrubber inlet where the salt brewing process takes place (Figure 3a). At this point, the gas flow velocity was on average approx. 20 m/s. The emitter cross-sectional area is 0.1018 m^2, which results in a gas volume flow of approximately 7300 m^3/h. Moreover, the measured brine temperature was in the range of 68–75 °C. Measurement point P2 is in the gas scrubber line (Figure 3b). The dust air-flow velocity was on average about 35 m/s, and the temperature was in the range of 30–34 °C. Production in the saltworks is carried out in a three-shift system, and each shift lasts 8 h. Due to the significant reduction in the production of salt assortments, the production equipment and dedusting systems operate continuously for an average of 10–12 days. After this period, there is a technical break until the assortment of salt is replenished. Currently, salt emission tests from the saltwork are carried out once a year in the fall and winter periods (November/December) and result directly from the desalination plant's work schedule. The air scrubber is used to clean the exhaust air from the salt dryer contaminated with salt dust. It is equipped with an external recirculation system with pumps. The rinsing liquid is condensate. Prior to startup of the scrubber, the position of the valves in the circulation piping system is checked. Before first use, remove the spray nozzle and thoroughly flush all water lines for 20 min at full pressure. Then, all filters installed in the piping system are cleaned. After 48 h of continuous operation, adjust all bolts on the flanges. Subsequent adjustments are made after a continuous two weeks of work. The mass flow and the condition of the liquid level are checked as part of the daily check. However, for the weekly check, the pump supply, water supply line, gas inlet, valves and fresh water supply are checked. The operation check of the scrubber can be carried out through the viewing holes with transparent surfaces. After a short start-up time of the water circuit—the time needed to fill the piping system—the scrubber is ready for operation. The filled scrubber reservoir serves as a water reservoir for the circulation pump, ensuring minimal NaCl emissions. During operation, the functioning of the high-level switch and keeping the overflow clean are checked. In addition, the concentration of NaCl in the circuit is checked periodically, and the fresh water supply is adjusted if necessary [25].

In the standard measuring system, the gravimetric dust meter is equipped with a cyclone, the essential elements of which are a filter and a tank. This structure of the cyclone allows the measurement of the respirable fraction precipitated on the filter and the remaining dust fraction deposited in the tank. In the case of industrial dusts, the obtained dustiness results are compared with the maximum permissible concentrations (NDS), which are used to determine the actual exposure of workers to dust hazard. In order to collect samples from the desalination plant at Wieliczka Salt Mine, AGH (University of Science and Technology) employees designed and manufactured a measuring probe with a salt-dust sampling selector (Figure 4a). Replacing the cyclone with a measuring probe resulted from the need to adjust the possibility of carrying out measurements, taking into account the nature of the installation structure. Dusty air, sucked in at the measurement site by the AS-50 aspirator, passes through the basic selector with a bag filter where the salt-dust fraction is deposited. Due to the high flow rate of dust-laden air, it was calculated that the constant flow rate in the aspirator was 50 l/min. For such high expenditure values, only a few dust meters can meet the assumed criteria; therefore, it was decided to use the AS-50 dust meter. The sampling time is constant and is 15 min. The weight of the dust deposited on the filter is determined as the weight increase of the filter weighed in the laboratory before and after sampling (Figure 4b). The applied bag filters were characterized by the high efficiency of retaining salt-dust grains and low air-flow resistance. The authors' research and analyses allowed us to conclude that for the designed measurement methodology, these will be propylene non-woven filters characterized by high chemical resistance to salt action and good medium permeability [26].

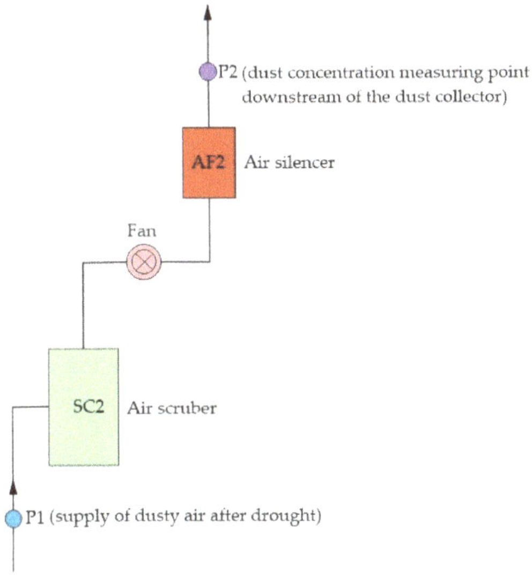

Figure 2. Scheme of distribution of measuring points in the saltwork of the "Wieliczka" salt mine.

Figure 3. Salt-dust concentration measurement points: (**a**) P1—in front of the scrubber; (**b**) P2—behind the scrubber.

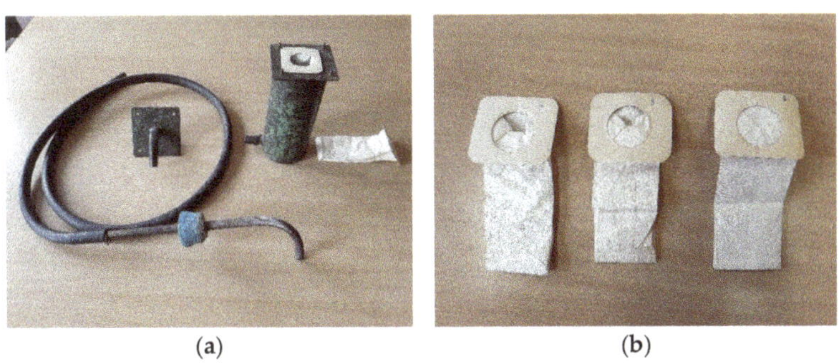

Figure 4. Salt-dust measuring instruments: (**a**) probe and selector; (**b**) measurement filters.

The saltwork in the Wieliczka Salt Mine was launched in 2003. The process of obtaining salt, which is a by-product, takes place as a result of desalination of water obtained from brine pumped out from underground workings. The water under very low pressure is heated and evaporates to leave pure evaporated salt. Measurements of salt-dust emissions discharged into the atmosphere have been carried out periodically, once a year by AGH (University of Science and Technology) employees since 2004. The average working time in the saltwork per month is from 10 to 12 days, because this is the amount of brine obtained from the drainage of workings, which is used as a contribution to the desalination installation. The research methodology includes six measurements of salt-dust mass increments on filters, including three before the inlet to the desalination installation and three at the outlet of the emitter. In order to reduce errors resulting from the diversity of concentrations in the mixture of dosed brine, three series of measurements are carried out each time. The results of the measurements from the last survey conducted in November 2021 are presented in Table 2.

Table 2. Dimensions of machines and devices with minimum movement distances.

No.	Measurement Point	Series	Mass Increase of Measurement Filters	Value of the Average Weight Gain of Dust	Gas Volume Taken	The Average Amount of Dust Concentration in the Gas
			Δm_i (g)	$\Delta m_n = \frac{1}{n}\sum_{i=1}^{n} \Delta m_i$	ω_i (m^3)	Z_i $\left(\frac{g}{m^3}\right)$
1	P1 (in front of the scrubber)	1	0.2513	0.2539	0.585	0.4340
2		2	0.2496			
3		3	0.2608			
4		1	0.2346	0.2365	0.585	0.4043
5		2	0.2292			
6		3	0.2458			
7		1	0.2463	0.2519	0.585	0.4306
8		2	0.2603			
9		3	0.2492			
10	P2 (behind the scrubber)	1	0.0057	0.0064	0.0585	0.0109
11		2	0.0071			
12		3	0.0063			
13		1	0.0048	0.0054	0.0585	0.0092
14		2	0.0055			
15		3	0.0059			
16		1	0.0051	0.0056	0.0585	0.0096
17		2	0.0061			
18		3	0.0057			

4. Discussion

Various dust control methods are frequently used in coal mines, among which the use of wet scrubbers has proven to be an effective technology to remove dust from air streams, rather than diluting or retaining dust [27]. The capture of airborne respirable coal dust by spraying water or wet scrubbers has been studied and developed over many decades as an engineering control to reduce dust exposure in coal mines and combat the pneumoconiosis of coal workers [28].

On the basis of annual tests in the saltwork in the Wieliczka Salt Mine, a list of changes in the operation of the desalination installation was made. In the analyzed period, the efficiency of the dedusting installation was in the range of 90.50–98.78%. The lowest value was recorded in 2009 and the highest in 2016 (Figure 5a–c). In the case of the amount of salt-dust emissions to the atmosphere, the value was 73.12–124.34 g/h. In this case, the highest value of emissions was obtained in 2012 and the lowest in 2021 (Figure 6a–c). A review of the records of the working time of the installation shows that on average it works about 180 days a year (50%). Throughout the entire period of measurements, the mean value of the efficiency was 95.99% with a standard deviation of σ = 0.57. In the case of the emission, the mean of the measurements was 99.51 g/h and the standard deviation σ = 6.93. When analyzing the changes in the efficiency of the installation, it can be noticed that in the years 2007–2009 there was a decrease in the dedusting value to approx. 90%. In the following years, there were regular increases in the efficiency of dedusting up to a value of approx. 98% in 2015. Since then, the operation of the systems has stabilized, which is confirmed by the test results until 2021. However, in the case of salt-dust emission into the atmosphere over the analyzed test period, it can be noticed that slight fluctuations with a slight downward trend have been observed in the last 5 years. The first period concerns the start-up of the installation and covers the years 2004–2006. During this period, the difference between the minimum and maximum values for the efficiency of the installation and salt-dust emissions was 5.7% and 28.35 g/h, respectively. The second period is related to the correction and adjustment of the operation of the installation, and it is the period from 2007 to 2012. In this period, the difference between the extreme values was reduced to the value of 5.45% and increased to the value of 41.14 g/h for efficiency and salt-dust emission, respectively. On the other hand, the third period is the stabilization of work, and it is counted from 2013 to the present (Figure 7a,b). This period is characterized by a significant improvement in efficiency because the difference between the extreme values is lower by 2.38% and 2.63%, respectively, for period II and I. In the case of salt-dust emission, the differences for the minimum and maximum values are higher by 3.64 g/h and 16.35 g/h for period II and I, respectively, which proves a significant improvement in the operation of the installation.

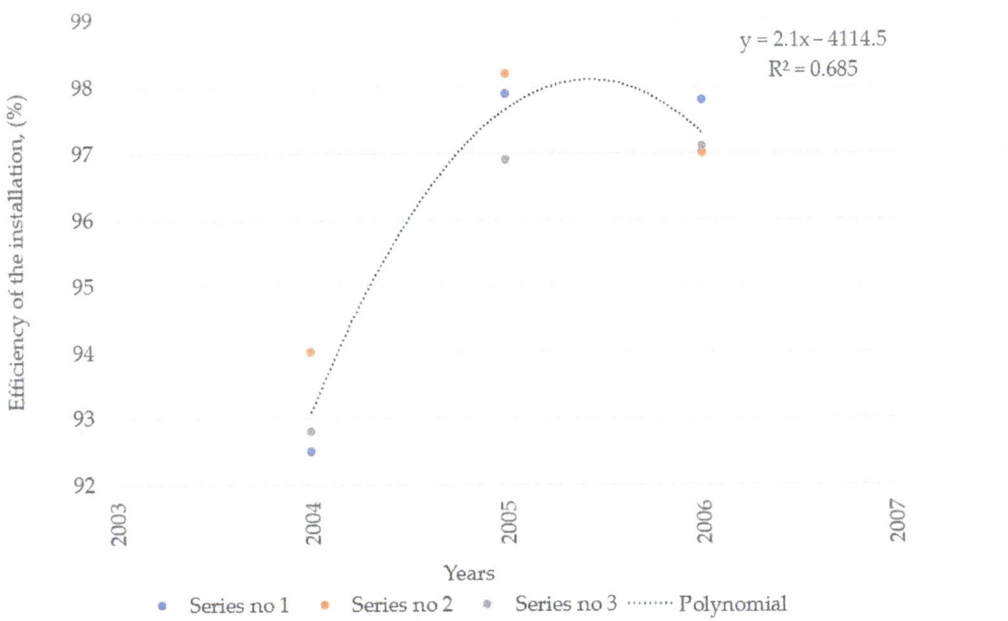

(a)

Figure 5. *Cont.*

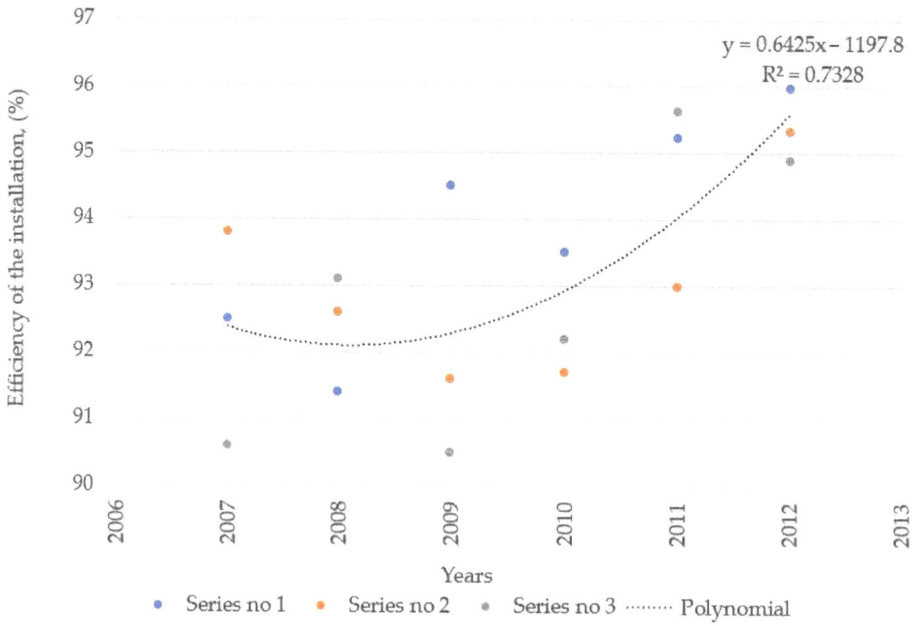

(**b**)

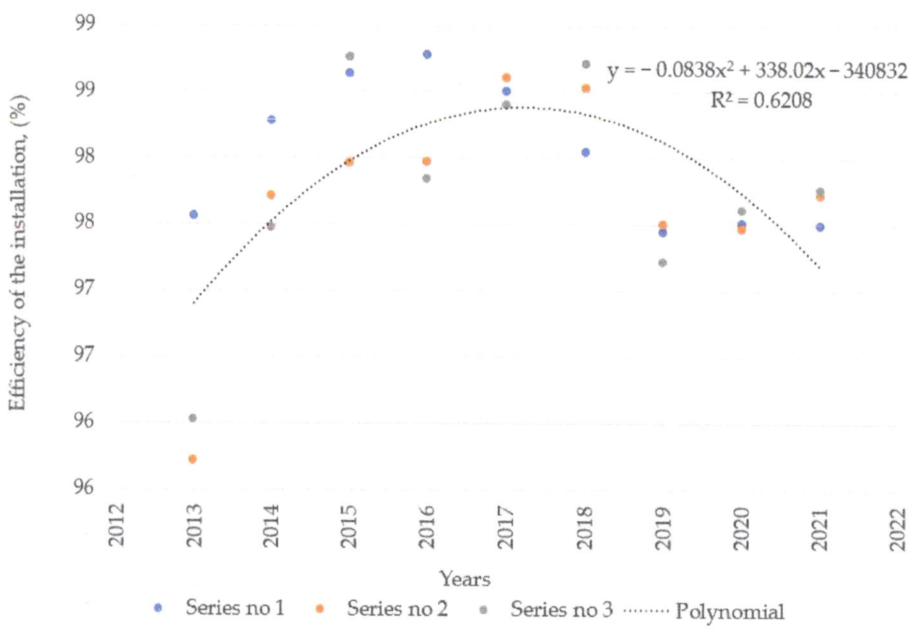

(**c**)

Figure 5. Efficiency of the dedusting installation in the Wieliczka Salt Mine in the years: (**a**) 2004–2006; (**b**) 2007–2012; (**c**) 2013–2021.

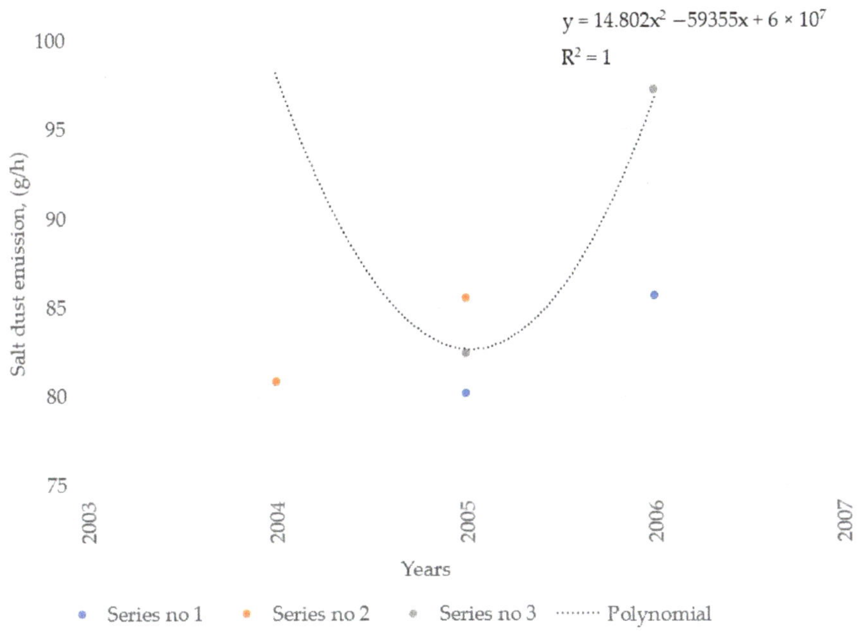

(a)

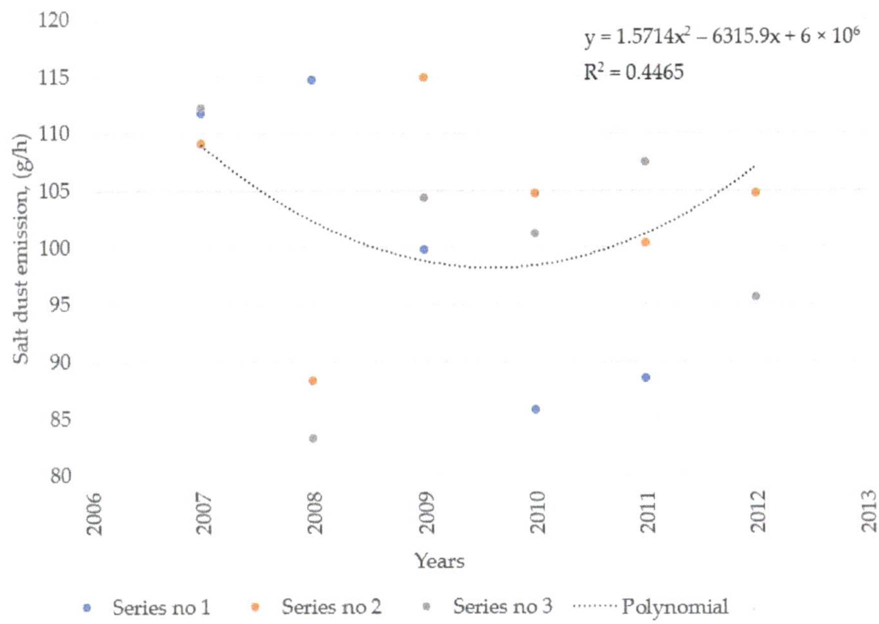

(b)

Figure 6. *Cont.*

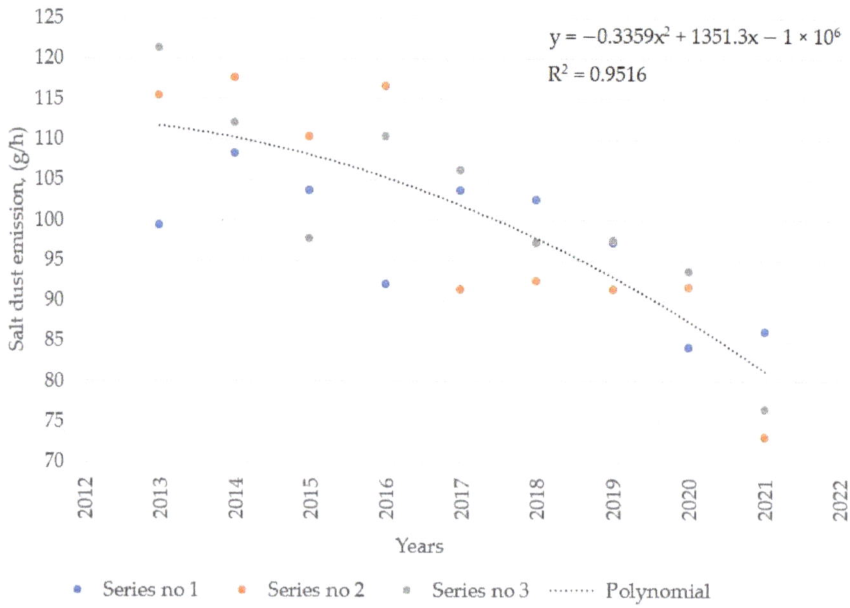

(c)

Figure 6. Salt-dust emissions in the Wieliczka Salt Mine in the years: (**a**) 2004–2006; (**b**) 2007–2012; (**c**) 2013–2021.

The efficiency of the dedusting installation presented in Figure 5b, in the adopted time intervals, showed an upward trend in comparison to Figure 5a,c. The reason for this was that in the analyzed period, the brine flowed in from various regions, which were then gradually liquidated. The first criterion for testing the salt-dust concentration is the provisions of the environmental protection law, which imposes an obligation on the enterprise to protect the earth. In the case of desalination plants, this is the prevention of contamination with risk-causing substances and the remigration of salinity due to the accumulation of soluble salts in the soil. In addition, the law provides information on the obligations to counteract adverse changes in the earth's surface by preventing and limiting the destruction of the land cover with vegetation [29]. Derogation from the regulations may be established industrial zones specified in spatial development plans as areas intended for production activities. The second criterion for carrying out measurements is the assessment of human exposure to the maximum permissible concentrations at the workplace. The concentrations of dust in the air at workplaces refer to the concentration of total dust, the concentration of the respirable fraction and the number of fibers above 5 μm [30]. However, in the case of salt dust, the regulations do not regulate the maximum permissible concentrations in the work environment, which confirms that it is not considered harmful to people. The factor that could influence its classification as harmful is the admixtures of free silica (SiO_2) or other elements, the content of which is often trace. Climate protection by improving air quality is part of the concept of the European Green Deal. Enterprises in the raw materials industry are obliged to take all measures to reduce the level of environmental pollution [31]. The activities carried out in the Wieliczka Salt Mine are in line with the trend of the proposed directions. Measurements of the chemical analysis carried out at the saltwork in the Wieliczka Salt Mine show that the pH of the brine fed to the scrubber is in the range from 6.9 to 8.1. On the other hand, the values of the individual elements are as follows: NaCl, Ca^{2+}, Mg^{2+}, SO_4^{2-},135; 0.98; 0.31; 3.86 g/L, respectively [32].

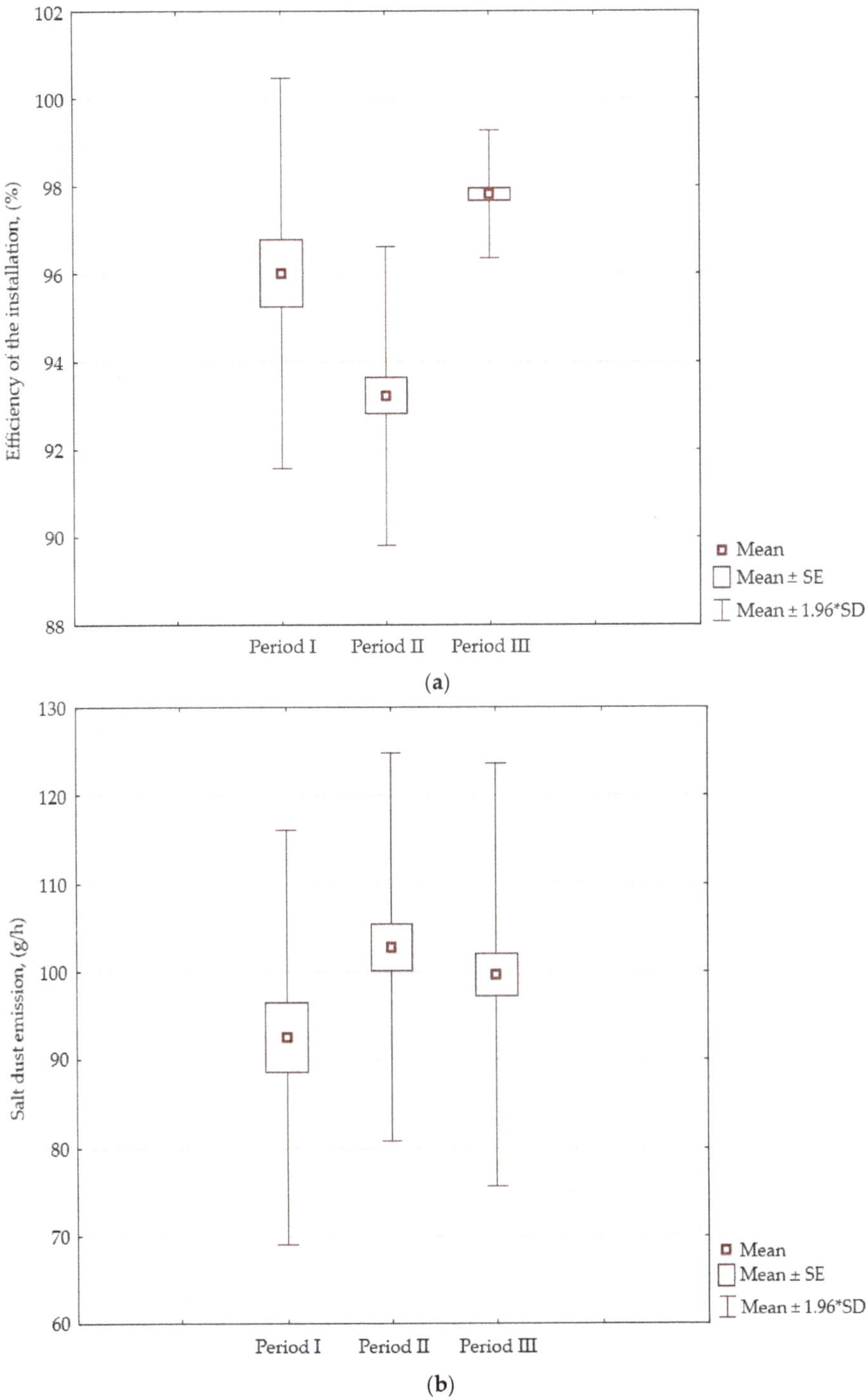

Figure 7. Three operating periods for the saltwork: (**a**) efficiency of the installation; (**b**) salt-dust emission; SE—standard error; SD—standard deviation.

5. Conclusions

The possibility of observing the emission of salt dust from the saltwork in the Wieliczka Salt Mine allows one to assess its impact on the environment and the exposure of people around the company. The fact that the tests are carried out cyclically and the results are analyzed for several years are of particular importance. In addition, the use of a repeatable test methodology as well as the use of individually designed measuring equipment allows for repeatability of results and minimization of errors. The research has been conducted for over a dozen years and covers almost the entire period of operation of the installation. On the basis of the obtained results, the following conclusions can be made:

- In the characteristics of the saltwork of the scrubber in the Wieliczka Salt Mine, three operating periods can be distinguished: start-up, correction and adjustment and stabilization of work. Comparing the individual periods with each other in terms of the maximum and minimum value, it can be concluded that: the efficiency of the installation in the third period is better by 2.63% and 2.38% compared to the first and the second period.
- The applied dedusting system is characterized by high efficiency in the range of 90.50–98.78%.

Salt-dust emission to the atmosphere is small and ranged from 73.12 g/h in 2021 to 124.34 g/h in 2012.

- The designed probe with the selector for the AS-50 gravimetric dust meter allows for measurements of dust in the desalination installation.
- Properly selected filters to work with the selector allow one to test the actual values of the salt-dust mass.

It is also worth noting that the Wieliczka Salt Mine, which is entered on the UNESCO World Heritage List, does not carry out any exploitation activities but still obtains salt in the saltwater utilization plant. Due to the amount of brine obtained, the saltwork operates 50% of the time during the year. A small annual emission of salt dust to the atmosphere causes a negligible or even marginal impact on the environment. In the future, the method of remote sensing, active or passive, can be used to compare the applied method. Such tests may allow the verification of the method used so far.

Author Contributions: Conceptualization, M.K. and K.S.; methodology, M.K. and K.S.; software, M.K.; validation, K.S. and M.K.; formal analysis, M.K. and K.S.; investigation, M.K.; resources, M.K.; data curation, M.K. and K.S.; writing—original draft preparation, M.K. and K.S.; writing—review and editing, M.K. and K.S.; visualization, M.K. and K.S.; supervision, M.K. and K.S.; project administration, M.K. and K.S.; funding acquisition, M.K. and K.S. All authors have read and agreed to the published version of the manuscript.

Funding: This research was prepared as part of AGH University of Science and Technology in Poland, scientific subsidy under number: 16.16.100.215.

Institutional Review Board Statement: Not applicable.

Informed Consent Statement: Not applicable.

Data Availability Statement: The data presented in this study are new and have not been previously published.

Conflicts of Interest: The authors declare no conflict of interest.

References

1. McClintock, M.A.; McDowell, W.H.; Gonzalez, G.; Schulz, M.; Pett-Ridge, J.C. African dust deposition in Puerto Rico: Analysis of a 20-year rainfall chemistry record and comparison with models. *Atmos. Environ.* **2019**, *216*, 116907. [CrossRef]
2. Puławska, A.; Manecki, M.; Flasza, M.; Styszko, K. Origin, distribution, and perspective health benefits of particulate matter in the air of underground salt mine: A case study from Bochnia, Poland. *Environ. Geochem. Health* **2021**, *43*, 3533–3556. [CrossRef] [PubMed]

3. Rogula-Kozłowska, W.; Kostrzon, M.; Rogula-Kopiec, P.; Badyda, A.J. Particulate Matter in the Air of the Underground Chamber Complex of the Wieliczka Salt Mine Health Resort. *Adv. Exp. Med. Biol.* **2017**, *955*, 9–18. [CrossRef]
4. Zhang, H.H.; Gu, W.J.; Li, Y.J.; Tang, M.J. Hygroscopic properties of sodium and potassium salts as related to saline mineral dusts and sea salt aerosols. *J. Environ. Sci.* **2020**, *95*, 65–72. [CrossRef] [PubMed]
5. Lutynski, A. Dust Hazards and their Control in Mechanical Processing Plants of Hard Coal Mines. *J. Pol. Miner. Eng. Soc.* **2021**, *1*, 13–18. [CrossRef]
6. Galbraith, J.H.; Hingston, F.J. Application of a directional Dust Gauge to measurement of impaction of atmospheric salt. *Atmos. Environ. Part A-Gen. Top.* **1991**, *25*, 2211–2221. [CrossRef]
7. Mottershead, D.; Gorbushina, A.; Lucas, G.; Wright, J. The influence of marine salts, aspect and microbes in the weathering of sandstone in two historic structures. *Build. Environ.* **2003**, *38*, 1193–1204. [CrossRef]
8. Semin, M.A.; Isaevich, A.G.; Zhikharev, S.Y. The Analysis of Potash Salt Dust Deposition in Roadways. *J. Min. Sci.* **2021**, *57*, 341–353. [CrossRef]
9. Korzeniowski, W.; Poborska-Młynarska, K.; Skrzypkowski, K. The idea of the recovery of municipal solid waste incineration (mswi) residues in Kłodawa Salt Mine SA by filling the excavations with self-solidifying mixtures. *Arch. Min. Sci.* **2018**, *63*, 553–565. [CrossRef]
10. Szlązak, N.; Obracaj, D. *Air Pollution in Underground Salt Mines*; Library of the School of Underground Exploitation Publishing House: Kraków, Poland, 2002; p. 5.
11. Kapusta, M. Impact of mining executive on improving occupational safety and health. *J. Pol. Miner. Eng. Soc.* **2017**, *18*, 183–193. [CrossRef]
12. Beuck, H.; Quass, U.; Klemm, O.; Kuhlbusch, T.A.J. Assessment of sea salt and mineral dust contributions to PM10 in NW Germany using tracer models and positive matrix factorization. *Atmos. Environ.* **2011**, *45*, 5812–5821. [CrossRef]
13. Boroughani, M.; Mohammadi, M.; Mirchooli, F.; Fiedler, S. Assessment of the impact of dust aerosols on crop and water loss in the Great Salt Desert in Iran. *Comput. Electron. Agric.* **2022**, *192*, 106605. [CrossRef]
14. Arenas-Díaz, F.; Fuentes, B.; Reyers, M.; Fiedler, S.; Böhm, C.; Campos, E.; Shao, Y.; Bol, R. Dust and aerosols in the Atacama Desert. *Earth-Sci. Rev.* **2022**, *226*, 103925. [CrossRef]
15. Ficker, H.; Gassner, H.; Achleitner, D.; Schabetsberger, R. Ectogenic Meromixis of Lake Hallstattersee, Austria Induced by Waste Water Intrusions from Salt Mining. *Water Air Soil Pollut.* **2011**, *218*, 109–120. [CrossRef]
16. Jain, K.R.; Edraki, M.; McIntyre, N. Controls of Wetting and Drying Cycles on Salt Leaching from Coal Mine Spoils. *Water Air Soil Pollut.* **2021**, *232*, 472. [CrossRef]
17. Tiwary, R.K. Environmental impact of coal mining on water regime and its management. *Water Air Soil Pollut.* **2001**, *132*, 185–199. [CrossRef]
18. Fooladi, M.; Ghadimi, F.; Zakariai, S.J.S.; Bonab, H.R. Influence of Physical and Chemical Material Properties on Mining Soil Erosion Processes Around Mineral Salts Company in Mighan playa, Arak, Iran. *J. Min. Environ.* **2021**, *12*, 725–741. [CrossRef]
19. Dhar, B.B. Environmental scenario of Indian mining-industry. In *Environmental Management, Geo Water and Engineering Aspects*; Chaudhary, R., Kumar, S., Eds.; Balkema: Rotterdam, The Netherlands, 1993; pp. 615–619.
20. *PN-Z-04008-7:2002/Az1:2004*; Air Purity Protection—Sampling—Principles of Air Sampling in the Work Environment and Interpretation of Results. Polish Committee for Standardization: Warsaw, Poland, 2004.
21. *PN-ISO 4225/Ak:1999*; Air Quality—General Issues—Terminolog. Polish Committee for Standardization: Warsaw, Poland, 1999.
22. *PN-91/Z-04030/05*; Air Purity Protection—Dust Content Tests—Total Dust Determination at Work Stations Using the Filtration-Weighing Method. Polish Committee for Standardization: Warsaw, Poland, 1991.
23. *PN-N-18001:2004*; Occupational Health and Safety Management Systems—Requirements. Polish Committee for Standardization: Warsaw, Poland, 2004.
24. Wacławik, J.; Cygankiewicz, J.; Knevhtel, J. *Climatic Conditions in Deep Mines*; Library of the School of Underground Mining Publishing House: Kraków, Poland, 1998; p. 21.
25. Scrubera. Scrubera HE 40 31 Operation and Maintenance Manual. In *Technical Documentation of the Wieliczka Salt Mine*; Wieliczka Salt Mine: Wieliczka, Poland, 2020. (In Polish)
26. Energos-Pro. Available online: http://www.energos-pro.pl/oferta/filtracja-procesowa/131-filtry-workowe (accessed on 31 May 2022).
27. Shengyong, H.; Yang, G.; Guorui, F.; Fei, H.; Changhe, L.; Jihua, L. Experimental study of the dust-removal performance of a wet scrubber. *Int. J. Coal Sci. Technol.* **2021**, *8*, 228–239. [CrossRef]
28. Organiscak, J.A.; Klima, S.S.; Pollock, D.E. Empirical engineering models for airborne respirable dust capture from water sprays and wet scrubbers. *Min. Eng.* **2018**, *70*, 50–57. [CrossRef]
29. Republic of Poland. Announcement of the Marshal of the Republic of Poland of September 29, 2021 on the publication of the consolidated text of the Act—Environmental Protection Law. *J. Laws* **2021**, *2021*, 1973.
30. Minister of Family, Labor and Social Policy. Regulation of the Minister of Family, Labor and Social Policy of June 12, 2018 on the maximum allowable concentrations and intensities of factors harmful to health in the work environment. *J. Laws* **2018**, *2018*, 1286.
31. Sukiennik, M.; Kapusta, M.; Bąk, P. Transformation of Corporate Culture in the Aspect of European Green Deal—Polish Raw Materials Industry. *J. Pol. Miner. Eng. Soc.* **2020**, *2*, 177–182. [CrossRef]
32. Wieliczka Salt Mine. *Technical Documentation of Wieliczka Salt Mine*; Wieliczka Salt Mine: Wieliczka, Poland, 2022; p. 5. (In Polish)

energies

MDPI

Article

Study on the Usefulness of Lean Management Tools and Techniques in Coal Mines in Poland

Marek Kęsek *, Paweł Bogacz and Marcin Migza

Faculty of Civil Engineering and Resource Management, AGH University of Krakow, 30-059 Kraków, Poland; bogacz@agh.edu.pl (P.B.); m.migza@gmail.com (M.M.)
* Correspondence: kesek@agh.edu.pl

Abstract: The article presents research on the possibilities of utilising the Lean Management method to increase the profitability of coal mining operations through cost rationalisation and production optimisation. The Lean Management method has its roots in the automotive industry, but its significant benefits can allow it to be transferred to other industries. Based on the analysis of global literature describing examples of its application in general mining, as well as the authors' own observations related to Polish hard coal mining, it was determined that the Lean Management method is not being utilised in this field. This article outlines the course and results of an original expert-mathematical study to assess the usefulness of specific Lean Management tools and techniques in hard coal mining. The study drew on knowledge obtained from selected, competent experts who demonstrated high levels of agreement in their assessments. Five core production areas were identified, and twelve widely used Lean Management tools and techniques were selected. The obtained assessments of the usefulness of Lean tools could serve as valuable guidance for mining management in the selection of methods for improving mining production in coal mines.

Keywords: mining industry; lean management; lean manufacturing; coal mining; expert mathematical study; expert concordance; profitability of production

Citation: Kęsek, M.; Bogacz, P.; Migza, M. Study on the Usefulness of Lean Management Tools and Techniques in Coal Mines in Poland. *Energies* **2023**, *16*, 7240. https://doi.org/10.3390/en16217240

Academic Editors: Sergey Zhironkin and Maxim Tyulenev

Received: 10 August 2023
Revised: 11 October 2023
Accepted: 19 October 2023
Published: 25 October 2023

1. Introduction

The EU climate policy pursued in recent years, particularly in the aspect of the European Green Deal, is unequivocally negative towards hard coal as an energy resource. A number of directives, strategies and regulations introduced aim to completely decarbonise the energy sector in the European Union by 2050. In the document entitled Energy Policy of Poland 2040 (PEP2040), published in 2021 [1], the share of coal in electricity generation in Poland in 2030 is projected to be between 37% and 56%, and in 2040 between 11% and 28% (depending on the price of CO_2 emission allowances).

On the other hand, Poland still has significant hard coal deposits. At the end of 2020, the total hard coal balance resources in Poland amounted to 64.4 billion tonnes, of which 28.4 billion tonnes were developed resources. Most of the resources are thermal coal deposits [2]. The above facts allow one to conclude, therefore, that Poland's energy sector is and will continue to be in the coming decades, largely based on hard coal. The Russian military aggression against Ukraine in February 2022, in addition to a number of other factors, has also had a significant impact on the hard coal sector and the European energy sector as a whole. The situation is dynamic and difficult to predict, but strong signals about the need for independence from Russian energy resources, coming from almost the entire European Union, indicate that energy security objectives may slow down the EU's decarbonisation plans. Confirmation of such a possibility can be found, inter alia, in the Assumptions to the Update of the Energy Policy of Poland until 2040 [3], prepared by the Ministry of Climate and Environment and adopted by the Polish Council of Ministers on 29 March 2022. The document indicates, inter alia, that Poland's updated energy policy must also take into account energy sovereignty, the use of domestic hard coal deposits may be periodically increased in situations where the energy security of the state is threatened, and the rate of reduction in the extraction and use of Polish coal may decrease slightly compared to previous scenarios [3]. PEP2040 indicates that in the long term, due to increasing environmental requirements and decreasing demand of the economy for fossil fuels, including hard coal, the role of these products in the economy will gradually decrease. During the transition period, the key task of mining companies is to continuously take measures to increase the efficiency of their operations as well as the competitiveness of their products [1].

As the first objective to increase the profitability of hard coal mining, the above-mentioned document mentions the rationalisation and optimisation of current operating costs and the sales system as well as the creation of stabilising mechanisms for periods of downturn [1]. One such solution could be the implementation of the Lean Management methodology. Unfortunately, there are few good examples of its application in Polish coal mines and above all, there is a lack of research on the usefulness of individual Lean tools in mining conditions.

The following article introduces the Lean Management method, which is currently one of the most popular process improvement methods in industry, presents concepts and examples of practical application of the method in the mining industry, describes the course and presents the results of a mathematical and expert study on the usefulness of Lean Management tools and techniques in the operating conditions of Polish coal mines.

2. Materials and Methods

2.1. Lean Management Method and Its Tools

Lean Management is a universal concept, especially in terms of the numerous successful implementations of this method in both the manufacturing and service sectors [4], and is one of the best-known and most widely used methods for the operational management of companies. It is also referred to as Lean Manufacturing, Lean Production and Lean for short [5]. Lean Management is defined in several ways in the literature: as a philosophy (e.g., [6,7]), a system (e.g., [8,9]), a concept (e.g., [5,10]), a methodology (e.g., [11,12]) or finally a method (e.g., [13–15]). The choice of the appropriate term should always be determined by the scope of the deliberations.

The Lean Management method has its roots and foundations in the TPS—the Toyota Production System (after the name of the automotive corporation in which this system developed over decades) [5]. On the other hand, the concept of Lean in the context of production management was first used in 1988 by J. Krafcik in his article, "Triumph of the Lean Production System" [8], with references precisely to the system used in Toyota factories. The idea of Lean Manufacturing was popularised by J. P. Womack, D. T. Jones and D. Roos. In 1990, they published the famous paper, "The Machine That Changed the World" [16], in which they pointed to the Toyota Production System as the first lean manufacturing system. The fathers of the Toyota Production System are considered to be Toyota president E. Toyoda and T. Ohno, director of one of Toyota's factories (and later vice-president) [17]. Toyota's ambition was to create an original Japanese production technique, adapted to the country and its culture and taking into account the new business environment. The Japanese did not force themselves to build everything from scratch but drew on American models [18]. Toyota was inspired by some of the solutions used in Ford's production system, but taken as a whole, the two systems were radically different. The key differences between the two are indicated in Table 1.

Table 1. Comparison of Ford and Toyota production systems [19].

Ford System	Toyota System
Producing large quantities of the same product—mass production	Production of small batches of differentiated products
Strip production—the direction to forward (earlier process determines the subsequent process)	Just in Time and Kanban—the direction "backwards" (the later process determines the earlier process)
Automation	Automation integrated with the human factor
Large production batches	Small production batches and rapid product changes
Large stocks, warehouses	Elimination of stocks—dynamic warehouses called supermarkets
Overproduction and defective products	Avoidance of overproduction and elimination of defective products
Planning	Response

Professor J. K. Liker, in his book *The Toyota Way* [20], outlined 14 principles that Toyota followed in implementing its production system. These principles are used for the implementation and application of Lean Management.

Liker also proposed the concept of the 'Toyota House' as a graphic representation of the Lean philosophy which is shown in Figure 1. The roof of the Toyota House represents the predominant goal of the organisation. It is based on two pillars: Just in Time, or precisely on time, and Jidoka, i.e., embedding quality into the production process. The foundations of the House of Toyota, and at the same time the entire Toyota Production System philosophy, are levelled production and process standardisation. Between the foundation and the roof, two forces at work lead to continuous improvement: a top-down one, i.e., motivating employees and a culture of teamwork, and a bottom-up one, i.e., eliminating waste.

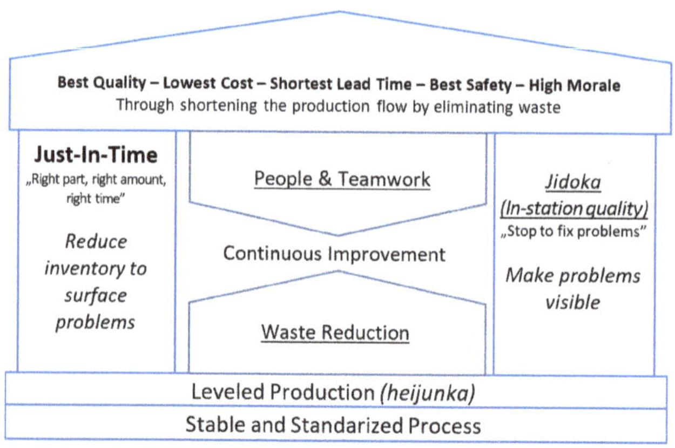

Figure 1. The Toyota House Model [21].

The success of TPS as a new approach to process management resulted in Toyota's rapid expansion in the global market and excellent financial performance, which attracted the attention of academics and practitioners alike—owners and managers of various companies around the world. As a result, the idea of lean manufacturing was quickly popularised [22]. Lean Management, based on the Toyota experience, places a very strong emphasis on eliminating waste in processes. In Lean, anything that does not add value from the customer's point of view is considered waste. One of the key assumptions of this method is to thoroughly analyse the processes (e.g., using value stream mapping—VSM) and to assign each activity to a group of value-added activities from the customer's point of view or to non-value-added activities [23]. Anything that does not add value in the eyes of the customer is, in the Lean method, waste (Japanese: muda). One of the founders of

Lean—T. Ohno—distinguished seven key process losses, commonly known as the '7 muda wastes' or '7 wastes of Lean' [24]: overproduction, waiting, transport, overprocessing, inventory, unnecessary movement and quality defects. The Lean Management method uses a number of tools and techniques to help the organisation reduce waste and add value for the customer, including [25]:

- 5S—A tool to better organise the workplace (and the whole factory) and to help maintain order. 5S is an acronym from the words: Selection; Systematics; Cleaning; Standardisation; Self-discipline (Japanese: Seiri, Seito, Seiso, Seiketsu, Shitsuke), (translated as 'sort', 'set in order', 'shine', 'standardise' and 'sustain');
- 5Why—The practice of asking the question "why?" several times at the origin of a problem in order to determine the root cause;
- Andon—Light (or sound) signalling on production lines for rapid notification of problems;
- Continuous flow—The continuous production and transfer (without downtime or storage) of individual pieces or small batches in a process. Production of only those items for which there is a current demand at the next workstation;
- Continuous Improvement—Continuous efforts to improve products, services or processes. It most often uses the PDCA and Kaizen approaches;
- Ishikawa diagram—A diagram that illustrates cause and effect relationships and separates causes from effects. Otherwise known as a fish diagram, a fault tree diagram, a cause-and-effect diagram,
- Spaghetti diagram—A technique for visualising the mobility path of workers, products, semi-finished products or materials. Most often, the drawing of movement paths is superimposed on the layout (diagram) of the plant;
- Flow Production—A way of organising mass production to minimise flow time through the process. Involves the standardisation of raw materials and parts, operation times and appropriate process layout and scheduling. It is otherwise referred to as flow production;
- Gemba—The place where the actual work is done, where value is added to the product or where the service is provided;
- Genchi Genbutsu—The principle of verifying an existing problem or event at the actual site of its origin, related to investigating the problem at its source;
- Heijunka—A technique of properly planning and balancing production to avoid downtime and stockpiling to level production;
- Hoshin Kanri—A strategic management principle aimed at integrating, unifying and organising the key elements of a company's strategy;
- Jidoka—Organisation of production systems to autonomously detect and eliminate errors and deviations from accepted standards. Most often associated with the separation of the machine from the operator and its autonomisation;
- Just in Time (JIT)—A production system for producing what is needed, at the right time and in the right quantity, to minimise wastage in the process;
- Kaikaku—Radical, revolutionary process improvement, often requiring investment. It is the opposite of, but also a synergistic complement to Kaizen;
- Kaizen—The continuous, gradual improvement of processes and organisations through the use of the small-steps method. Also linked to the employee idea system;
- Kanban—A Pull system production control tool using cards, representing an internal order for a component;
- One Piece Flow (OPF)—A process organisation that allows components (semi-finished products) to be processed and transferred through the process one piece at a time. It involves a transport and production batch equal to 1, otherwise understood as a one-piece flow;
- Overall Equipment Effectiveness (OEE)—An indicator of the total use of machinery. It is the product of machinery availability, actual productivity and production quality;
- Plan, Do, Check, Act (PDCA)—An algorithm for conducting improvement activities in the spirit of Continuous Improvement: Plan; Do; Check; Apply;

- Poka-yoke—An approach to preventing inadvertent errors. It is understood as any tool that prevents an error/quality defect from occurring in a process, independent of the operator. It is usually uncomplicated, cheap and quick to prepare;
- Pull system—A production system in which the start of a job at a station is initiated by a demand at the next station in the process. Orders are generated in the process from the last step (from the end) backwards. It is referred to as a suction system;
- Single Minute Exchange of Die (SMED)—A technique for minimising machine changeover times;
- Supermarket—An inter-operational warehouse that supports self-controlled production. It allows retrieval and replenishment of stock only for a certain number of components, thus preventing overproduction. It is an important component of the suction system;
- Total Productive Maintenance (TPM)—A maintenance management technique aimed at ensuring maximum availability of equipment. It strives to achieve zero breakdowns, zero shortages and zero accidents at work;
- Visual Management (VM)—A set of techniques and ways to visualise problems, events, information and objectives. An improvement in communication and process management.

2.2. Concepts for Using Lean Management in Mining

The global mining industry is increasingly looking for proven organisational solutions to reduce mining and mineral processing costs. These solutions are often imported from other industries. Lean management is an example of a method that is receiving increasing recognition in this sector [4].

The Lean method has emerged and developed over decades mainly in the automotive production environment. Despite numerous examples of applications in various other processes and industries, it is the automotive production halls that should be considered its 'natural environment'. Lean requires full, consistent and methodical implementation and adaptation of tools and techniques to local conditions [26]. As the production conditions in mines differ significantly from those of an automotive company, the implementation of the method in the conditions of a mining company must, therefore, differ from that of an automotive company.

A comparison of conditions in the mining industry with those in the automotive industry in the context of the possibility of implementing Lean was presented in the works of K. Dunstan [27] and J. Helman [28]. The authors conducted their analysis mainly from an ore mining perspective, and the results of their analysis are presented in Tables 2 and 3.

The authors of both papers agreed on a number of important differences between the two industries, including the volatile, unstable working conditions in mining versus the stable, safe conditions of the production halls and the built-in need for a push system in the mining sector.

Table 2. Comparison of the specifics of the raw materials and minerals industry and Automotive [27].

Raw Materials and Minerals Industry	Automotive Sector
A smelter or refinery cannot be stopped hence the Push system is built into the process	The assembly line can be stopped, so it is possible to create a suction system
Continuous production, 24 h a day	Discrete production (in units), often in cycles of less than 1 day
Generates considerable dust	Low dust
Demanding working environment	Stable, secure working conditions
Variable working environment	Stable working environment
Remote locations	Large centres
Impact of the weather	In-house conditions
High variability of material availability	Material availability under control
Geographically dispersed teams	Relatively small factories
The molten metal has a short shelf life before it solidifies	Components suitable for long-term storage and warehousing

Table 3. Comparison of the specificities of the mining industry and the automotive industry [28].

Mining Industry	Automotive Industry
The work of customers cannot be stopped thus, a push system is in operation at the mine	The assembly line can be stopped, so transformation to a suction system is possible
Continuous production	Production in cycles
Unstable/variable working conditions	Stable working conditions
Variable working environment	Permanent working environment
Geological hazards may halt production	No threat to production
High variability of material availability	Material availability under control
Large spread of workplaces (up to several kilometres)	Working in a relatively small factory
Other industrial companies are customers of the mines	Sales of products mainly to individual customers

The possibility of replacing the push system with a Pull system (Pull system) in mining was discussed in more detail by J. Helman in another publication [29], demonstrating that there is no substantive justification for attempting to apply a Pull system to strictly mining processes. Nevertheless, elements of the Pull system can be implemented in areas of a mine where there is a flow of different types of materials [29]. At this point, it is also worth citing the work of M. Sukiennik and P. Bak [30], which considers the possibility of implementing Lean in the broader power industry. The paper lists important elements that differentiate this industry from most industrial sectors [30]:

- Process continuity, which is required in the power industry but not in typical companies;
- Stability of the working environment—in the power industry, this is basically nonexistent;
- Diversification of suppliers—virtually impossible in the power industry;
- The need for specialised staff in the power industry is present in many areas of operation, which makes staff turnover difficult and contributes to problems in a situation of demographic decline;
- In the power industry, companies are closely linked to other entities, which is not the case in typical industries;
- The need to maintain a certain level of production that determines the actions of power companies.

The article referenced above highlights the cultural aspect of Lean implementation outlines the elements that can contribute to implementation failure, and proposes three main steps for implementing Lean in the power industry [30]:

- Initial implementation;
- Proper implementation;
- Developing a culture of continuous improvement.

J. Korski, in his work [31], points to Lean Management as the peak current system approach to production process management aimed at customer satisfaction and economic efficiency. The article also comprehensively describes the possibilities and conditions for applying this approach to mining companies.

In the work of V. Mikhalchenko and Y. Rubanik extensively discussed the five principles of Lean Management in the context of Russian coal mining, concluding that the implementation of Lean principles is possible, but will require a fundamental change in the basic principles of design and management of industrial coal mining systems. A qualitative increase in the efficiency of resource use through the implementation of a new approach to shaping production systems on Lean principles will contribute to a significant increase in the competitiveness of mining enterprises in modern economic conditions [32].

For the sake of balance, it is worth citing a more sceptical paper by S. Haugen, who, based on her mining experience, discusses in detail the 14 principles of the Toyota Production System in the context of metal mining. The paper emphasises that the implementation of continuous flow and the reduction of inventory buffers brings problems to light, motivating improvement, but before buffers are eliminated, it is important to ensure that we are able to deal with new problems that arise because of this [33].

S. Haugen points out that a stable, predictable and repeatable production process is a prerequisite for Lean principles or tools to work properly, and mining processes are not such. Furthermore, Lean contributes to lower production costs by increasing flexibility and speed (and not just eliminating waste). If this effect is not desirable for a mining company (e.g., due to continuous high market demand and low value of capital frozen in work-in-progress), Lean may not be a good solution [33].

Several other works, in addition to presenting a general approach to the application of Lean in mining, propose a more detailed analysis—at the level of specific tools. In an article discussing the possibilities of implementing Lean in underground ore mining [34], A. R. Wijaya, R. Kumar and U. Kumar gave examples regarding the occurrence of eight Lean losses in the mining industry and analysed the feasibility of implementing selected Lean principles and tools. The results are presented in Tables 4 and 5.

Table 4. Incidence of eight losses in mining [34].

Loss	Occurrence in Mining
Waiting	Waiting for dust to be removed after blasting operations, downtime of transport equipment.
Overproduction	Overproduction is not a problem for the mining industry, as the market is stable, and it can be said that it can always absorb any amount of production [sic!].
Quality (repairs)	Repairs in mining often are the result of the very nature of the changing working environment, but it is possible to increase the quality, for example, by standardising procedures.
Unnecessary movement	Loss related to the manner in which work is carried out. It can be caused by a lack of a formal, standardised approach to training new operators.
Redundant processing	Oversized tunnels due to the instability of the rock casing, but also to the operator's inexperience.
Storage	Stockpiling resulting from production downtime (breakdowns). A loss that is difficult to combat in mining, as stockpiles are often treated as a positive thing here.
Transport	Losses due to suboptimal choice of means of transport, place of collection, size of equipment.
Untapped potential	Under-utilisation of operators' time, as well as productivity and capacity losses.

Table 5. Feasibility of implementing selected Lean principles and tools in mining [34].

Principle/Tool	Application in Mining
Standardisation	Often made considerably more difficult due to the varying conditions of the working environment, but possible to implement in many areas, such as anchoring quality control.
Total Productive Maintenance (TPM)	In practice, due to the remote location of the work site, the operator is encouraged to perform simple maintenance, but often operators lack the required knowledge and skills to do this correctly.
5S and Visual Management	Difficult, due to frequent subcontracting in the industry. Subcontractors do not engage in 5S and VM implementation as this is not part of the contract.
Just In Time (JIT)	One of the weakest points in the implementation of the Lean concept in mining research into the individual adaptation of JIT to mining conditions is needed.
Jidoka	The separation (and collaboration) of machine and humans is an important element of change in mining, seeking to increase automation.
Respect for the people	The need for a structured approach to training due to the high workforce.

In addition, J. Helman, in the article cited previously [28], presents an assessment regarding the adaptability of exemplary Lean Management tools in mining, as shown in Table 6.

Table 6. Adaptability of Lean Management tools in mining [28].

Tool	Adaptability	Examples of Implementation Areas
Just In Time	Yes	Order system and all warehouses.
One Piece Flow	Not explicitly	Machine and operator flow diagrams, cross training.
Total Productive Maintenance	Yes	Vehicles, conveyors and other machinery.
5S	Yes	Warehouses, tool rooms and other areas where materials and equipment are stored.
Kanban	Yes	Storage facilities in the Heavy Machinery Chamber, engine room, and shaft bottom.
Heijunka	Yes	All sites use Kanban cards.
Continuous improvement	Yes	All miners, foremen, etc.

J. C. Yingling, R. B. Detty and J. J. Sottile describe in detail the Lean concept and the possible benefits of its potential application in US coal mines. The authors point to the possible application of the following Lean principles and tools in specific areas of the mine, among others [35]:

- TPM;
- SMED;
- Pull system;
- Standardisation/standardised work;
- Continuous flow;
- JIT;
- Kanban;
- PDCA;
- Kaizen.

As indicated in the above-cited article [35], many of the elements of the SMED tool are already applicable to longwall changeover processes. Also described are tools and principles (e.g., flexibility, production levelling, flow design), which, according to the authors, may be impossible or difficult to translate directly to mining conditions [35].

In their article, A. Bator and A. Paluchniak [36] proposed the use of one of the main elements of the Lean method—the 5S tool (extended by the sixth S, which stands for Safety) to increase work safety in mines. The paper takes a closer look at 5S and points to the possibility of its application in the Polish coal mining industry in order to avoid many accidents, especially those of a fatal nature.

In 2016, J. Brodny and K. Stecuła pointed out the validity of applying the TPM concept to reduce costs in coal mines by increasing the efficiency of machine utilisation. They described the most important TPM indicator (and at the same time, one of the most important in the whole Lean method)—the OEE, i.e., the machine utilisation efficiency index. The paper proposes a method of obtaining the data for calculating the indicator and presents the results of an analysis regarding 16 work shifts for a longwall complex consisting of a shearer, a longwall conveyor, a face conveyor and a crusher. According to the authors' calculations, the effective utilisation of the entire machinery set during the study period was less than 54%. Even having taken into account the specifics of mining production, this result was considered far too low compared to other production industries. As the authors themselves pointed out, it is difficult to assess the results obtained against other mining companies due to the lack of reliable data to which they could be compared [37].

A similar concept of using the OEE indicator in measuring the efficiency of a longwall complex was presented in detail by R. Polak [38]. The paper also presents an analysis of data from over 850 failures and determines the distributions of two other important Total Productive Maintenance indicators: MTBF (Mean Time Between Failures) and MTTR (Mean Time To Repair).

The vast majority of publications present the authors' own view on the feasibility of implementing Lean in mining, based on their knowledge of the issue and the literature, but rarely supported by the results of wider research or documented implementation trials. For this reason, the following section discusses examples of practical applications for the Lean Management method in the Polish and global mining industry.

2.3. Examples of Lean Management Method Application in the Polish and Global Mining Industry

The literature presents numerous examples of mining companies that have attempted to implement the Lean method. Alcoa was probably the first raw materials company to use the Toyota Production System and developed its own version, called the Alcoa Business System (ABS) [39]. According to the International Quality and Productivity Centre (IQPC), since the combination of two business improvement strategies, Lean and Six Sigma, began to permeate the Australian mining industry, there has been strong evidence demonstrating that overall process improvement has been achieved, resulting in faster and higher levels of return on investment [40].

A. Klippel, C. Petter and J. Antunes, in a 2008 article [41], reported on two cases when ore mining process mapping based on the Lean concept was applied. The examples cited involved the use of value stream mapping to classify activities as value-adding or non-value-adding (needed and unnecessary from the customer's point of view). Specific actions were taken to eliminate non-value-adding activities and minimise waste, resulting in significant benefits in increased productivity, reduced costs and improved workplace safety [41].

The Rio Tinto Group began implementing Lean in aluminium ore mining in 2004 as a complement to its Six Sigma improvement programme. Subsequently, the project was expanded to include copper ore mining as well as coal and iron ore operations, mainly in Australia. Building on Rio Tinto's understanding of Lean ('the continuous removal of waste'), K. Dunstan, B. Lavin and R. Sanford proposed that Lean in practice consists of [27]:

- Involving cell leaders;
- Asking employees to respect the agreed standards for their work;
- Enabling employees to write their own standards and improve them;
- Visual presentation of key production performance data (visual management);
- Enabling shopfloor staff to make data-driven operational decisions;
- Organising operational and maintenance staff into production teams;
- Applying the business improvement toolkit.

The Rio Tinto Group named its Lean Six Sigma programme 'IPT' (from Improving Performance Together) and applied it to all its business units [42].

Rio Tinto Alcan used Six Sigma and Lean Manufacturing tools to streamline processes and increase productivity. In 2008 alone, savings from the use of Lean Management and Six Sigma at Rio Tinto Alcan amounted to more than $28 million [43].

BHP Billiton ran a similar programme called Business Excellence. Several of the group's oil sands operations in Canada have begun to apply Lean and Six Sigma principles and methods to reduce defects and wastage in the production operations of the bitumen materials creation process [39].

Quadra FNX Mining (now KGHM International) had its own programme, called 'Rising the Bar'. The programme was based on the appropriate application of Lean Six Sigma tools, starting with value stream mapping for the entire process, which establishes a benchmark for performance using standard definitions and KPIs (Key Performance Indicators) [39].

The Diavik diamond mine also focused resources on the Lean Six Sigma business improvement system. The method at Diavik aimed to improve the efficiency of operations by removing duplicate or redundant activities and automating processes as much as possible. The focus was on rethinking processes in depth and defining how they can be done in a way that is better, faster and cheaper [44]. All departments at the Diavik diamond mine were required to find ways to reduce costs. The automation of underground processes and the Lean Six Sigma initiative are examples of sustainable operations at Diavik. A success story was that ideas to reduce cost intensity were generated by employees and subcontractors at all levels of the organisation. Total savings for the first six months of 2013 amounted to $6 million [44].

The above-mentioned implementations at Rio Tinto, BHP Billiton and Diavik are examples of using a combination of Lean Management and Six Sigma methods. Lean tends to focus on day-to-day small improvements, while Six Sigma focuses on large-scale, often months-long optimisation projects that bring leaps in productivity, quality improvements and cost reductions. Six Sigma, despite placing the utmost importance on hard statistical and analytical tools, often reaches for Lean Management tools in implementation projects. Lean itself, on the other hand, fares somewhat less well with highly automated processes. The combination of both concepts often creates a synergy effect [45].

A paper by Castillo et al., published in 2015, presented the results of a study on the impact of Lean Management implementation on labour outcomes in three copper mines in Chile. The study used a survey questionnaire administered to area managers. The survey form asked about changes in the following areas [46]:

- Design effectiveness (disruption, physical work progress, programme reliability, productivity, efficiency);
- Organisational effectiveness (teamwork, participation, communication, commitment and learning).

The study indicated a statistically significant improvement in both design and organisational efficiency in all three mines studied.

In the underground coal mining industry, a good example of the application of Lean was found in the German mines of RAG AG, which are no longer in operation due to depleted resources. Measures to eliminate waste and standardise processes began there in 2000, under the slogan 'Zero accidents and 100 per cent added value'. Lean processing guidelines for this programme are [47]:

- Creating value from the customers' point of view;
- Look at the entire value stream;
- Pull principle for processes;
- Synchronised flow;
- Seek perfection.

RAG adapted many existing Lean tools to the mining context and developed its own specific tools for mining processes. Two examples of this transfer were the SMED tools and the PDCA cycle. RAG called these the 'Boxenstopp' and the APUC cycle, or Analysis, Planning, Implementation, Controlling (German: Analyse, Planung, Umsetzung, Controlling) [47].

In 2000, RAG started a pilot implementation of Lean in the business processes of one of its mines. The necessary Lean tools (e.g., 5S, Visual Management, Boxenstopp/SMED, Pull system) were successively implemented into the daily practice of the employees. Following the initial successes of the implementation of Lean and its tools, implementations were also launched at the group's other mines. Subsequently, the implementation was extended to support processes such as machine maintenance and logistics, as well as administration and management processes [48].

RAG has also defined its Lean principles to be applied in process development by employees and management [48]:

- Create customer value, which means stable, efficient processes without waste;
- Look at the whole value stream, not just its elements;
- Pull principle, the customer or the next stage of the process initiates the work in the position before;

- Implement a continuous improvement process.

In RAG production, the most frequently used tools were visual management (VM), SMED and the Pull system. An important aspect was the interaction between production and the other departments. The idea of the seven Lean losses was presented and discussed at many meetings, workshops and training courses. This enabled employees to look at processes from a broader perspective, which helped to identify losses and propose solutions to problems. The search for the seven losses in the processes became part of daily work and part of continuous improvement [48].

While analysing the above examples, it can be seen that mining companies tailor their Lean Management implementation programme to their needs, often using an individual name for their programme [49].

In the article "Study on Coal Lean Mining Theory and Practice", Z. Liu cited the example of one brigade (in an underground coal mine in China, using the pillar system) where, using the Lean approach, it was possible to reduce the operational time of performing activities by 77 min (16.5%) [50].

In their article, E. M. Ade and V. S. Deshpande described the basic definitions and principles of Lean Management and gave an example of the implementation of an overhead rail to transport miners to the face in one coal mine in India as an example of Lean Thinking [51].

An example of a similar improvement in the spirit of Lean from Poland was given by A. Bator et al. [52], as well as P. Bogacz and M. Migza [49]. In the Mysłowice-Wesoła mine, a belt conveyor used only for transporting people was implemented, which made it possible to significantly increase the availability of workers at the actual work site [53] following [52]. A similar, albeit on a larger scale (a longwall with the longest panel length in the world) implementation took place at LW 'Bogdanka' [54]. Another example given by the above-cited authors is the significant increase in the utilisation of machinery and equipment at the Silesia mine, mainly through the implementation of a 24/7 working system, but also through the introduction of planned maintenance and increased flexibility of the workforce—each brigade is universal in terms of qualifications and must be able to, both, mine and perform basic rebuilding [55] after [52]. The aforementioned article points out that the factors that most affect the efficiency of mines are the cycle time of the operations being carried out, the productivity of the machines and the use of human resources, and that all these factors can be optimised by applying Lean methodologies. For this to be successful, as a first step, mine management should develop a training programme to develop awareness of Lean Manufacturing among mine employees [52].

The need for Lean training, but mainly among managers of coal mines, was also indicated by D. Sztajerska and M. Bogdański. According to the authors, without equipping managers (at various levels of the organisation) with knowledge of modern process management concepts, mainly Lean Management and Lean Manufacturing, it will not be possible to improve processes in a mine [56].

The cited article by these authors also describes implemented improvements in the maintenance process of one of the Polish Mining Group S.A mines. One of the implemented solutions was a standard failure card, which is shown in Figure 2. The authors indicate that on the basis of the implemented failure cards, it is planned to perform root cause analysis using, among others, the Ishikawa Diagram and the 5Why tool [56].

The best-documented examples, including the adaptation and implementation of Lean elements in Polish underground mining, refer to the implementation at KGHM Polska Miedź S.A. as part of the project Adaptation and Implementation of Lean Methodology in Copper Mines.

The following areas of the above project have been designated [57]:

- "Lean Mining" area;
- The area Kaizen—Employee Ideas;
- The "TPM" area;
- The area "Process Approach";

- The area "Modelling and simulation".

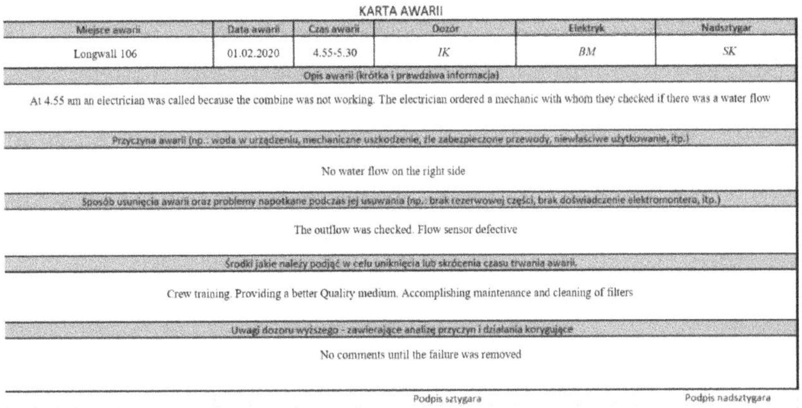
KARTA AWARII

Miejsce awarii	Data awarii	Czas awarii	Dozór	Elektryk	Nadsztygar
Longwall 106	01.02.2020	4.55-5.30	*IK*	*BM*	*SK*
Opis awarii (krótka i prawdziwa informacja)					
At 4.55 am an electrician was called because the combine was not working. The electrician ordered a mechanic with whom they checked if there was a water flow					
Przyczyna awarii (np.: woda w urządzeniu, mechaniczne uszkodzenie, źle zabezpieczone przewody, niewłaściwe użytkowanie, itp.)					
No water flow on the right side					
Sposób usunięcia awarii oraz problemy napotkane podczas jej usuwania (np.: brak rezerwowej części, brak doświadczenie elektromontera, itp.)					
The outflow was checked. Flow sensor defective					
Środki jakie należy podjąć w celu uniknięcia lub skrócenia czasu trwania awarii.					
Crew training. Providing a better Quality medium. Accomplishing maintenance and cleaning of filters					
Uwagi dozoru wyższego - zawierające analizę przyczyn i działania korygujące					
No comments until the failure was removed					

Podpis sztygara — Podpis nadsztygara

Figure 2. Completed equipment failure card at a coal mine [56].

Methods with potential for being applied in the mining industry in the areas identified above are Total Productive Maintenance, 5S, standardisation, Kaizen and PDCA [57]. The implementation of Lean elements at KGHM Polska Miedź S.A. took place in both the mining and processing parts.

The effects of TPM implementation in O/ZWR KGHM Polska Miedź S.A. include [58]:

- Elimination of the tube-and-chain conveyor for the crusher dust extraction system, thereby achieving zero breakdowns and saving 6.0 kW per shift;
- Reduction in downtime by 11%, including the planned downtime by 8% and unplanned downtime by 45%;
- Suction piping improvements—threefold increase in failure-free operation.

Further work utilising the concept and specific Lean Management tools at KGHM Polska Miedź S.A. included value stream mapping of the current and future state for the shift foreman's workflow (scope: loading, hauling, anchoring and drilling) [59], reduction of administrative documentation maintained by a foreman and a pilot programme to implement the Kaizen concept. An employee idea submission programme was introduced, and some staff were trained in problem-solving methodologies: FMEA and Ishikawa Diagram. The pilot programme resulted in 29 Kaizen ideas [59].

The example of KGHM Polska Miedź proves that the implementation of Lean tools in underground mining is not only possible but can also bring real benefits. The examples cited regarding the application of Lean tools in a copper mine do not show any characteristics that could negate the possibility of application in other underground mines, such as coal mines [4]. Examples of successful Lean implementations in mining show that it is an attractive development path [60].

Rational implementation of Lean Management tools in coal mines can result in significant savings. For this to be possible, an individual approach of implementers and Lean experts is required, i.e., not copying ready-made solutions from other industries, but expert selection (and adaptation) of Lean tools to the conditions of the mine. On the part of mine management, a greater openness to organisational change and a strong commitment to the implementation process itself is required [4].

An analysis of the world literature on the possibilities of implementing Lean in the mining industry, with examples of implementation, can also be found in works [61–63].

As indicated in the above analysis, there are examples in the literature of implementing Lean Management elements in the Polish coal mining industry, but unfortunately, they are few and usually concern a limited scope of work. The cited cases of success in implementing Lean in the global coal mining industry, as well as a good example of implementing elements of this method in Polish underground ore mining, suggest that the implementation of Lean could be very beneficial for Polish coal mines. Unfortunately, there is a lack of research indicating which Lean Management tools and techniques could be most useful in this sector.

2.4. Expert Mathematical Study

In order to investigate the suitability of Lean Management tools and techniques in coal mines and thus to confirm the theses and individual examples of Lean implementations cited above, the authors of this article used an expert-mathematical method.

The expert-mathematical method, also known as the method of competent judges [64], remains an effective method for forecasting and scientifically solving complex tasks. Its essence consists of the use of data obtained as a result of a scientifically justified procedure of collecting, systematising and analysing information from experts in a given field [65]. It is a method using, among other things, elements of the Delphi method, group evaluation of experts, analysis of experts' competence and concordance analysis (concordance of experts' evaluations by Kendall's W coefficient).

2.4.1. Lean Management Method and Its Tools

The work began with the design of a research sheet in line with the requirements of the expert-mathematical method. Its most important function was to collect expert assessments of the usefulness of Lean tools and techniques in the conditions of Polish coal mines.

In the next stage, five basic production areas of Polish coal mines were determined (own elaboration based on [66]), for which the usefulness of individual Lean Management tools and techniques was to be determined:

- Accessibility works;
- Preparatory works;
- Exploitation of the deposit;
- Logistics (haulage of excavated material and transport of materials);
- Processing of hard coal.

Twelve widely used tools and techniques of the Lean Management method were then selected:

- 5S;
- TPM—Total Productive Maintenance
- Ishikawa diagram;
- Spaghetti Diagram;
- JIT—Just in time;
- Kanban;
- VSM—Value Stream Mapping;
- Mistake proofing (Poka-yoke);
- Employee Suggestion Programme (Kaizen);
- OPF—One Piece Flow;
- SMED—Single Minute Exchange of Die;
- VM—Visual management.

The spreadsheet prepared by the authors allowed for an assessment of the usefulness of each of the 12 tools in each of the five areas of mine operation on a scale from 0 (no usefulness) to 10 (very high usefulness), for a total of 60 questions.

An additional function of the sheet was the possibility for the experts to self-assess their knowledge regarding each of these five production areas and the Lean Management method, as well as the sources of their knowledge. This structure of the sheet allowed for an analysis of the experts' competence in each of the five areas and in terms of the Lean method.

The sheet was created in the Google Forms tool and consisted of a foreword and two main parts: an expert self-assessment and an assessment of the usefulness of Lean tools at the mine.

Table 7 shows the structure of the expert self-assessment sheet for the first of the following six sections:

Table 7. Study on the usefulness of Lean Management tools in the Polish coal mining industry—part 1—experts' self-assessment.

Question	Scale Answers		
Please Specify Your Knowledge of the Issue of Accessibility Works	**From 0 to 10 ***		
Please indicate the level of influence of the various sources of knowledge on your knowledge of the issue regarding access works:	High	Medium	Low
Theoretical analysis Practical experience Knowledge of national and international literature Knowledge of the state of the issue abroad Personal intuition			

* 0 = No knowledge of the tool, 10 = Very good knowledge of the tool.

Section 1—Lean in Accessibility Works;
Section 2—Lean in Preparatory Works;
Section 3—Lean in the Field Operation;
Section 4—Lean in Logistics (Ore Haulage and Material Transport);
Section 5—Lean in Coal Preparation;
Section 6—Knowledge of Lean Management.

Table 8 shows the structure of the Lean tool suitability assessment sheet at the mine for the first of the twelve tools (5S, Total Productive Maintenance, Ishikawa Diagram, Spaghetti Diagram, Just in Time, Kanban, Value Stream Mapping, Mistake proofing—Poka-yoke, Employee Suggestion Program—Kaizen, OPF—One Piece Flow, Single Minute Exchange of Die, Visual Management).

Table 8. Study on the usefulness of Lean Management tools in the Polish coal mining industry—part 2—assessment of the usefulness of Lean tools (example for 5S tool).

Question	Scale Answers
The 5S Tool	**From 0 to 10 ***
Please evaluate the usefulness of the 5S tool for accessibility works	...
Please evaluate the usefulness of the 5S tool for preparatory work	...
Please evaluate the usefulness of the 5S tool for the deposit exploitation processes	...
Please evaluate the usefulness of the 5S tool for logistics	...
Please assess the usefulness of the 5S tool for coal preparation processes	...

* From 'Lack of knowledge of the subject' to 'Very good knowledge of the subject'.

2.4.2. Selection of Experts

The study adopted a multi-step expert qualification system. The first step was the selection of an initial group of experts, who were drawn from practitioners in the mining industry and from a group of academics. For academics, the following qualification criteria were adopted:

- Authorship or co-authorship of scientific papers in the field of broadly defined management in hard coal mining in Poland;
- Possession of at least a doctoral degree;
- For industrial workers, on the other hand, the following criteria were adopted:

- Working in a Polish coal mine;
- Senior management (survey sheets were mainly addressed to mine directors and technical directors of mine operations).

In step two, a competency analysis of the experts who took part in the study was carried out based on a self-assessment and an analysis of the experts' sources of knowledge. This part asked, among other things, about their knowledge regarding the operation of selected production areas of coal mines and about Lean Management.

Step three was to analyse the concordance of the experts' ratings. The most commonly used measure for this purpose is the concordance coefficient (Kendall's W coefficient), proposed by Kendall and Smith [67]. For a concordance analysis to be statistically significant, a certain minimum number of experts is required, depending, among other things, on the number of questions (factors assessed in the study).

The minimum number of experts is calculated from the relationship defined by Formula (1) below [68] after [65]:

$$N_E = \frac{f_\beta(b-1)}{(\gamma+1)(b-1)W_0} \tag{1}$$

where:

N_E—minimum number of experts required;
f_β $(b-1)$—the quantiles of the distribution χ^2 corresponding to the confidence level β and the number of degrees of freedom $b-1$;
b—the number of factors assessed;
γ—assumed accuracy in assessing concordance (Kendall's W);
W_0—the minimum limit value of the concordance coefficient (Kendall's W).

The multi-step approach reduces the number of experts whose knowledge is used in the study, which, on the one hand, may result in an insufficient number of experts, but on the other hand, ensures that those who have gone through all three steps of the algorithm have the necessary knowledge and competence.

2.4.3. Assessment of Experts' Competence

The expert team should be competent and composed of high-calibre specialists with (moderately) compatible views. In addition to objective indicators of competence, such as, for example, the number of scientific publications, academic title held, the number of years worked, or position held. Self-evaluation can be an important element. The practice has shown that an expert's self-assessment of his or her relative competence in various fields can be relatively well correlated with his or her actual proficiency in those fields [69] after [70].

In the survey sheet prepared according to the method of competent judges, in addition to the main subject of the experts' assessment, there is also a section on the expert's self-assessment (on a scale from 0 to 10) [71] and sources of knowledge. This allows for the subsequent elimination of expert judgements that, for example, do not feel competent in a particular part of the subject under study.

Additional verification of the expert's self-assessment is done by analysing the sources of knowledge held by the expert and determining the competence degree coefficient. The expert's degree of competence coefficient is calculated from the relationship described by Equation (2) [70,71] et al.:

$$K_k = \frac{k_z + k_a}{2} \tag{2}$$

where:

K_k—expert competence coefficient;
K_z—coefficient of knowledge of the problem in question on a point scale multiplied by 0.1, obtained from the self-assessment;

K_a—the argumentation coefficient obtained by adding up the scores from the master table (shown in Table 9), as indicated by the experts in the analysis of the impact of knowledge sources on the knowledge of the issue.

Table 9. Benchmark table for the determination of the argumentation coefficient k_a [70] i [71].

Source of Argumentation (Expert Knowledge)	Degree of Influence of the Source on the Expert's Assessment		
	High	Medium	Low
Theoretical analysis	0.30	0.20	0.10
Practical experience	0.50	0.40	0.20
Knowledge of national and international literature	0.10	0.10	0.10
Knowledge of the state of the issue abroad	0.05	0.05	0.05
Personal intuition	0.05	0.05	0.05

It is worth noting that practical experience and theoretical analysis are the most important elements when calculating the argumentation coefficient. The level of knowledge of the literature, the state of the issue abroad and, finally, the expert's own intuition has no influence on the value of the argumentation coefficient. Thus, an expert with a low influence of practical experience and no theoretical research on the issue will obtain the lowest possible argumentation coefficient value. Such a result may not be sufficient to qualify this expert's knowledge for further research, even they rated their own knowledge of the issue highly. This methodology favours experts with a high impact of theoretical analysis and practical experience.

The higher the value of the competence index is taken as a limit value, the more proficient experts' knowledge will be used in the study, eliminating the knowledge of less competent persons. As a consequence, however, it is possible to reject people who are, in fact, high-level experts but have assessed their knowledge too harshly. Another consequence may be that the number of experts whose knowledge will be used in the study is too low.

As an example of a limiting competence factor to qualify an expert for a study, some authors give a value of $K_k = 0.5$ [70]. Others consider this value as "low" and 0.8 as "medium" [65]. In light of the above, $K_k = 0.7$ appears to be a safe limit value, indicating sufficient competence of the expert taking part in the study.

2.4.4. Assessment of Experts' Compliance

There are three main sources of disagreement in opinions on a given subject of evaluation. The first is the low competence of the group of evaluators (the competence of the group as a whole rather than of individual members is more relevant here). The second reason for disagreement is related to an improperly organised evaluation process. The third source of disagreement is an ill-defined object of evaluation [72].

The finding of non-conformity in the assessments should result in action being taken to eliminate the causes of non-conformity or, where the causes are not remediable, refraining from formulating an overall assessment [72].

In order to assess the concordance of the experts' assessments, the results of their assessments should be transformed into the form of rank orderings (preference series).

Spearman's rank correlation coefficient ρ or Kendall's rank correlation coefficient τ is used to measure the ordering consistency between two preference series. When comparing more than two rank series, Kendall's W concordance coefficient is the most commonly used measure for assessing preference congruence [72].

The measurement of concordance is reduced to the construction of the W coefficient, in which the numerator expresses a value reflecting the degree of actual links between the preference series (S) and the denominator, an analogous value calculated for the situation of full concordance of rank orderings, i.e., the maximum possible S_{max} [72], which is carried out using Formula (3):

$$W = \frac{S}{S_{max}} = \frac{\sum_{j=1}^{n} \left(R_j - \frac{m(n+1)}{2}\right)^2.}{\frac{1}{12} m^2 (n^3 - n)} \tag{3}$$

where:

S—the degree of actual linkage between preferential series;
S_{max}—the maximum achievable degree of linkage between preferential series;
R_j—the sum of ranks for the jth object;
n—the number of assessed objects;
m—the number of expert assessors.

If the results of the expert assessments produce series with tied ranks, the averaged rank method is necessary to use the concordance coefficient in measuring concordance. The method involves averaging the tied ranks so that they produce a series analogous to the series in the strong order. This transformation results in an ordering whose sum of the ranks is equal to (n (n + 1))/2, i.e., equal to the sum of the analogous series with unrelated ranks [72].

To determine the value of the concordance coefficient for tied ranks, in addition to averaging the ranks, it is also necessary to make an adjustment in the denominator of the concordance coefficient (S_{max}). For this purpose, for ranks with tied ranks, the value of Ti is calculated according to Formula (4):

$$T_i = \frac{1}{12} \sum_{j=1}^{k} (t_j^3 - t_j) \tag{4}$$

where:

k—the number of groups having the same rank ($j = 1, 2, \ldots, k$) in the i-th row;
t_j—the number of identical tied ranks in a given group.

When all m series have tied ranks, the value of T is determined from the following Equation (5):

$$T = \sum_{i=1}^{m} T_i \tag{5}$$

where:

T—correction for tied ranks occurring in all ranks.

In the case of full concordance between m series, the tied ranks refer to the same objects, so the magnitude of T is multiplied by m. Finally, the concordance coefficient for the case with tied ranks is determined by the following Equation (6):

$$W = \frac{S}{S_{max} - mT} = \frac{\sum_{j=1}^{n} \left(R_j - \frac{m(n+1)}{2}\right)^2}{\frac{1}{12} m^2 (n^3 - n) - mT} \tag{6}$$

The Kendall's W concordance coefficient calculated in this way takes a level on a scale from 0 to 1. The degrees of concordance for the W coefficient are prioritized as follows [71]:

- Sufficient—$W \in$ <0.20; 0.40>;
- Good—$W \in$ <0.41; 0.60>;
- Plus good—$W \in$ <0.60; 0.80>;
- Very good—$W \in$ <0.81; 0.95>;
- Ideal—$W \in$ <0.96; 1.00>.

The values obtained should be further checked with an appropriate significance test. Assuming the independence of the experts, we can consider that the occurrence of a particular rank ordering is as likely as any other. On this basis, the distribution S can be identified. For a given m and n, there are $n!^m$ all possible rank orderings. For low values of m and n, arrays of the actual probability distribution of obtaining a particular value of a given S have been developed.

For $n > 7$, a satisfactory approximation of the true S distribution is the chi-square distribution is given. For robust rankings, the value of the χ statistic$_r{}^2$ is calculated from the Equation (7):

$$\chi_r^2 = m(n-1)W = \frac{S}{\frac{1}{12}mn(n+1)} \quad (7)$$

For tied ranks, this value is calculated according to the Formula (8):

$$\chi_r^2 = m(n-1)W = \frac{S}{\frac{1}{12}mn(n+1) - \frac{1}{n-1}T} \quad (8)$$

Testing the statistical significance of the concordance coefficient involves the null hypothesis H_0, that the rank series under study are uncorrelated. The null hypothesis is rejected if the value of χ_r^2, calculated according to the above relationship, is equal to or higher than the value of χ_α^2, as read from the chi-square distribution tables for $n-1$ degrees of freedom (df) and for the assumed significance level α.

The rejection of the null hypothesis means that the concordance coefficient W determined between the expert assessments is not random and can therefore, be used in further research work.

3. Results

The expert survey described methodologically in Section 2.4 was conducted between 1 March 2021 and 15 April 2021. The sheets were distributed electronically (by email). Each email contained introductory content, a brief description of the purpose and scope of the study, a request to participate in the research survey and a link to the survey form.

The form was anonymous and easy to complete by indicating the answer sets of single-choice questions.

According to the authors, the anonymity of the experts encouraged independent thinking and allowed them to express their opinions freely without fear of criticism. The lack of contact between experts, on the other hand, prevented one expert from influencing another and prevented the study from being dominated by one person or part of the respondents [65].

A group meeting with experts was also not considered due to the restrictions in place at the time due to the prevailing COVID-19 pandemic.

Based on the criteria described in the previous section, a group of 110 experts was selected and sent an email with a link to the form. Among the representatives of science, there were 59 experts representing scientific and research institutions and 51 associated with coal mines operating within the territory of Poland.

A total of 54 responses were received, giving a response rate of 49%. It is worth noting here that, although there are surveys that achieve results above 70% or even 80% [65], in many surveys conducted via the Internet, the rate is much lower. It is even indicated that the average level is 5–7% [65]. The 49% obtained was therefore considered by the authors to be a sufficient result. All answers were complete (the design of the sheet prevented partial answers—all questions were compulsory).

3.1. Results of the Assessment of Experts' Competence

The survey sheet targeted experts who should be characterised by a wide range of competencies within the coal mining industry, but not every high-calibre expert has a similar level of knowledge of such disparate production areas as preparatory works, mining or processing. In addition, for the expert's assessments to be useful, he or she should also have relevant knowledge of the Lean Management method. For this reason, the experts were asked to self-assess separately for each of the five main production areas of the coal mines as well as their knowledge of the Lean Management method and to indicate the sources of their knowledge. The value of the competence coefficients for each expert in each area was then calculated. A level of 0.7 was adopted as the level of the competence coefficient, allowing the expert to be considered sufficiently competent. This level of the coefficient was adopted as the minimum limit value qualifying the experts for the study.

For each of the five designated production areas of the mine, experts with the minimum required competence factor value in the area and in the Lean Management method were selected. This made it possible to designate between 14 and 16 experts (depending on the area) whose assessments were used in the remainder of the study. Out of a total of 54 experts who completed the survey form, only 19 met the requirements in at least one production area of the mine and in knowledge of the Lean Management method.

The reason for rejecting as many as 32 forms was insufficient knowledge of the Lean method. In order to competently assess the applicability of Lean Management tools to a mine's production area, it is essential to have expert knowledge of the method. This points to the need to popularise Lean Management among mining experts, especially in view of the numerous successful applications in other industries.

To check if the number of experts was sufficient, the relationship described by Equation (1) was used, and the minimum number of experts was calculated for the following parameters:

- The number of factors $b = 12$ (12 Lean tools);
- In line with practice [74], the value of γ was assumed to be 0.01;
- In line with practice [71], a minimum sufficient value for the Kendall concordance coefficient $W_0 = 0.2$;
- Confidence level $\beta = 0.95$.

For each of the mine production areas, the minimum number of experts calculated in the above manner is 9, while the number of experts who were qualified as sufficiently competent is between 14 and 16, depending on the mine area. For each area, the minimum number of experts has therefore been reached.

3.2. Results of Expert Concordance

In the next step, a check was made on the level of concordance regarding the qualified experts. In the case under consideration, there are series with related ranks. This is primarily due to the very purpose of the survey. In order to obtain ranks with a strong order, it would have been necessary to ask the experts to rank the individual Lean tools in order from least to the most useful in a given production area of the mine. Such an approach would not have answered the key questions of whether and how useful Lean tools could be in particular production areas of mines. After all, in a ranking of strengths, even a tool that would be indicated as the best among the 12 proposed would not necessarily, according to the expert, be suitable for implementation. Similarly, in the expert's opinion, when all tools are suitable for implementation in a particular area of the mine, a strong ranking would not convey this information. Further expansion of the survey sheet would be necessary.

In the current form of the spreadsheet, each expert could independently rate the usefulness of each of the 12 tools in each area on a scale from 0 to 10. The resulting tool usefulness ratings can be ranked from least useful to most useful, but the results should be treated as ranks with tied ranks, as more than one tool could receive the same rating from a given expert—and thus be ranked ex aequo in the same place in the ranking. In the present case, this even had to happen, as there are 12 evaluated objects (Lean tools), and the possible scoring has only 11 items (from 0 to 10).

If the results of the experts' evaluations produce ranks with tied ranks, these should be transformed into averaged ranks, as detailed in Section 2.4.4. The transformation was done for each expert in each area. After the transformation of the ranks, the mutual positions of the individual tools in the 'ranking' do not change. However, the sum of the averaged ranks changes and, after the transformation, is equal to the sum of the ranks in the strong ranking (i.e., in the case studied, the sum of consecutive numbers from 1 to 12, i.e., 78). This makes it possible to determine the value of the concordance coefficient, unencumbered by the error resulting from the unequal sum of ranks.

An additional advantage of such a transformation is that it allows a more accurate comparison of expert assessments of different stringency. For example, an expert using only ratings between 0 and 5 in practice may, in fact, rate very similarly to an expert using ratings between 4 and 10 (as long as the lowest, average and highest ratings are for the same tools). Of course, in the final analysis of the usefulness of the tools in question, it is the individual experts' "absolute" numerical assessment of the usefulness of the tool itself that is most important.

Table 10 shows an example of the averaged ranks (transformed from expert assessments) for assessing the application of Lean tools in accessibility works.

Table 10. Averaged ranks of ratings given by competent experts to Lean tools for accessing works.

No. Expert	5S	TPM	Diagram Ishikawa	Diagram Spaghetti	JIT	Kanban	VSM	Poka-Yoke	Kaizen	OPF	SMED	VM
3	8.0	8.0	8.0	8.0	2.0	4.0	2.0	12.0	8.0	8.0	8.0	2.0
5	12.0	2.0	9.5	6.5	4.0	6.5	4.0	9.5	9.5	1.0	4.0	9.5
7	10.0	11.5	11.5	5.5	5.5	5.5	9.0	5.5	5.5	1.5	5.5	1.5
9	12.0	10.0	10.0	4.5	2.0	4.5	10.0	7.5	7.5	1.0	4.5	4.5
14	10.5	10.5	10.5	10.5	1.0	2.0	6.0	6.0	6.0	3.5	8.0	3.5
42	7.5	7.5	7.5	2.5	11.5	2.5	7.5	11.5	7.5	2.5	7.5	2.5
43	11.5	11.5	9.0	6.0	6.0	4.0	6.0	3.0	9.0	1.5	1.5	9.0
47	9.0	9.0	9.0	9.0	9.0	9.0	5.0	3.5	1.5	3.5	9.0	1.5
48	7.5	10.0	12.0	10.0	5.5	3.5	10.0	3.5	7.5	1.5	5.5	1.5
49	8.5	8.5	2.5	4.0	8.5	8.5	8.5	8.5	8.5	1.0	2.5	8.5
50	9.5	9.5	9.5	5.0	9.5	4.0	6.0	2.0	9.5	3.0	1.0	9.5
52	10.5	10.5	10.5	3.5	6.0	2.0	10.5	6.0	8.0	1.0	6.0	3.5
53	10.0	10.0	12.0	4.5	8.0	4.5	6.5	6.5	10.0	1.0	2.0	3.0
54	9.5	9.5	6.5	4.5	4.5	3.0	1.0	9.5	12.0	2.0	6.5	9.5

To facilitate interpretation of the table, a colour scale was used, where the colour of each cell represents the value of the averaged rank given by the m-th expert to the n-th tool. The scale has 12 colours (from dark red for the lowest possible averaged rank value of 1, through shades of red, orange, yellow and green to dark green representing the highest possible averaged rank value of 12).

To determine the value of the concordance coefficient for the tied ranks, in addition to averaging the ranks, it is also necessary to make a T correction in the denominator of the concordance coefficient. In the study carried out, all m series for each of the mine areas studied have tied ranks. In this case, the T-value was calculated using Formula (5). The calculations were carried out separately for each study area.

The next step was to calculate the value of Kendall's W concordance coefficient, using Formula (6), and the value of the mean rank correlation coefficient p_{av} Spearman for each of the mine's production areas. The results of the above calculations are shown in Table 11.

The value of the calculated concordance coefficient W ranged from 0.41 for the accessibility works to 0.46 for the exploitation of the deposit. Referring to the scale for the W coefficient described in Section 2.4.4, the results obtained show good concordance between the expert assessors for each of the mine areas examined.

On the basis of the obtained values of the concordance coefficient W and Spearman's rank correlation, it should be considered that the concordance between the experts is sufficient to make generalised judgements and decisions based on them. However, these values should be checked with an appropriate significance test beforehand.

Table 11. Kendall's W coefficients and Spearman's p_{av} values for the qualified groups of experts in each area.

Mine Production Area	Number of Qualified Experts m	Number of Factors (Tools) Assessed n	Kendall Factor Value W	Mean Spearman Rank Correlation Coefficient Value p_{av}
Accessibility works	14	12	0.41	0.36
Preparatory works	15	12	0.45	0.41
Exploitation of the deposit	16	12	0.46	0.43
Logistics	15	12	0.43	0.39
Hard coal preparation	14	12	0.43	0.38

For n > 7 (and, in the case under study, n = 12), a satisfactory approximation of the true S distribution is a chi-square distribution.

Testing the statistical significance of the coefficient of concordance involves posing the null hypothesis (H_0) that the rank series under study are not related. In the case under test, the null hypothesis is rejected if the calculated value of χ_r^2 for the tied ranks is equal to or higher than the value of χ_α^2, as read from the chi-square distribution tables for $n-1$ degrees of freedom (df) and for an assumed significance level $\alpha = 0.05$.

The calculated χ values$_r{}^2$ and the readings from the tables [?] of the χ values$_\alpha{}^2$ for the expert assessments studied are given in Table 12.

Table 12. Calculated χ values$_r{}^2$ and table read χ values$_\alpha{}^2$.

Mine Production Area	Calculated Value χ_r^2	Array Value Reading χ_α^2
Accessibility works	62,779	19,675
Preparatory works	73,674	19,675
Exploitation of the deposit	81,439	19,675
Logistics	70,357	19,675
Hard coal preparation	65,899	19,675

On the basis of the above data for each of the mine's production areas, the null hypothesis stating that the series under study are not related to each other must be rejected. It can therefore be assumed that the concordance W coefficients determined between the experts' assessments are not random—they are therefore statistically significant. The evaluations obtained from the selected, competent experts can therefore be used to formulate generalised evaluations and decisions.

3.3. Empirical Results of the Expert Study

As shown in Sections 3.1 and 3.2, a sufficient number of competent experts participated in the survey, and their responses should be considered to be in agreement, allowing them to be used in further research.

Table 13 shows the averaged results of the assessments of the usefulness of the various Lean tools in each area of the mine operation.

Based on the analysis of the results in Table 13, it can be seen that in all areas, the 5S tool was considered the most useful by the experts. Among the tools whose averaged usefulness was rated above 8 were Total Productive Maintenance and Ishikawa Diagram.

Similarly, the lowest-rated tool was consistently One-Piece Flow, with an average rating of just 3.7 for the usefulness of this tool across the different areas of mining operations. The next lowest-rated tools were SMED and Kanban, although for these tools, the rating was more dependent on the area of mining operations.

Table 13. Averaged results of suitability assessments.

Tool	Accessibility Works	Preparatory Works	Exploitation of the Deposit	Logistics	Hard Coal Preparation
5S	8.0	8.5	8.7	9.1	9.2
TPM	7.9	8.3	8.6	8.7	9.1
Ishikawa diagram	8.0	8.0	8.1	8.1	8.4
Spaghetti diagram	6.4	6.6	6.3	7.1	7.4
JIT	5.9	5.9	6.1	7.1	6.6
Kanban	5.1	5.2	5.4	6.6	5.7
VSM	6.2	6.0	5.9	6.1	6.2
Poka-yoke	5.6	5.4	6.3	6.1	6.1
Kaizen	6.9	6.6	6.8	7.3	6.8
OPF	3.1	3.5	3.6	4.0	4.4
SMED	4.9	4.4	4.4	4.9	5.6
VM	5.5	5.3	5.8	6	7.1

In addition to the mean scores, basic statistics were determined for all results (an example of the coal preparation area is shown in Table 14) and presented in the form of empirical columnar distributions (shown in Table 15).

Table 14. Basic statistics for assessing (on a scale of 0 to 10) the usefulness of applying Lean Management tools in hard coal preparation.

Tool	The Number of Assessments m	Arithmetic Mean $\bar{a}$	Median Me	Dominant D	Stretch R	Deviation S
5S	14	9.2	9	10	3	0.9
TPM	14	9.1	9	10	3	1.0
Ishikawa diagram	14	8.4	9	8	5	1.4
Spaghetti diagram	14	7.4	8	9	6	1.9
JIT	14	6.6	7	5	10	2.5
Kanban	14	5.7	6	9	10	2.9
VSM	14	6.2	7	9	8	2.4
Poka-yoke	14	6.1	8	8	10	3.5
Kaizen	14	6.8	8	10	9	3.2
OPF	14	4.4	5	0	9	2.9
SMED	14	5.6	6	7	9	2.8
VM	14	7.1	8	10	9	2.9

Table 15. Distributions of empirically obtained expert evaluations (each bar represents the number of evaluations obtained, and the horizontal axis represents the rating on a scale from 0 to 10).

Tool	Accessibility Works	Preparatory Works	Exploitation of the Deposit	Logistics	Hard Coal Preparation
5S	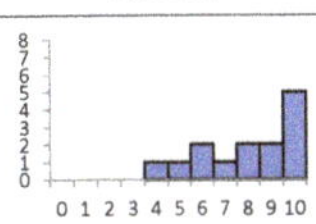	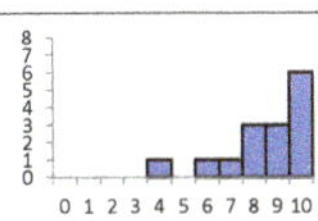	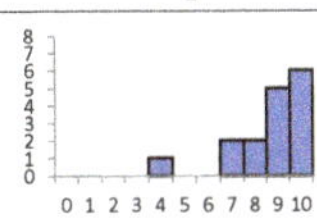	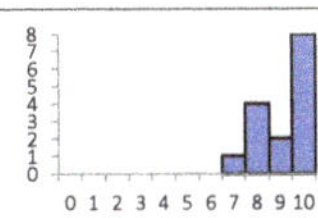	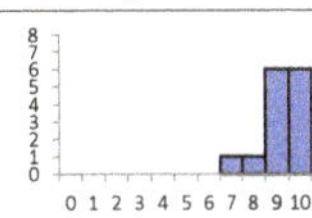
TPM	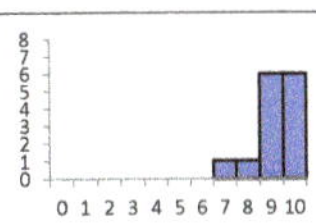	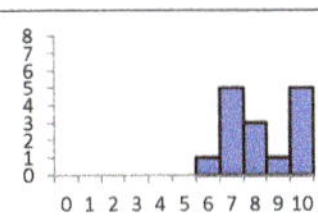	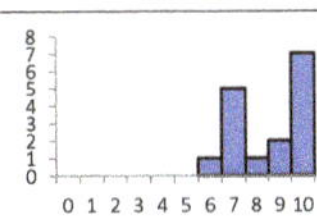	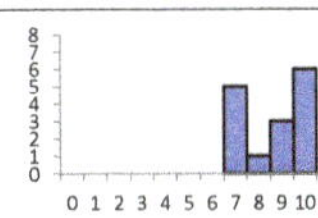	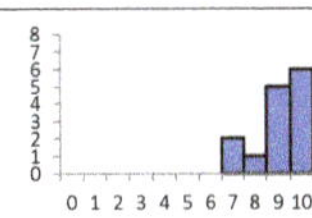

Table 15. *Cont.*

Tool	Accessibility Works	Preparatory Works	Exploitation of the Deposit	Logistics	Hard Coal Preparation
Ishikawa diagram	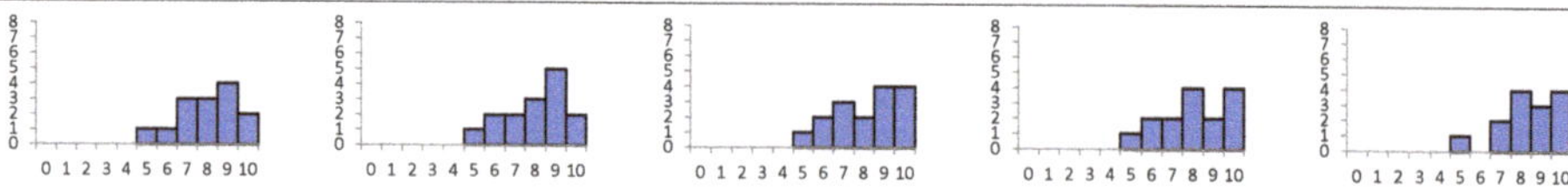				
Spaghetti diagram	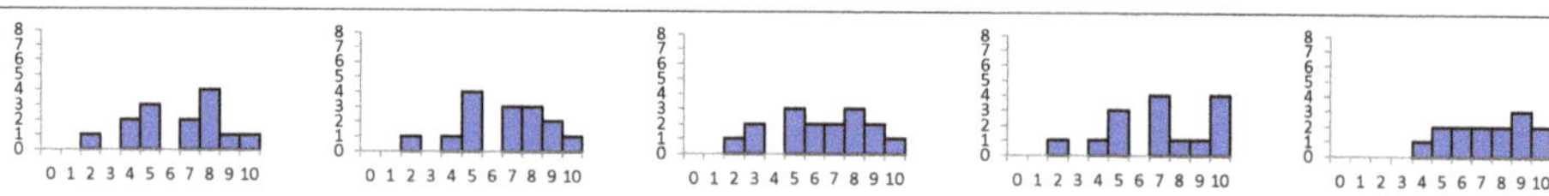				
JIT	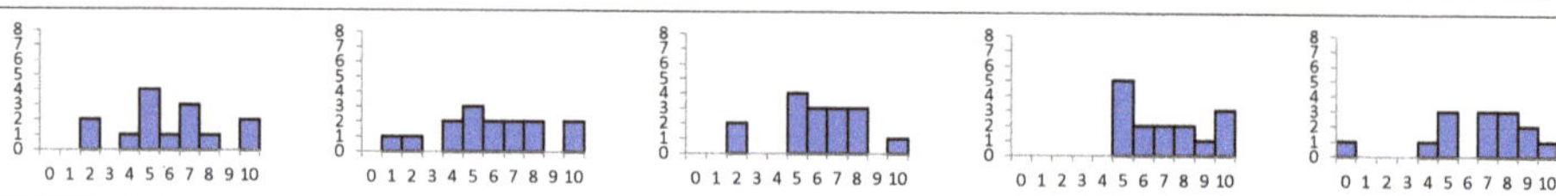				
Kanban	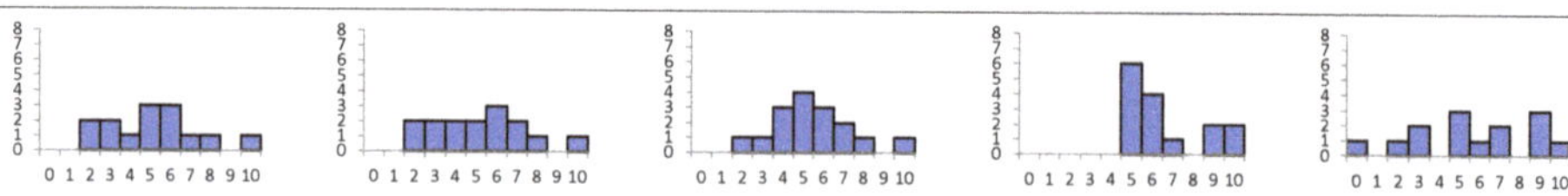				
VSM	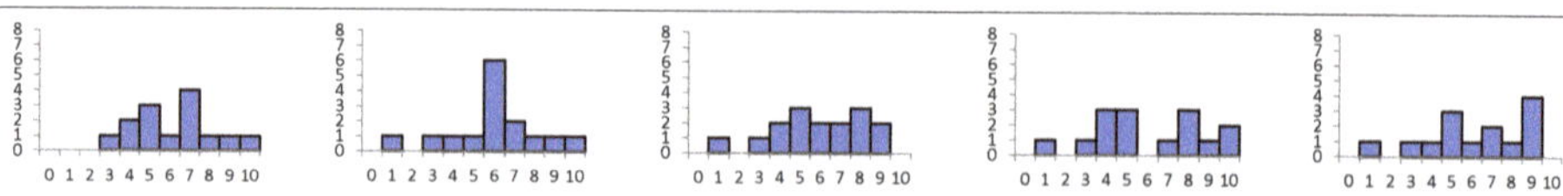				
Poka-yoke	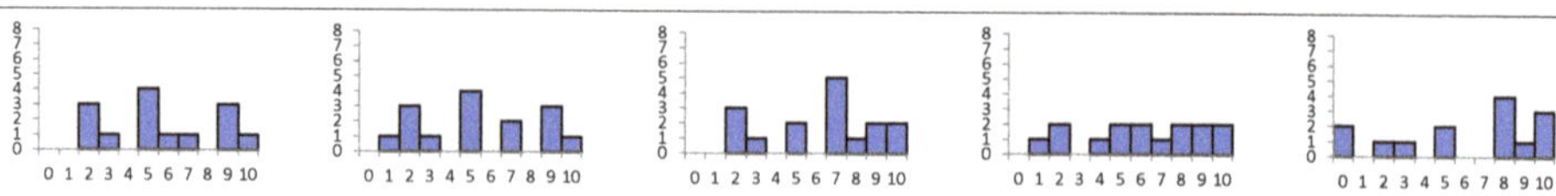				
Kaizen	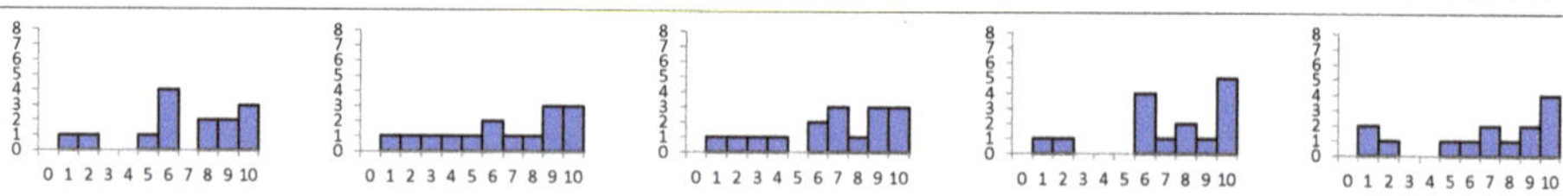				
OPF	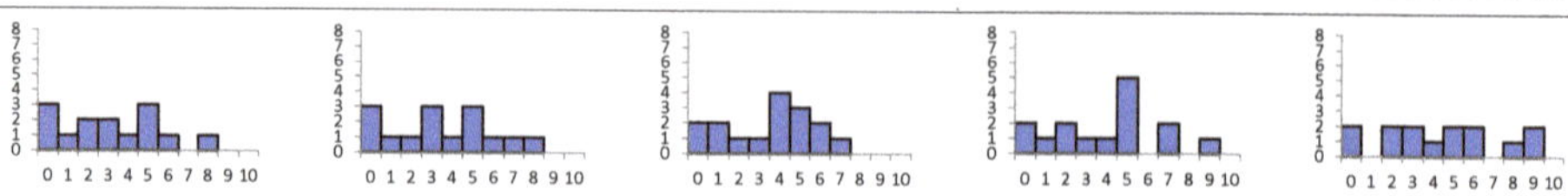				
SMED	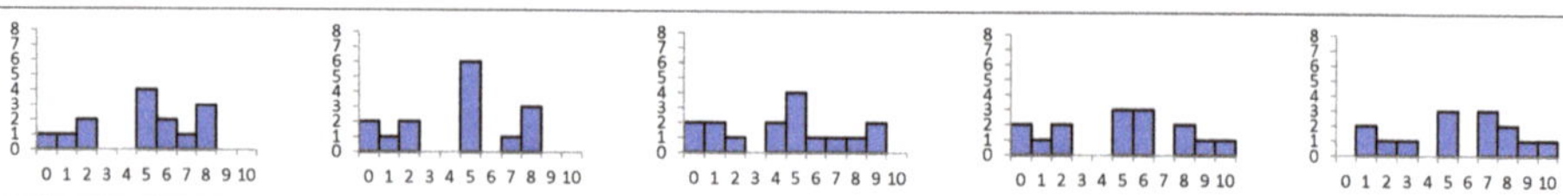				
VM	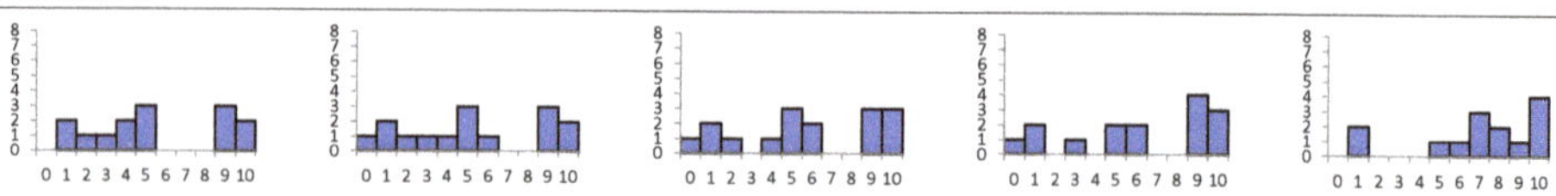				

4. Discussion and Conclusions

The literature examples cited in the paper indicating the applicability of individual elements of the Lean Management method in mining conditions are mainly based on the opinions and experience of the authors of the individual publications cited and on individual practical implementations by some mining companies. However, there is certainly a lack of articles presenting the results of empirical studies of this issue, going in the direction of systematising knowledge on the usability as well as the implementation algorithm of individual Lean tools in mining. For this reason, the following article presents the course and results of the author's expert-mathematical study on the assessment of the usefulness of individual Lean Management tools and techniques in a coal mine. The study made use of the knowledge obtained from selected, competent experts, who showed a high level of agreement in their assessments.

They indicated the highest usefulness of the 5S, Total Productive Maintenance (TPM), and Ishikawa Diagram tools. It is worth noting that the experts highlighted the high utility of these tools in all areas of the mine's operation, with slightly (a few per cent) lower relative levels compared to providing works. If we consider the methodological guidance of Lean Manufacturing, it should be noted that both 5S and TPM belong to tools used for managing production factors, specifically, space and equipment and machinery, respectively. This points to the greatest utility of such Lean tools, as well as production factors themselves, in the coal mining industry. Directly following the Lean tools used for standardisation and optimisation of production factors, experts identified, in terms of the highest utility, the Ishikawa Diagram. This tool is used for identifying the root causes of problems and is one of the quality tools belonging to the group of problem-solving tools. According to the authors of this publication, this indicates a high need for the application of solutions related to identifying the causes of defects and subsequently eliminating them. This is an evident reference to the third principle of Lean Manufacturing, which is the principle of identifying root causes [23]. It may also be related, although it is not a final conclusion but rather a supposition, to the connection of this tool with the previously mentioned 5S and TPM. Especially in the case of the latter, it seems very likely, considering the approach to error elimination through the identification and elimination of their causes, which takes place in Step 4 and Step 5 of TPM [58].

Analyzing the values of assessments provided by experts regarding the other Lean Manufacturing tools, it is noteworthy that their indicated level of usefulness is significantly lower. It is approximately 25–30% lower in this regard compared to 5S, TPM, and the Ishikawa Diagram. This is symptomatic considering the fact that among these tools are those related to logistics, material flow, and the product stream itself. Significantly lower, by over 40%, is the assessment of the usefulness of SMED, and over 70% lower is the usefulness of OPF. This is surprising, especially in the case of the SMED tool, which, like TPM, relates to the machinery park, although it is focused on changeovers rather than maintenance. This can be explained to some extent by the connection of SMED with maximising flow capacity.

Further interesting insights were provided by basic statistical calculations regarding the usefulness of applying Lean tools. It is worth noting, in this regard, the relatively low values of dispersion characteristics, especially the range, for the highest and lowest-rated tools, as well as the highest levels of these characteristics for tools related to maximising the operational efficiency of product flow, mainly JIT, Kanban, and Poka-Yoke. This indicates a lack of consensus among experts regarding these tools.

The assessments on the usefulness of Lean tools presented in the previous chapter can become a useful guideline for mine managers in the selection of methods to improve mine production.

It is worth noting that, despite the good agreement shown between the experts, for most of the surveyed tools, it is possible to observe a considerable variability in the ratings obtained. In addition, it is important to recognise that the scores obtained are subjective—for one expert, a 7 may mean high usefulness, while for another, it may mean medium usefulness at best. On the basis of the research conducted so far, it is not possible to draw a clear line of demarcation between tools of low usefulness and tools of medium or high usefulness, as this boundary may run slightly differently for each expert.

The next step to consider seems to be the selection of an appropriate inference method in order to determine the tools that should be applied in particular areas of the mine's operation in order to assist managers in deciding whether to implement a particular tool in a particular area.

Analysing the available methods of inference and decision support, it seems that a method that can be used to solve the above research problem is the method of inference using fuzzy logic rules.

The next planned stage of the research is to determine a theoretical, fuzzy model of the inference system. Then, this model can be fed with the empirical results obtained in the above study to build a model and algorithm for the application and implementation, respectively, of the Lean Management method in coal mines.

Author Contributions: Conceptualisation, M.K., P.B. and M.M.; methodology, M.K., P.B. and M.M.; validation, M.K., P.B. and M.M.; formal analysis, M.K., P.B. and M.M.; investigation, M.K., P.B. and M.M.; resources, M.K., P.B. and M.M.; writing—original draft preparation, M.K., P.B. and M.M.; writing—review and editing, M.K., P.B. and M.M.; visualisation, M.K., P.B. and M.M.; supervision, M.K., P.B. and M.M. All authors have read and agreed to the published version of the manuscript.

Funding: This paper presents the results of research conducted at AGH University of Krakow no. 16.16.100.215.

Data Availability Statement: No new data supporting reported results were created.

Conflicts of Interest: The authors declare no conflict of interest.

References

1. *Polityka Energetyczna Polski 2040*; Ministerstwo Klimatu i Środowiska: Warszawa, Poland, 2021.
2. Szuflicki, M.; Malon, A.; Tymiński, M. *Bilans Zasobów złóż Kopalin w Polsce wg Stanu na 31 XII 2020 r.*; Państwowy Instytut Geologiczny—Państwowy Instytut Badawczy: Warszawa, Poland, 2021.
3. Założenia do Aktualizacji Polityki Energetycznej Polski do 2040 r. (PEP2040). In *Wzmacnianie Bezpieczeństwa i Niezależności Energetycznej*; Ministerstwo Klimatu i Środowiska: Warszawa, Poland, 2022.
4. Migza, M.; Bogacz, P. Możliwość wykorzystania narzędzi Lean Management w przedsiębiorstwach sektora górnictwa podziemnego w Polsce. *Przegląd Górniczy* **2015**, *71*, 58–61.
5. Walentynowicz, P. *Zakres Zastosowania Lean Management w Przedsiębiorstwach Produkcyjnych—Wyniki Badań Empirycznych, in*; Knosala, R., Ed.; Innowacje w zarządzaniu i inżynierii produkcji; Oficyna Wydawnicza Polskiego Towarzystwa Zarządzania Produkcją: Opole, Poland, 2013; pp. 407–418.
6. Bożek, M.; Handzelewicz, A. Determinanty efektywnego wdrożenia filozofii Lean Maufacturing. *Probl. Jakości* **2012**, *44*, 13–20.
7. Leite, H.D.R.; Vieira, G.E. Lean philosophy and its applications in the service industry: A review of the current knowledge. *Production* **2015**, *25*, 529–541. [CrossRef]
8. Krafcik, J. Triumph of the Lean Production System. *Sloan Manag. Rev.* **1988**, *30*, 41–52.
9. Martyniak, Z. System lean management. *Organ. Kier.* **1998**, *1*, 21–29.
10. Parkes, A. Lean Management. Genesis. *Management* **2015**, *19*, 106–121. [CrossRef]
11. Hakim, H. Not just for cars: Lean methodology. *Nurs. Manag.* **2014**, *45*, 39–43. [CrossRef] [PubMed]
12. Kimsey, D.B. Lean Methodology in Health Care. *AORN J.* **2010**, *92*, 53–60. [CrossRef]
13. Cholewicka-Goździk, K. Metoda LEAN—doskonalenie procesów i produktów: Wokół książki Jamesa P. Womack'a i Daniela T. Jones'a. *Probl. Jakości* **2001**, *1*, 21–25.
14. Jóźwiakowski, P. Lean Management—Metoda racjonalnego zarządzania produkcją, Zeszyty Naukowe DWSPiT. *Stud. Z Nauk. Tech.* **2015**, *4*, 33–46.
15. Lipecki, J. Lean management jako metoda restrukturyzacji zarządzania. *Ekon. Organ. Przedsiębiorstwa* **1998**, *8*, 12–15.
16. Womack, J.P.; Jones, D.T.; Roos, D. *The Machine That Changed the World: The Story of Lean Production*; Rawson Associates: New York, NY, USA, 1990.
17. Toyota Production System. Toyota Motor Corporation. Available online: https://global.toyota/en/company/vision-and-philosophy/production-system/ (accessed on 19 February 2022).
18. Dekier, Ł. *Zastosowanie Systemu Sugestii Przedsiębiorstwach Zarządzanych Zgodnie z Metodą Lean Management*; Rozprawa doktorska; Uniwersytet Ekonomiczny w Poznaniu: Poznań, Poland, 2017.
19. Jakonis, A. Kulturowe uwarunkowania Lean management. *Przedsiębiorczość I Zarządzanie* **2011**, *12*, 29–55.
20. Liker, J.K. *Droga Toyoty*; MT Biznes: Warszawa, Poland, 2005.
21. Liker, J.K.; Morgan, J.M. The Toyota Way in Services: The Case of Lean Product Development. *Acad. Manag. Perspect.* **2006**, *20*, 5–20. [CrossRef]
22. Lisiński, M.; Ostrowski, B. *Lean Management w Restrukturyzacji Przedsiębiorstwa*; Wydawnistwo Antykwa: Kraków, Poland, 2006.
23. Womack, J.P.; Jones, D.T. *Lean Thinking*; Simon & Schuster: New York, NY, USA, 2003.
24. Ohno, T. Toyota Production System. In *Beyond Large-Scale Production*; Productivity Press: New York, NY, USA, 1988.
25. Marchwinski, C.; Shook, J.; Schroeder, A. *Lean Lexicon. 4.0*; The Lean Enterprise Institute: Cambridge, MA, USA, 2008.
26. Kubis, N. Narzędzia Lean Management. *Zagadnienia Tech.-Ekon.* **2005**, *50*, 291–303.

27. Dunstan, K.; Lavin, B.; Sanford, R. The application of lean manufacturing in a mining environment. In *Proceedings International Mine Management Conference, Melbourne, Australia, 16–18 October 2006*; The Australasian Institute: Melbourne, Australia, 2006; pp. 145–157.
28. Helman, J. Analysis of the potentials of adapting elements of Lean Methodology to the unstable conditions in the mining industry. *AGH J. Min. Geoengin.* **2012**, *36*, 151–157.
29. Rosienkiewicz, M.; Helman, J. *Koncepcja Zastosowania Elementów Systemu Ssącego w Przemyśle Wydobywczym, w*; Knosala, R., Ed.; Innowacje w Zarządzaniu i Inżynierii Produkcji; Oficyna Wydawnicza Polskiego Towarzystwa Zarządzania Produkcją: Opole, Poland, 2014; pp. 646–657.
30. Sukiennik, M.; Bąk, P. Applying Lean Management Solutions in the Context of the Organisational Culture of Energy Sector Enterprises. *Inżynieria Miner.* **2018**, *20*, 117–122.
31. Korski, J. Lean Management w przedsiębiorstwie górniczym. *Wiadomości Górnicze* **2011**, *62*, 23–27.
32. Mikhalchenko, V.; Rubanik, Y. Application of the "lean thinking" concept to the analysis of coal mining region's sustainable development objective. In *E3S Web of Conferences, Proceedings of the First Interregional Conference "Sustainable Development of Eurasian Mining Regions" (SDEMR-2019), Kemerovo, Poland, 25–27 November 2019*; EDP Sciences: Les Ulis, France, 2019; Volume 134, p. 03001.
33. Haugen, S. *Lean Mining*; Mineralproduksjon: London, UK, 2013; pp. B21–B40.
34. Wijaya, A.R.; Kumar, R.; Kumar, U. Implementing Lean Principle into Mining Industry—Issues and Challenges. In Proceedings of the 18th International Synposium on Mine Planning and Equipment Selection, Banff, Alberta, Canada, 16–19 November 2009; Volume 16.
35. Yingling, J.C.; Detty, R.B.; Sottile, J.J. Lean Manufacturing principles and their applicability to the mining industry. *Min. Res. Eng.* **2000**, *9*, 215–238. [CrossRef]
36. Bator, A.; Paluchniak, A. Wykorzystanie metody 5S do poprawy bezpieczeństwa pracy. *Przegląd Górniczy* **2013**, *69*, 7–10.
37. Brodny, J.; Stecuła, K. *Analiza Efektywności Wykorzystania Zestawu Maszyn Górniczych, w*; Knosala, R., Ed.; Innowacje w Zarządzaniu i Inżynierii Produkcji; Oficyna Wydawnicza Polskiego Towarzystwa Zarządzania Produkcją: Opole, Poland, 2016; pp. 419–421.
38. Polak, R. Adaptacja kluczowych miar efektywności strategii TPM w warunkach kopalni węgla kamiennego. *Zesz. Nauk. Inst. Gospod. Surowcami Miner. I Energią Pol. Akad. Nauk.* **2014**, *87*, 49–68.
39. Turnbull, G.K. *The Alcoa Business System: Pathway to Performance*; Alcoa Inc.: Pittsburgh, PA, USA, 2003.
40. Mottola, L.; Scoble, M.; Lipsett, G. Machine Monitoring and Automation as Enablers of Lean. In Proceedings of the Second International Future Mining Conference, Sydney, NSW, Australia, 22–23 November 2011; pp. 81–86.
41. Klippel, A.; Petter, C.; Antunes, J. Lean management implementation in mining industries. In *Proceedings Dyna Conference 2008*; Universidad Nacional de Colombia: Bogota, Columbia, 2008; pp. 81–89.
42. Rio Tinto. Annual Report. 2010. Available online: http://www.riotinto.com/ (accessed on 21 February 2022).
43. Gladstone Sustainable Development Report 2008. Our People, Our Operations, Our Community, Boyne Smelters Limited, Rio Tinto Alcan Yarwun. 2008. Available online: https://www.yumpu.com/en/document/view/52658581/gladstone-sustainable-development-report-2008 (accessed on 11 February 2022).
44. Diavik Dialogue Newsletter. Volume 16, 2nd Quarter 2013, Diavik Diamond Mines, Rio Tinto, Yellowknife 2013. Available online: https://www.miningnorth.com/_rsc/site-content/library/diamondmining/2013-10-30_Diavik_Ten_Year_Milestones_PUB_FINAL.pdf (accessed on 10 February 2021).
45. Kęsek, M.; Bogacz, P.; Migza, M. The application of Lean Management and Six Sigma tools in global mining enterprises, 2nd International Conference on the Sustainable Energy and Environmental Development. *IOP Conf. Ser. Earth Environ. Sci.* **2019**, *214*, 012090. [CrossRef]
46. Castillo, G.; Alarcon, L.F.; Gonzalez, V.A. Implementing Lean Production in Copper Mining Development Projects: Case Study. *J. Constr. Eng.* **2015**, *141*, 05014013. [CrossRef]
47. Löchte, J.; Langhanki, B. *Zero Accidents and 100 Percent Value Added—Utopian Challenging, Aachen International Mining Symposia 2015*; RWTH Aachen University: Aachen, Germany, 2015; pp. 27–28.
48. RAG. *Aktiengesellschaft 2016*; RAG AG: Aachen, Germany, 2016.
49. Bogacz, P.; Migza, M. Zastosowanie Lean Six Sigma w doskonaleniu procesów produkcyjnych w przemyśle wydobywczym. *Inżynieria Miner.* **2016**, *17*, 23–29.
50. Liu, Z. Study on Coal Lean Mining Theory and Practice. *Adv. Mater. Res.* **2013**, *605*, 538–541. [CrossRef]
51. Ade Er, M.; Deshpande, V.S. Lean Manufacturing and Productivity Improvement in Coal in Mining Industry. *Int. J. Eng. Res. Dev.* **2012**, *2*, 35–43.
52. Bator, A.; Fuksa, D.; Kęsek, M.; Ślósarz, M. Zarządzanie produkcją odchudzoną - kierunki działań dla poprawy funkcjonowania kopalń. *Przegląd Górniczy* **2015**, *71*, 8–10.
53. Górnictwo: Jazda na Taśmie w Kopalni Mysłowice-Wesoła, Portal Górniczy NETTG.pl. 2015. Available online: https://nettg.pl/gornictwo/128262/gornictwo-jazda-na-tasmie-w-kopalni-myslowice-wesola/set/page/2 (accessed on 13 February 2022).
54. Bogacz, P.; Cieślik, Ł.; Osowski, D.; Kochaj, P. Analysis of the scope for reducing the level of energy consumption of crew transport in an underground mining plant using a conveyor belt system mining plant. *Energies* **2022**, *15*, 7691. [CrossRef]
55. Gałązka, W. PG Silesia: System 24/7 Daje Wydajność i Zatrudnienie, Portal Górniczy NETTG.pl. 2014. Available online: https://nettg.pl/gornictwo/121579/pg-silesia-system-24-7-daje-wydajnosc-i-zatrudnienie (accessed on 13 February 2022).

56. Sztajerska, D.; Bogdański, M. Improvement of maintenance process in a coal mine—Case study, Zeszyty Naukowe. *Organ. Zarządzanie/Politech. Śląska* **2021**, *150*, 279–298.
57. Sobol-Wojciechowska, J.; Szwancyber, Ł.; Zaremba, L. Implementacja metodologii Lean w warunkach KGHM „Polska Miedź" SA—Poprawa efektywności i innowacyjności przedsiębiorstwa wydobywczego. *Wiadomości Górnicze* **2013**, *64*, 399–405.
58. Konieczny, A.; Kidoń, M.; Kanikuła, T. TPM to kształtowanie postaw, a nie konserwacja maszyn—Dwuletnie doświadczenia O/ZWR KGHM Polska Miedź, Lean Enterprise Institute Polska. 2014. Available online: htttp://www.lean.org.pl (accessed on 13 February 2022).
59. Burduk, A.; Chlebus, T.; Helman, J.; Kowalski, A.; Rosienkiewicz, M.; Stefaniak, P. Adaptacja wybranych metod Lean Manufacturing do wybranych warunków przemysłu wydobywczego. *Napędy Sterow.* **2014**, *16*, 106–112.
60. Migza, M.; Bogacz, P. Lean Thinking in Mining Industry. In *Problemy Nedropol'Zovaniâ: Meždunarodnyj Forum-Konkurs Molodyh Učenyh: 22–24 Aprelâ 2015*; Nacional'nyj Mineral'no-Syr'evoj Universitet «Gornyj»: Sankt-Peterburg, Russia, 2015; p. 214.
61. Flynn, J.R.; Vlok, J.P. Lean Approaches in Asset Management within the Mining Industry. In Proceedings of the 2014 World Congress on Engineering Asset Management, Pretoria, South Africa, 28–31 October 2014; Springer: Cham, Switzerland, 2015; pp. 101–118.
62. Lööw, J.; Johansson, J. An overview of Lean Production and its application in mining. In Proceedings of the Aachen International Mining Symposia: Mineral Resources and Mine Development, Aachen, Germany, 27–28 May 2015; RWTH Aachen University, Institute of Mining Engineering: Aachen, Germany, 2015; pp. 121–136.
63. Seifullina, A.; Er, A.; Nadeem, S.P.; Garza-Reyes, J.A.; Kumar, V. A Lean Implementation Framework for the Mining Industry. *IFAC-Pap.* **2018**, *51*, 1149–1154. [CrossRef]
64. Gorbaniuk, O. Wykorzystanie Procedury Sędziów Kompetentnych w Naukach Społecznych i Możliwość jej Oceny Psychometrycznej za Pomocą Narzędzi Dostepnych w Statistica, StatSoft Polska. Available online: https://media.statsoft.pl/pdf/czytelnia/wykorzystywanie_procedury_sedziow_kompetentnych.pdf (accessed on 10 February 2022).
65. Zając, S.; Izdebski, W.; Skudlarski, J. Metoda ekspercko-matematyczna jako narzędzie wspomagające prognozowanie i naukowe rozwiązywanie skomplikowanych zadań, Prace Naukowo-Dydaktyczne PWSZ im. S. *Pigonia Krośnie* **2015**, 331–355.
66. Turek, M. *Podstawy Podziemnej Eksploatacji Pokładów Węgla Kamiennego*; Główny Instytut Górnictwa: Katowice, Poland, 2010.
67. Kendall, M.G.; Smith, B.B. The Problem of m Rankings. *Ann. Math. Stat.* **1939**, *10*, 275–287. [CrossRef]
68. Masiuk, A. *Wpływ Profilaktyki Eksploatacyjnej na Efektywność Produkcji mleka, Rozprawa Habilitacyjna*; Fundacja Rozwój SGGW: Warszawa, Poland, 1998.
69. Męczyńska, A. *Metoda Heurystyczna—Grupowa Ocena Ekspertów w Zastosowaniu do Analizy Procesów, Produktów, w*; Knosala, R., Ed.; Komputerowo zintegrowane zarządzanie; WNT: Warszawa, Poland, 1999.
70. Grabowska, J. Grupowa ocena ekspertów do identyfikacji czynników kształtujących system informacji logistycznej. Założenia i opis metody, wybór czynników i dobór ekspertów. *Zesz. Nauk. Politech. Śląskiej. Ser. Organ. Zarządzanie* **2015**, *78*, 155–165.
71. Mieszaniec, J. Identyfikacja Kluczowych Składników Potencjału Konkurencyjności Kopalń Węgla Kamiennego. Ph.D. Thesis, Akademia Górniczo-Hutnicza im, Stanisława Staszica w Krakowie, Kraków, Poland, 2002.
72. Cabała, P. Zastosowanie współczynnika konkordancji w pomiarze zgodności ocen ekspertów. *Przegląd Stat.* **2010**, *2–3*, 36–52.
73. Skudlarski, J. Hierarchia ważności czynników techniczno-organizacyjnych w odniesieniu do efektywności obsługi serwisowej ciągników rolniczych. *Acta Sci. Pol. Tech. Agrar.* **2005**, *4*, 49–57. [CrossRef]
74. Wieczorek, M. *Statystyka Lubię to! Zbiór zadań*, 1st ed.; SGH: Warsaw, Poland, 2013.

MDPI
St. Alban-Anlage 66
4052 Basel
Switzerland
www.mdpi.com

Energies Editorial Office
E-mail: energies@mdpi.com
www.mdpi.com/journal/energies

www.ingramcontent.com/pod-product-compliance
Lightning Source LLC
Chambersburg PA
CBHW041946260926
51850CB00018B/1622
9783725808052